Developmental Approaches to Human Evolution

Developmental Approaches to Human Evolution

Edited by

Julia C. Boughner and Campbell Rolian

WILEY Blackwell

Published by John Wiley & Sons, Inc., Hoboken, New Jersey
Published simultaneously in Canada

For general information on our other products and services or for technical support, please contact our Customer Care Department within the United States at (800) 762-2974, outside the United States at (317) 572-3993 or fax (317) 572-4002.

Wiley also publishes its books in a variety of electronic formats. Some content that appears in print may not be available in electronic formats. For more information about Wiley products, visit our website at www.wiley.com.

Library of Congress Cataloging-in-Publication data applied for

ISBN: 9781118524688

Cover image: Janka Dharmasena © iStock / Getty Images Plus, Ralf Hettler/iStockphoto, luismmolina/iStockphoto.

Printed in Singapore by C.O.S. Printers Pte Ltd

10 9 8 7 6 5 4 3 2 1

Contents

Contributors

Julia C. Boughner
Department of Anatomy and Cell Biology
College of Medicine
University of Saskatchewan
Saskatoon, SK, Canada

Anne Buchanan
Department of Anthropology
Pennsylvania State University
University Park, PA, USA

Terence D. Capellini
Department of Human Evolutionary Biology
Harvard University
Cambridge, MA, USA

Christine J. Charvet
Behavioral and Evolutionary Neuroscience Group
Department of Psychology
Cornell University
Ithaca, NY, USA

Bernard Crespi
Department of Biological Sciences
Simon Fraser University
Burnaby, BC, Canada

Rui Diogo
Department of Anatomy
Howard University College of Medicine
Washington, DC, USA

Mireia Esparza
Secció d'Antropologia
Departament de Biologia Animal
Universitat de Barcelona
Barcelona, Spain

Barbara L. Finlay
Behavioral and Evolutionary Neuroscience Group
Department of Psychology
Cornell University
Ithaca, NY, USA

Philipp Gunz
Department of Human Evolution
Max-Planck-Institute for
Evolutionary Anthropology
Leipzig, Germany

Benedikt Hallgrímsson
McCaig Institute for Bone and Joint
Research
Alberta Children's Hospital
Research Institute
Department of Cell Biology and
Anatomy
Faculty of Medicine
University of Calgary
Calgary, Canada

Rodrigo S. Lacruz
Department of Basic Science and
Craniofacial Biology
New York University College of
Dentistry
New York, NY, USA

Carles Lalueza-Fox
Institut de Biologia Evolutiva
CSIC-Universitat Pompeu Fabra
Barcelona, Spain

Emma Leach
Department of Biological Sciences
Simon Fraser University
Burnaby, BC, Canada

Neus Martínez-Abadías
Centre for Genomic Regulation (CRG)
The Barcelona Institute of Science
and Technology, Dr. Aiguader
Universitat Pompeu Fabra (UPF)
Secció d'Antropologia
Departament de Biologia Animal
Universitat de Barcelona
Barcelona, Spain

Philip L. Reno
Department of Anthropology
The Pennsylvania State University
University Park, PA, USA

Campbell Rolian
Faculty of Veterinary Medicine
University of Calgary
Calgary, AB, Canada

Torstein Sjøvold
Osteologiska Enheten
Stockholms Universitet
Stockholm, Sweden

Kenneth Weiss
Departments of Anthropology
and Biology
Pennsylvania State University
University Park, PA, USA

Bernard Wood
Center for the Advanced Study of
Hominid Paleobiology
Department of Anthropology
George Washington University
Washington, DC, USA

Nathan M. Young
Department of Orthopaedic Surgery
University of California
San Francisco, CA, USA

Foreword: Humans from Embryos

If there ever have been any large-scale transformations of the study of evolution, then there have been two recent ones, both precipitating towards the end of the 20th century: the first is the establishment of a rigorous framework for phylogenetic and comparative studies, and the second is the origination of developmental approaches to evolutionary biology. The latter depended to some extent on the success of the former, in the sense that developmental inferences about the mechanisms of evolutionary change require one to use phylogenies and the comparative method to contextualize the mechanistic results from different species. Given the experimental bent of the study of developmental evolution, it is not surprising that, for much of its history, devo-evo was a largely zoological and botanical enterprise, focusing on *Drosophila* wing pigmentation, Nematode vulva development, and the origin of limbs and flowers, among other "model characters." Human evolution was not a natural proving ground for devo-evo due to the ethical and practical strictures on any work with humans and other primates. But a determined cadre of biological anthropologists worked towards changing that, making human evolution an important application of developmental evolutionary thinking and research. How then is it possible to learn about human and primate developmental evolution without executing on primate species breeding experiments, transgenic techniques, and interfering with primate embryonic development? The answer is given here in the exciting collection of chapters on "Developmental Approaches to Human Evolution."

Surveying the contributions to this volume, it seems there are two principal work-arounds to overcome the lack of opportunity for direct experimentation in primates. One could be called *homology-based mechanistic inferences*. The idea is to take advantage of mechanistic knowledge obtained from model organisms, mostly "the" model mammal, mouse. The assumption is that homologous developmental mechanisms are also relevant in primates, and thus evolutionary changes in the orthologous genes in the human lineage are taken as evidence for a corresponding developmental evolutionary change in human evolution. These inferences are often scaffolded with comparative genomic data showing that relevant genes or *cis* regulatory elements, related to the trait of interest, have changed in the human lineage. The other approach could be called *theoretical or computational morphology*. It combines phenotypic and variational studies with theoretical models. These models can be abstract genetic ideas about the structure of the genotype-phenotype map or specific mechanistic models of developmental processes. These approaches are surprisingly powerful in generating overarching explanations of patterns of morphological and even paleontological disparity.

Neither of these approaches will suffice to make anthropology a source of novel insights into the principles and mechanisms of developmental evolution *per se* in the same way as experimental systems can. But nevertheless, the inclusion of evolutionary developmental data and principles will continue to lead and already has led to a deeper and richer understanding of how the human species arose and how our human uniqueness resulted from the same material substrate as the rest of biological diversity. As illustrated in this volume, one area where evolutionary developmental anthropology has a leg up over most model organism research is its relevance and access to issues of cognitive evolutionary change. It is an exciting time to be a biological anthropologist!

Günter P. Wagner
Yale University

Chapter 1
Introduction to Evo-Devo-Anthro

Campbell Rolian[1] and Julia C. Boughner[2]
[1] *Faculty of Veterinary Medicine, University of Calgary, Calgary, AB, Canada*
[2] *Department of Anatomy and Cell Biology, College of Medicine, University of Saskatchewan, Saskatoon, SK, Canada*

Evolutionary developmental biology, or evo-devo, is a relatively young branch of biology concerned with how and why organismal development matters to evolution. Evo-devo encompasses a range of unified research questions and empirical approaches that can be grouped into two complementary research areas (Laublicher 2007). The first area concerns the process of organismal development. This research uses molecular tools such as studies of gene and protein expression patterns to understand how *processes* of organismal development have evolved and produced phenotypic diversification at macroevolutionary scales. The second approach focuses on the role that process plays in structuring the *pattern* of heritable phenotypic variation among individuals. This approach relies on quantitative genetic theory and morphometric tools to measure developmentally determined patterns of phenotypic variation, typically at the level of populations, and to understand how these patterns have biased or constrained the rate and direction of evolutionary change within and between species (Raff 2000).

In the past couple of decades, the types of research questions that evo-devo addresses have also become of great interest to biological anthropologists. The discipline is gaining traction among researchers interested in the role(s) played by organismal development in the evolution of uniquely human, and non-human primate, traits. This volume aims to provide an overview of past and ongoing research in evo-devo specifically as it applies to the study of human and primate evolution – Evolutionary Developmental Anthropology (EDA, or Evo-Devo-Anthro). In this

Developmental Approaches to Human Evolution, First Edition. Edited by
Julia C. Boughner and Campbell Rolian.

introductory chapter, we begin with a brief survey of the origins and principal discoveries of evolutionary developmental biology. We then discuss the emergence of evo-devo in anthropology in the context of past research at the interface of human/primate development and evolution. Finally, we summarize the current state of affairs in the growing field of evo-devo anthropology, and highlight a number of knowledge gaps which are promising avenues for future research in this field.

A Brief History of Evo-Devo

As this volume attests, many different approaches to studying the reciprocal interactions of development and evolution fall under the broad umbrella of evolutionary developmental biology. As a result, finding consensus on a single definition of evo-devo that describes what the field is, and what it seeks to accomplish, can be challenging. The lack of consensus stems from two distinct goals: studying process at macro- and microevolutionary scales. These goals are driven by the desire to understand the broad developmental processes that drive the evolution and diversification of form among species and higher taxonomic levels, versus the lower level (but likely similar) processes that pattern the structure of phenotypic variation among individuals within populations, as the fuel for natural selection. This lack of consensus on what evo-devo is also stems from the fact that, although it is in some ways a "new" discipline (Carroll 2005), its roots run deep. The study of organismal development, evolutionary processes, and their complex interactions is at least 150 years old, dating back to 19th-century evolutionary embryologists such as Ernst Haeckel and Francis Balfour, and to Charles Darwin himself (Hall 1999). These pioneers focused on the comparative study of embryology as a window into organismal development, and were particularly interested in what these processes could reveal across taxa about phylogenetic relationships and the evolution of specific traits with a shared evolutionary origin but different morphologies and functions (i.e., homologies such as the hands of dolphins, bats, and humans) (Hall 1999; Maienschein 2007).

Many of these early studies in comparative embryology were concerned with comparing *patterns* of growth and development within and among species. These now-classic works inferred that differences in ontogenetic patterns must account for variation within populations, but especially morphological divergence among taxa in deep time. Evolutionary embryologists were less concerned with lower level biological *processes* (i.e., cellular dynamics) that would explain described developmental *patterns* across all vertebrates. This was largely a practical issue: describing macroscopic changes in vertebrate fetal development between taxa was considerably simpler than documenting changes in the spatial relationships of cells and tissues during morphogenesis. Productivity in this area of study has since increased dramatically with the advent and benefit of modern molecular tools.

In contrast to other fields in biology such as population genetics, progress in embryology for much of the 20th century was relatively slow, in part due to the technical challenges associated with studying embryonic development, to the extent that the discipline contributed relatively little to the modern evolutionary synthesis of the 1940s (Carroll 2005; Maienschein 2007). The Modern Synthesis united several disciplines studying evolutionary biology from different angles, particularly population genetics and paleontology (Mayr and Provine 1998). It suggested that the evolution of quantitative traits is gradual, and occurs through mechanisms consistent with Mendelian genetics, namely through small genetic changes that produce continuous (i.e., bell-curved) variation within populations, which can then be acted upon by selective forces. Proponents further argued correctly that this process, occurring at the population level (microevolution), could be extrapolated to higher taxonomic levels and longer timescales to explain macroevolutionary patterns.

Despite the realization that development interposes itself between genes and phenotypes, and hence likely influences the transition from one to the other, the study of embryology was not revived after the synthesis. Rather, the synthesis served to affirm the primacy of genes and phenotypes in determining evolutionary change, relegating organismal development, not to mention the field of epigenetics (*sensu* Waddington, Jamniczky *et al.* 2010) to a secondary, less important, process linking genes to phenotypes. For several decades following the synthesis, organismal development continued to be viewed as a black box, something that was "hopelessly complex and would involve entirely different explanations for different types of animals" (Carroll 2005:6). However, the reasons for ignoring organismal development in the study of the evolution of animal form, in particular early embryological events such as morphogenesis, were not entirely philosophical. Considerable practical obstacles in developmental biology remained: although genes were now seen as primary drivers of evolutionary change, prior to the late 1970s no one had successfully identified and localized genes that *determine* animal form, let alone how changes in their structure or function could lead to the *evolution* of animal form and function.

Breakthroughs in developmental biology were finally achieved in the late 1970s and 1980s, first in fruit flies, and eventually in vertebrates. These breakthroughs were rooted in technological innovations in molecular genetics: especially the ability to identify, localize, and manipulate genes physically; in particular to visualize their expression patterns; and to relate these to temporal and spatial effects on the development of organismal form (for example through gene inactivation) (Anderson and Ingham 2003). Homeotic genes were among the first genes shown to control key aspects of development, and some consider their discovery to mark the birth of evo-devo (Hall 1999; Carroll 2005). Homeotic genes regulate segmental patterning in the metazoan embryo. Early analyses revealed that loss-of-function mutations in these genes in *Drosophila* caused segmental identity shifts (homeotic transformations), where one segment along the embryo's anterior-posterior axis would take on the likeness of an adjacent segment (Lewis 1978). Homeotic genes act as transcription factors, proteins that regulate the transcriptional activity of other genes (Mallo *et al.* 2010).

Soon after their discovery in *Drosophila,* similar genes with similar tasks in embryonic patterning were uncovered in vertebrates, including humans (Tournierlasserve *et al.* 1989; Krumlauf 1994). These discoveries led to a fundamental evo-devo concept: the developmental genetic toolkit (Carroll *et al.* 2005). Toolkits describe subsets of genes that specify animal form during embryological development. Toolkit genes are distinct from those involved in the routine functions of all cells (housekeeping genes,

Zhu *et al.* 2008), and those that are uniquely expressed in differentiated cell types. Toolkit genes belong to signalling pathways, many acting as transcription factors that regulate the activities of other genes that specify cell fates and/or establishing spatial and temporal expression patterns during morphogenesis. The same toolkits are recruited several times during an individual's development, and contribute to the development of vastly different structures. As a result, toolkit genes have pleiotropic effects on the phenotype, a fact amply demonstrated by disrupting the function of any of these genes (Goodman and Scambler 2001; Goodman 2002).

Developmental toolkits are also highly conserved – down to the level of the nucleotide sequence – across divergent animal phyla. This sequence conservation provided evidence that homology at the phenotypic level, which played such an important role in 19th-century evolutionary embryology, was reflected in homology at the genetic level. Put differently, morphological structures in related species look similar not only because they were inherited from a common ancestor with a similar character trait, but more specifically because these species inherited homologous developmental genetic toolkit(s) specifying these traits. Highly conserved toolkit genes were even found to specify anatomical structures that evolutionary embryologists would have identified as being analogous rather than homologous, such as the eyes of fruit flies and vertebrates (Mark *et al.* 1997). In other words, even though certain traits between vertebrates and invertebrates may not seem homologous in an anatomical sense, they are, with some cautions (Hall 2007), genetically and developmentally homologous via inheritance of the same toolkits from a last common ancestor deep in evolutionary time (Shubin *et al.* 1997).

The existence of seemingly universal developmental toolkits implies that morphology evolves not because of differences in the structure or complement of toolkit genes, but rather because of how and when they are expressed, in other words because of gene regulatory differences among taxa (Prud'homme *et al.* 2007). This idea had already been proposed by Mary-Claire King and A. C. Wilson in a seminal paper in 1975 (King and Wilson 1975). They compared the sequence homology at a macromolecular (i.e., amino acids among proteins) in *Pan* and *Homo*, highlighting a distinction between the well-known organismal differences of these sister taxa, and the high degree of conservation in the amino acid sequence of dozens of proteins primarily related to cell function. They concluded that the striking differences in biology between these taxa could not be explained by differences in the sequence and function of protein-coding genes, and were therefore due to mutations in the non-coding, regulatory regions that control the expression of these genes.

In the age of evo-devo, King and Wilson's observation was extended to anatomical structures. The discovery of homeotic genes and highly conserved developmental toolkits led to the realization that even the striking morphological differences between a human and a chimpanzee were due to differences in the regulation of the same/homologous toolkit genes. This idea is now commonly known as the "*cis*-regulatory hypothesis" (Carroll 2008; Wittkopp and Kalay 2012). *Cis*-regulatory elements (CREs) are non-coding DNA regions located on the same DNA strand as the toolkit gene(s) they regulate. CREs typically bind transcription factors, modulating the expression of their targets in a context-specific manner (Wittkopp and Kalay 2012).

Thus CREs enable the same toolkits to be redeployed at different times and in different developmental contexts, for example in different types of tissues or body segments. This plasticity allows morphology to evolve via mutations in CREs rather than in their protein-coding targets. Mutations in CREs not only preserve the structural integrity of the toolkit genes, they also mitigate the potential pleiotropic effects of mutations in the coding sequence of the toolkit gene(s) they regulate. The importance of CREs at macroevolutionary scales is still a matter of debate (Hoekstra and Coyne 2007; Lynch and Wagner 2008; Wagner and Lynch 2008). Still, genetic change in CREs remains a strong candidate for enabling microevolutionary change; that is, producing continuous variation in quantitative traits at the level of the population.

Evolution, Development, and Anthropology

The study of primate development and ontogeny has a long history in biological anthropology. In the late 19th century, European anatomists contributed a number of studies on prenatal and juvenile specimens of apes, monkeys, and even humans (reviewed in Schultz 1926). Many were based on the description of anatomy in primate fetuses, often obtained by chance from zoos or tropical expeditions. Although some were quite detailed (Deniker 1885), most tended to focus on a few readily measured, external characteristics such as body mass and head circumference (Figure 1.1), or more qualitative descriptions of the face, cranium, and soft anatomy. Detailed quantitative analyses were hampered by small sample sizes, which rarely exceeded half a dozen. Such small samples made it difficult to draw conclusions regarding variation among individuals, or even whether the few individuals described were representative of their species. Moreover, samples rarely included longitudinal ontogenetic

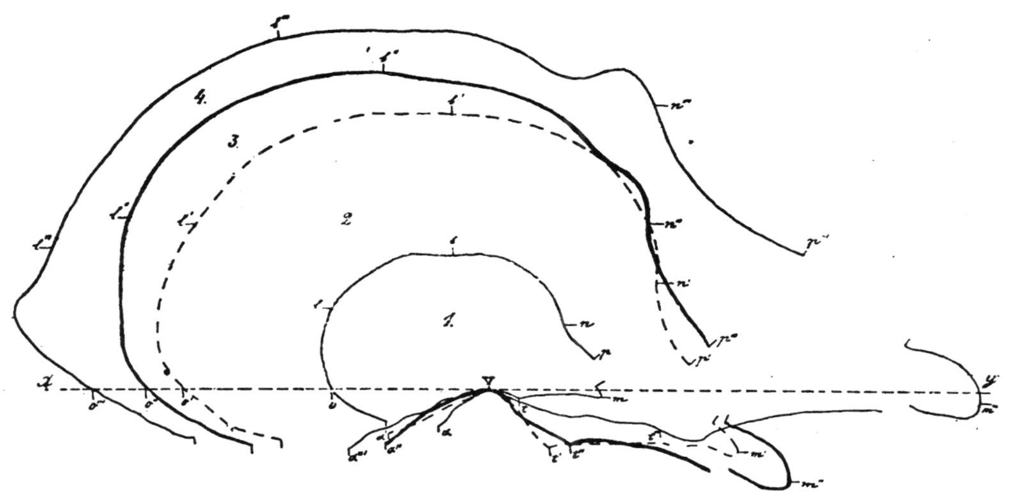

Figure 1.1. Figure from Deniker's 1885 study on cranial growth in hominoids. The figure depicts cranial shape changes over ontogeny in gorillas, starting with a fetus (1, center) to near-adult crania (4, outermost).

series, precluding a temporal dimension to the study of primate development. Most importantly, however, despite the prior publication of both Charles Darwin's *On the Origin of Species* and *The Descent of Man*, few of these studies were conducted in an explicitly evolutionary framework.

Still, these early studies made important contributions to the increasingly intertwined fields of human/primate development and evolution. These comparative studies were some of the earliest to illustrate key evolutionary developmental concepts, such as homology and heterochrony, as they apply to primates and humans. Taken together, they revealed that many interspecific differences in morphology observed in adults first appear prenatally, during fetal growth or even earlier in embryonic patterning. Beyond simply affirming a link between embryology, ontogeny, and evolution, this relatively simple observation led to the recognition that phenotypic divergence among primates, even among closely related taxa such as hominoids, is driven by organismal development, including not only observable differences in postnatal growth rates, but also earlier stages of embryonic and fetal development.

These studies laid the foundation for the more systematic and quantitative comparative ontogenetic studies of European anatomists in the first half of the 20th century, culminating in the work of Adolph Schultz. Between about 1920 and 1960, Schultz published close to 50 articles dealing with growth, development, and variability in different regions of the body, including the head, teeth, vertebral column, limbs, and internal organs in dozens of primate taxa (Howells 1977; Wood 1996). The importance of his contributions is not only in their quantitative nature, but also because, by the same token, they provided a detailed account of variation and variability in morphology within and among primates. Schultz published the bulk of this work before the Modern Synthesis, and long before the advent of molecular embryology. He thus lacked a framework for explaining how, mechanistically, the various parts of the vertebrate body are patterned, grow, and interact, and more importantly, how heritable variation in these complex traits arises. Nonetheless, his contributions highlight the fact that much can be learned about primate evolutionary diversification by studying patterns of growth and development among closely related taxa, and presage the importance of studying phenotypic variation and variability (i.e., the propensity of a trait to vary) for understanding the relationship between organismal development and evolution.

Following in the tradition of evolutionary embryologists and Adolph Schultz, much of the evo-devo research in anthropology in the second half of the 20th century continued to focus on comparative patterns of embryonic, fetal, and postnatal ontogeny within and among taxa. Specifically, many studies addressed heterochrony, or how differences in the timing and rate of development between different parts of the body, and/or between the same parts in different taxa, could explain the evolution of size and shape differences across species (i.e., allometry, Gould 1975, 1977; Alberch *et al.* 1979; Shea 1981, 1983; Shea and Bailey 1996; Minugh-Purvis and McNamara 2002; Mitteroecker *et al.* 2004). Heterochronic shifts are readily tractable with cross-sectional ontogenetic samples, even when the non-adult sample is limited to a few individuals or a few developmental stages. This is especially useful for studying primate evo-devo, because of the relative paucity of prenatal collections of primates.

Moreover, comparative studies of heterochrony, ontogenetic allometry, and evolutionary shifts in life history events using extant primates are useful for understanding the evolution of development as documented in the hominin fossil record, even when single juvenile samples are recovered (e.g., Alemseged *et al.* 2006). This presents an inherent advantage over a developmental genetics approach, which is challenging in a primate framework (see below).

Integration, Modularity, and Evolvability in Biological Anthropology

In parallel with comparative studies of allometry, heterochrony, and other ontogenetic processes in primates, the second half of the 20th century also saw the application of concepts such as morphological integration and modularity to the study of primate evolutionary developmental biology. Morphological integration and modularity describe the interconnectedness of different structures in the vertebrate body (Wagner 1995; Cheverud 1996a; Wagner and Altenberg 1996). These concepts are based on two simple observations: (1) the body of a complex organism is organized into discrete, internally consistent structures that share a common developmental origin or function (i.e., modules), (2) these modules do not develop independently, but rather must interact and grow in such a way that they preserve the proper function of the whole organism throughout its ontogeny (i.e., integration). From a quantitative perspective, modularity and integration are typically assessed in a hierarchical fashion through the magnitude of phenotypic covariation among traits. Specifically, sets of traits that covary more strongly in size and shape with each other than they do with other traits form a module. At the level of the whole organism, some of these modules interact with each other, for example due to their physical proximity, and grow in a coordinated fashion in response to changes in the overall size of the organism (i.e., growth, Hallgrímsson *et al.* 2002). In this sense, the whole organism is also integrated. A good example can be found in the mammalian skull: the cranial base, face, and neurocranium are neighboring structures that form independent modules with different developmental origins and functions, yet the growth of all three is coordinated even while the function of each part (e.g., mastication, sensory input) is preserved (e.g., Hallgrímsson *et al.* 2004, 2007; Bastir and Rosas 2005, 2009; Mitteroecker and Bookstein 2008).

The idea of modularity and integration traces its origins as far back as Georges Cuvier and Charles Darwin, both of whom recognized that body parts are functionally and developmentally correlated, to the extent that "when slight variations in one part occur, and are accumulated through natural selection, other parts become modified" (Darwin 1859:147). The idea that organisms are integrated entities made of correlated parts remained largely qualitative, until 1948, when Everett Olson and Robert Miller published their seminal volume *Morphological Integration* (Olson and Miller 1958). Olson and Miller proposed that both developmental and functional interactions were important sources of correlation among body parts, contributing towards building "integrated" organisms in which these parts function in concert with all others. Olson and Miller also developed the first quantitative methods, mostly based on statistical correlation, for empirically identifying groups of phenotypic traits that are more strongly integrated on the basis of shared developmental and/or functional factors. One of the examples they used were the teeth of owl monkeys (*Aotus*), revealing stronger patterns of integration (correlation) within teeth than, for example, between the lower and upper molars.

Since then, there have been many publications dealing with morphological integration in primates, beginning with the pioneering work of James Cheverud on the monkey cranium (Cheverud 1982, 1996b). To date, integration and modularity has been studied most extensively in the primate cranium (reviewed in Mitteroecker and Bookstein 2008) and dentition (reviewed in Grieco *et al.* 2013), but also increasingly in the postcranial skeleton (e.g., Young 2004; Young and Hallgrímsson 2005; Rolian 2009; Young *et al.* 2010; Lewton 2012). Many of these studies assess shape (co)variation using geometric morphometric tools, and determine the extent to which these patterns constrain or facilitate evolutionary change in skeletal form (evolvability, Rolian *et al.* 2010; Young *et al.* 2010; Grabowski 2011). These types of studies formulate and test models of integration/modularity based on genetic and developmental processes that, at least in theory, affect the strength of correlations among parts (e.g., pleiotropy); and in this sense, they are studies in evolutionary developmental anthropology. It is important to emphasize, however, that they are not empirical analyses of these genetic and developmental processes *per se*. Instead, most rely on existing developmental evidence from model organisms to test hypotheses regarding the outcome of development on patterns of integration, and what these mean for the evolvability of the primate and hominin skeleton (Rolian 2014).

Developmental Genetics: Next Steps and New Frontiers in Evo-Devo-Anthro

The field of evo-devo is just over 30 years old, and that of vertebrate evo-devo as applied to primates about half that age (e.g., Weiss and Buchanan, this volume). Within this timeframe, and due in no small part to technological breakthroughs in rapid and high-throughput genomics and developmental biology research tools, substantial progress has been made in understanding how processes of organismal development relate to the patterns of macroevolutionary diversity across vertebrate species. In terms of understanding mammalian evo-devo, the primary animal model system remains the very well characterized, now-universal laboratory rodent (e.g., Boughner, Lacruz, Martínez-Abadías *et al.*, Reno, this volume). The success of mouse (and other rodent models such as rats and voles), is due to the relative ease with which these small mammals can be housed and manipulated experimentally, including the benefit of relatively short gestational periods of less than a month and larger litters of about 8–12 pups.

Experimental developmental biology work of the sort routinely done using rodent models is largely impractical in primates, not least because of longer gestational periods, single-births, greater costs of housing, and ethical challenges in handling these larger animals. In this respect, EDA invariably has limitations in terms of collecting data directly from the developmental genetic processes that regulate primate ontogeny and on which selection can act. The news is not all bad, however, and there is one very good reason to continue to use rodent models for work in EDA: developmental processes in rodents are to a high degree homologous with those of other mammals, including primates (Rolian 2014; Reno, this volume). This homology encompasses morphological modularity in the skull and postcranial skeleton (e.g., Hallgrímsson *et al.* 2002, 2004; Willmore *et al.* 2009) and in the dentition (Chai *et al.* 2000; Jernvall and Thesleff 2012), to the extent that it is tenable to generate and use, for example, mouse mutants that replicate macroevolutionary change evidenced in the hominin fossil record to model the developmental-genetic processes that underlie phenotypic variation in primates living and extinct (Martínez-Abadías *et al.*, this volume). Thus it is possible to meaningfully address fundamental questions in paleoanthropology even if the data are collected from experimental model systems other than primates.

Yet even distantly related model organisms, such as yeast, share with vertebrates, including humans, fundamental cellular processes (e.g., chromatin assembly, Harkness 2005) that help build our understanding of these most basic units – cells – of body tissues and parts including organism maintenance, ageing, and life history. Also, non-mammalian models offer various unique advantages in terms of experimental techniques and options. For example, in mouse it is virtually impossible to perturb development *in utero* without sacrificing both the pregnant mother and her prenatal offspring, or time consuming and costly to create new mouse mutants via genetic engineering approaches. In contrast, the chick/avian model is ideally suited to prenatal *in ovo* surgeries and grafts; and not only does the zebrafish model offer transparent embryos as literal windows into development, but also fish morphogenesis can be perturbed quickly and effectively by adding varying amounts of mutagens to the tank water. Thus, using a variety of model organisms (i.e., yeast, worm, chick, frog, and zebrafish) to describe and experimentally manipulate highly conserved low-level developmental-genetic processes is a feasible and valid solution to gain new mechanistic insights into primate ontogeny and evolution. As other model organisms are adopted and characterized alongside new molecular genetic and developmental biology techniques, the options available to evo-devo anthropologists will no doubt increase as well.

Cross-discipline work is also important to get traction in research areas that to date have proven more intangible. As the degree of cognition and culture appears to distinguish anatomically modern *Homo sapiens* among other primates living and fossil, new frontiers in EDA could tackle the evolutionary-developmental mechanisms underlying behavioral traits such as complex language and abstract thought (Charvet and Finlay, Crespi and Leach, Lalueza-Fox, this volume). Also, the developmental-genetic basis of primate life history demands further attention as an absolutely longer life history period, due in no small part to a protracted childhood, appears to be another trait that is specific, if not unique, to *Homo sapiens* (Gunz, this volume). While the evolutionary patterns of life history variation among primates are now well established, how evolutionary changes in life history are genetically and developmentally regulated remain fascinating and underexplored topics in EDA. In this and all EDA contexts, knowledge of the fossil record and the major transitions it documents is paramount to frame, and then properly test, the right hypotheses.

A related challenge is how to accurately model ontogeny in fossil primates using what is known, or might yet be gleaned from, a relatively few existing fossil specimens. At first glance, the fossil record seems ill-suited to studies of evolution and development, as McNulty writes, "for the simple reason that fossils neither evolve nor develop" (McNulty 2012, p.488). In other words, we can't do developmental genetics or embryology with vertebrate fossils; and except perhaps for our most recent Neanderthal relatives (Green *et al.* 2010), evolutionary genomics in fossil hominins are also impossible. To this, we may add a further complication: with few exceptions (e.g., the Pleistocene site of Sima de los Huesos, De Castro *et al.* 2004; Rosas and Bastir 2004; Gomez-Robles and Polly 2012), hominin fossil samples are generally too small or too distant in time and space to derive reliable "population" estimates of phenotypic (co)variation patterns required for the types of integration and modularity analyses described above. Despite these limitations, there is still an important role for fossils in human evo-devo studies. Although fossil samples are small, more often than not consisting of single data points, they can serve as "yardsticks" by which to test or validate evolutionary hypotheses derived from developmental analyses in neontological taxa (Thewissen *et al.* 2012). For example, fossils can be used to: confirm if different patterns of integration observed among *living* primate taxa have influenced the evolutionary sequence of skeletal changes in our past (Young *et al.* 2010); or test development-based hypotheses about adaptive versus non-adaptive origins of derived skeletal features in hominins (Ackermann and Cheverud 2004; Rolian *et al.* 2010); and, more generally, validate models of life history and ontogenetic evolution within hominoids and hominids (e.g., Bromage 1989; Ackermann and Krovitz 2000; Dean *et al.* 2001; McNulty 2012).

Given the practical difficulties of working with primates in an evo-devo context, biological anthropologists can also explore the utility of *in silico* experiments to test predictions about developmental-evolutionary change during the morphogenesis of a body part (e.g., teeth, limbs), including over longer periods of time in a population (see, e.g., Salazar-Ciudad and Marin-Riera 2013; Rolian, this volume). These increasingly sophisticated computation-driven insights should help build useful theories about phenotypic and ontogenetic plasticity, and subsequently phenotypic evolution, in extant and extinct primates. Also, compared to many developmental genetic methods, classic approaches of studying phenotypic evolvability, integration, and variation using skeletal specimens (e.g., evolutionary quantitative genetics; geometric morphometrics; and other measures of morphological modules and their change) is straightforward and inexpensive with access to large collections, and thus sample populations that more likely capture normal ranges of variation. In this sense, these types of studies will likely continue to be fruitful avenues for research in EDA. These skeleton-based studies can also reasonably be applied to the fossil record, and – no less importantly – fossils used appropriately as yardsticks by which to gauge theories of integration.

As the power of *in vivo*, *in vitro*, and *in silico* techniques increases, another important step to advance EDA is not only for anthropologists to build expertise in developmental biology but also to talk and work directly beside developmental biologists, who continue to cultivate new experimental animal models and *in vivo* and *in vitro* techniques as well as transgenic and other genetic engineering approaches. Further, the unique ethos and methodologies of each field are driven by fundamental questions that reflect a collective curiosity about why species, as well as individuals, look different from each other; as well as why pathologies (i.e., extremes of phenotypic variation) occur and how they can be diminished or prevented.

Lastly, the field of Evolutionary Developmental Biology itself continues to change, witnessed, for example, by a recent push to incorporate ecological principles, as well as epigenetic influences and mechanical stimuli, into contemporary models of organismal development (Gilbert and Epel 2009; Jamniczky *et al.* 2010; Mammoto and Ingber 2010; Abouheif *et al.* 2014). These new directions and their unique breakthroughs will surely feed into EDA research; and potentially vice versa. For instance, persistent "black boxes" in our understanding include how natural selection (i.e., survival in a particular ecological context) shapes developmental process to effect evolutionary change; and how developmental processes beyond the genetic code constrain evolutionary potential. Beyond these broad questions, there also remain proximate questions regarding primate evolution and development for which we have few answers. This is undoubtedly a long list, and we mention but a few topics here. For example, considering the amount of data already available for other regions of the vertebrate skeleton (e.g., limbs and skulls), it is remarkable that we know relatively little regarding developmental evolution in the axial skeleton (Burke *et al.* 1995). In light of differences in vertebral patterning among hominoids and within the hominin lineage, using mouse models to understand, for example, the underlying mechanisms of homeotic changes in the axial skeleton could give useful insight into important events in human evolution such as the transition to bipedalism (Pilbeam 2004; Williams 2012; Young and Capellini, this volume). Another research area that remains underexplored in EDA, as in developmental biology more broadly, concerns soft tissues (Diogo and Wood, this volume). Little is known regarding the developmental evolution of soft tissues such as muscles, sensory tissues, and the integument, despite the tremendous amount of phenotypic diversity across primates in these traits (Hamrick 2003; Diogo *et al.* 2014). Thus we make the case that now, more than ever, those working or training to work in evolutionary developmental anthropology cast their intellectual nets widely to capture and then make use of as great a wealth of new methods, insights, and inspirations as possible. The end goal of this book is to stimulate interest and build momentum in the burgeoning field of Evolutionary Developmental Anthropology by sharing current EDA work to highlight scientific ground covered thus far as well as suggest new and needed forays into uncharted research territories.

References

Abouheif, E., Fave, M. J., Ibarraran-Viniegra, A. S. *et al.* 2014. Eco-evo-devo: The time has come. *Ecological Genomics: Ecology and the Evolution of Genes and Genomes* 781:107–125.

Ackermann, R. R., and Cheverud, J. M. 2004. Detecting genetic drift versus selection in human evolution. *Proceedings of the National Academy of Sciences, USA* 101:17946–17951.

Ackermann, R. R., and Krovitz, G. E. 2000. Morphometric analysis of craniofacial shape and growth patterns in *Australopithecus africanus*. *American Journal of Physical Anthropology* Suppl. 30:91.

Alberch, P., Gould, S. J., Oster, G. F., and Wake, D. B. 1979. Size and shape in ontogeny and phylogeny. *Paleobiology* 5:296–317.

Alemseged, Z., Spoor, F., Kimbel, W. H. *et al.* 2006. A juvenile early hominin skeleton from dikika, ethiopia. *Nature* 443:296–301.

Anderson, K. V., and Ingham, P. W. 2003. The transformation of the model organism: A decade of developmental genetics. *Nature Genetics* 33:285–293.

Bastir, M., and Rosas, A. 2005. Hierarchical nature of morphological integration and modularity in the human posterior face. *American Journal of Physical Anthropology* 128:26–34.

Bastir, M., and Rosas, A. 2009. Mosaic evolution of the basicranium in homo and its relation to modular development. *Evolutionary Biology* 36:57–70.

Bromage, T. G. 1989. Ontogeny of the early hominid face. *Journal of Human Evolution* 18:751–773.

Burke, A. C., Nelson, C. E., Morgan, B. A., and Tabin, C. 1995. *Hox* genes and the evolution of vertebrate axial morphology. *Development* 121:333–346.

Carroll, S. B. 2005. *Endless forms most beautiful: The new science of evo devo and the making of the animal kingdom*. New York: Norton.

Carroll, S. B. 2008. Evo-devo and an expanding evolutionary synthesis: A genetic theory of morphological evolution. *Cell* 134:25–36.

Carroll, S. B., Grenier, J. K., and Weatherbee, S. D. 2005. *From DNA to diversity: Molecular genetics and the evolution of animal design*. Oxford: Blackwell.

Chai, Y., Jiang, X. B., Ito, Y. *et al.* 2000. Fate of the mammalian cranial neural crest during tooth and mandibular morphogenesis. *Development* 127:1671–1679.

Cheverud, J. M. 1982. Phenotypic, genetic, and environmental morphological integration in the cranium. *Evolution* 36:499–516.

Cheverud, J. M. 1996a. Developmental integration and the evolution of pleiotropy. *American Zoologist* 36:44–50.

Cheverud, J. M. 1996b. Quantitative genetic analysis of cranial morphology in the cotton-top (*Saguinus oedipus*) and saddle-back (*S. fuscicollis*) tamarins. *Journal of Evolutionary Biology* 9:5–42.

Darwin, C. 1859. *On the origin of species by means of natural selection*. London: J. Murray.

De Castro, J. M. B., Martinon-Torres, M., Carbonell, E. *et al.* 2004. The atapuerca sites and their contribution to the knowledge of human evolution in europe. *Evolutionary Anthropology* 13:25–41.

Dean, C., Leakey, M. G., Reid, D. *et al.* 2001. Growth processes in teeth distinguish modern humans from *Homo erectus* and earlier hominins. *Nature* 414:628–631.

Deniker, J. 1885. Le développement du crâne chez le gorille. *Bulletins de la Société d'Anthropologie de Paris, IIIe Série,* 8:703–714.

Diogo, R., Molnar, J. L., and Smith, T. D. 2014. The anatomy and ontogeny of the head, neck, pectoral, and upper limb muscles of *Lemur catta* and *Propithecus coquereli* (primates): Discussion on the parallelism between ontogeny and phylogeny and implications for evolutionary and developmental biology. *Anatomical Record-Advances in Integrative Anatomy and Evolutionary Biology* 297:1435–1453.

Gilbert, S. F., and Epel, D. 2009. *Ecological developmental biology: Integrating epigenetics, medicine, and evolution*. Sunderland, MA: Sinauer Associates.

Gomez-Robles, A., and Polly, P. D. 2012. Morphological integration in the hominin dentition: Evolutionary, developmental, and functional factors. *Evolution* 66:1024–1043.

Goodman, F. R. 2002. Limb malformations and the human *Hox* genes. *American Journal of Medical Genetics* 112:256–265.

Goodman, F. R., and Scambler, P. J. 2001. Human *Hox* gene mutations. *Clinical Genetics* 59:1–11.
Gould, S. J. 1975. Allometry in primates, with emphasis on scaling and evolution of brain. *Contributions to Primatology* 5:244–292.
Gould, S. J. 1977. *Ontogeny and phylogeny*. Cambridge, MA: Belknap Press of Harvard University Press.
Grabowski, M. W. 2011. Morphological integration and correlated evolution in the hominin pelvis. *American Journal of Physical Anthropology* 144:146–146.
Green, R. E., Krause, J., Briggs, A. W. *et al.* 2010. A draft sequence of the neandertal genome. *Science* 328:710–722.
Grieco, T. M., Rizk, O. T., and Hlusko, L. J. 2013. A modular framework characterizes micro- and macroevolution of old world monkey dentitions. *Evolution* 67:241–259.
Hall, B. K. 1999. *Evolutionary developmental biology*. London: Chapman & Hall.
Hall, B. K. 2007. Homoplasy and homology: Dichotomy or continuum? *Journal of Human Evolution* 52:473–479.
Hallgrímsson, B., Lieberman, D. E., Liu, W., Ford-Hutchinson, A. F., and Jirik, F. R. 2007. Epigenetic interactions and the structure of phenotypic variation in the cranium. *Evolution & Development* 9:76–91.
Hallgrímsson, B., Willmore, K., Dorval, C., and Cooper, D. M. L. 2004. Craniofacial variability and modularity in macaques and mice. *Journal of Experimental Zoology Part B: Molecular and Developmental Evolution* 302B:207–225.
Hallgrímsson, B., Willmore, K., and Hall, B. K. 2002. Canalization, developmental stability, and morphological integration in primate limbs. *Yearbook of Physical Anthropology* 45:131–158.
Hamrick, M. W. 2003. Evolution and development of mammalian limb integumentary structures. *Journal of Experimental Zoology Part B: Molecular and Developmental Evolution* 298B:152–163.
Harkness, T. A. A. 2005. Chromatin assembly from yeast to man: Conserved factors and conserved molecular mechanisms. *Current Genomics* 6:227–240.
Hoekstra, H. E., and Coyne, J. A. 2007. The locus of evolution: Evo devo and the genetics of adaptation. *Evolution* 61:995–1016.
Howells, W. W. 1977. Schultz, A. 1891–1976. *American Journal of Physical Anthropology* 46:191–195.
Jamniczky, H. A., Boughner, J. C., Gonzalez, P. N. *et al.* 2010. Mapping the epigenetic landscape: Rediscovering Waddington in the post-genomic age. *Integrative and Comparative Biology* 50:E82–E82.
Jernvall, J., and Thesleff, I. 2012. Tooth shape formation and tooth renewal: Evolving with the same signals. *Development* 139:3487–3497.
King, M. C., and Wilson, A. C. 1975. Evolution at two levels in humans and chimpanzees. *Science* 188:107–116.
Krumlauf, R. 1994. *Hox* genes in vertebrate development. *Cell* 78:191–201.
Laublicher, M. D. 2007. Does history recapitulate itself? Epistemological reflections on the origins of evolutionary developmental biology. *In:* M. D. Laublicher and J. Maienschein (eds) *From embryology to evo-devo: A history of developmental evolution.* Cambridge, MA: MIT Press.
Lewis, E. B. 1978. Gene complex controlling segmentation in *Drosophila*. *Nature* 276:565–570.
Lewton, K. L. 2012. Evolvability of the primate pelvic girdle. *Evolutionary Biology* 39:126–139.
Lynch, V. J., and Wagner, G. P. 2008. Resurrecting the role of transcription factor change in developmental evolution. *Evolution* 62:2131–2154.

Maienschein, J. 2007. To evo-devo through cells, embryos and morphogenesis. *In:* M. D. Laublicher and J. Maienschein (eds) *From embryology to evo-devo: A history of developmental evolution*. Cambridge, MA: MIT Press.

Mallo, M., Wellik, D. M., and Deschamps, J. 2010. *Hox* genes and regional patterning of the vertebrate body plan. *Developmental Biology* 344:7–15.

Mammoto, T., and Ingber, D. E. 2010. Mechanical control of tissue and organ development. *Development* 137:1407–1420.

Mark, M., Rijli, F. M., and Chambon, P. 1997. Homeobox genes in embryogenesis and pathogenesis. *Pediatric Research* 42:421–429.

Mayr, E., and Provine, W. B. 1998. *The evolutionary synthesis: Perspectives on the unification of biology*. Cambridge, MA: Harvard University Press.

Mcnulty, K. P. 2012. Evolutionary development in *Australopithecus africanus*. *Evolutionary Biology* 39:488–498.

Minugh-Purvis, N., and Mcnamara, K. 2002. *Human evolution through developmental change*. Baltimore: Johns Hopkins University Press.

Mitteroecker, P., and Bookstein, F. 2008. The evolutionary role of modularity and integration in the hominoid cranium. *Evolution* 62:943–958.

Mitteroecker, P., Gunz, P., Bernhard, M. *et al.* 2004. Comparison of cranial ontogenetic trajectories among great apes and humans. *Journal of Human Evolution* 46:679–697.

Olson, E. C., and Miller, R. L. 1958. *Morphological integration*. Chicago: University of Chicago Press.

Pilbeam, D. 2004. The anthropoid postcranial axial skeleton: Comments on development, variation, and evolution. *Journal of Experimental Zoology Part B: Molecular and Developmental Evolution* 302B:241–267.

Prud'homme, B., Gompel, N., and Carroll, S. B. 2007. Emerging principles of regulatory evolution. *Proceedings of the National Academy of Sciences, USA* 104:8605–8612.

Raff, R. A. 2000. Evo-devo: The evolution of a new discipline. *Nature Reviews Genetics* 1:74–79.

Rolian, C. 2009. Integration and evolvability in primate hands and feet. *Evolutionary Biology* 36:100–117.

Rolian, C. 2014. Genes, development, and evolvability in primate evolution. *Evolutionary Anthropology* 23:93–104.

Rolian, C., Lieberman, D. E., and Hallgrímsson, B. 2010. The coevolution of human hands and feet. *Evolution* 64:1558–1568.

Rosas, A., and Bastir, M. 2004. Geometric morphometric analysis of allometric variation in the mandibular morphology of the hominids of Atapuerca, Sima de los Huesos site. *Anatomical Record Part A: Discoveries in Molecular Cellular and Evolutionary Biology* 278A:551–560.

Salazar-Ciudad, I., and Marin-Riera, M. 2013. Adaptive dynamics under development-based genotype-phenotype maps. *Nature* 497:361–364.

Schultz, A. 1926. Fetal growth of man and other primates. *Quarterly Review of Biology* 1:465–521.

Shea, B. T. 1981. Relative growth of the limbs and trunk in the African apes. *American Journal of Physical Anthropology* 56:179–201.

Shea, B. T. 1983. Allometry and heterochrony in the African apes. *American Journal of Physical Anthropology* 62:275–289.

Shea, B. T., and Bailey, R. C. 1996. Allometry and adaptation of body proportions and stature in African pygmies. *American Journal of Physical Anthropology* 100:311–340.

Shubin, N., Tabin, C., and Carroll, S. 1997. Fossils, genes and the evolution of animal limbs. *Nature* 388:639–648.

Thewissen, J. G. M., Cooper, L. N., and Behringer, R. R. 2012. Developmental biology enriches paleontology. *Journal of Vertebrate Paleontology* 32:1223–1234.

Tournierlasserve, E., Odenwald, W. F., Garbern, J. *et al.* 1989. Remarkable intron and exon sequence conservation in human and mouse homeobox *Hox* 1.3 genes. *Molecular and Cellular Biology* 9:2273–2278.

Wagner, G. P. 1995. Adaptation and the modular design of organisms. *Advances in Artificial Life* 929:317–328.

Wagner, G. P., and Altenberg, L. 1996. Complex adaptations and the evolution of evolvability. *Evolution* 50:967–976.

Wagner, G. P., and Lynch, V. J. 2008. The gene regulatory logic of transcription factor evolution. *Trends in Ecology & Evolution* 23:377–385.

Williams, S. A. 2012. Variation in anthropoid vertebral formulae: Implications for homology and homoplasy in hominoid evolution. *Journal of Experimental Zoology Part B: Molecular and Developmental Evolution,* 318B:134–147.

Willmore, K. E., Roseman, C. C., Rogers, J. *et al.* 2009. Comparison of mandibular phenotypic and genetic integration between baboon and mouse. *Evolutionary Biology* 36:19–36.

Wittkopp, P. J., and Kalay, G. 2012. *Cis*-regulatory elements: Molecular mechanisms and evolutionary processes underlying divergence. *Nature Reviews Genetics* 13:59–69.

Wood, B. 1996. Hominid palaeobiology: Have studies of comparative development come of age? *American Journal of Physical Anthropology* 99:9–15.

Young, N. M. 2004. Modularity and integration in the hominoid scapula. *Journal of Experimental Zoology Part B: Molecular and Developmental Evolution* 302B:226–240.

Young, N. M., and Hallgrímsson, B. 2005. Serial homology and the evolution of mammalian limb covariation structure. *Evolution* 59:2691–2704.

Young, N. M., Wagner, G. P., and Hallgrímsson, B. 2010. Development and the evolvability of human limbs. *Proceedings of the National Academy of Sciences, USA* 107:3400–3405.

Zhu, J., He, F. H., Hu, S. N., and Yu, J. 2008. On the nature of human housekeeping genes. *Trends in Genetics* 24:481–484.

Chapter 2

Chondrocranial Growth, Developmental Integration and Evolvability in the Human Skull

Neus Martínez-Abadías[1,2,3], Mireia Esparza[3], Torstein Sjøvold[4] and Benedikt Hallgrímsson[5]

[1] *Centre for Genomic Regulation (CRG), The Barcelona Institute of Science and Technology, Dr. Aiguader, Barcelona, Spain*

[2] *Universitat Pompeu Fabra (UPF), Barcelona, Spain*

[3] *Secció d'Antropologia, Departament de Biologia Animal, Universitat de Barcelona, Barcelona, Spain*

[4] *Osteologiska Enheten, Stockholms Universitet, Stockholm, Sweden*

[5] *McCaig Institute for Bone and Joint Research, Alberta Children's Hospital Research Institute, Department of Cell Biology and Anatomy, Faculty of Medicine, University of Calgary, Calgary, Canada*

Introduction

The vertebrate skull exhibits a remarkable morphological diversity and the human skull represents one of the more interesting examples. Human skulls are unusual among mammals due to the combination of a large neurocranium, small face, and highly flexed cranial base. To understand the evolution of this characteristic craniofacial shape, it is also necessary to understand its developmental basis. The skull is both integrated and modular, which profoundly influences how natural selection can act on craniofacial shape variation to produce evolutionary changes in the shape of the head (Hallgrímsson *et al.* 2007a; Hallgrímsson and Lieberman 2008; Lieberman 2011). Previous work using mouse mutants with perturbed craniofacial development has shown that changes to brain and chondrocranial growth produce patterns of highly integrated change in craniofacial shape (Hallgrímsson *et al.* 2009). In this chapter, we use a combination of mouse mutants in which the growth of the chondrocranium is perturbed and a human cranial collection with known relatedness among individuals to investigate how this central developmental determinant of craniofacial covariation structure influences the evolvability of the human skull.

Selection acts on phenotypic variation, but evolutionary change is dependent on the underlying genetic variation and the mapping of phenotypic variation to genetic variation (Lynch and Walsh 1998; Hansen and Houle 2008). Quantitative genetic tools can be used to predict how complex phenotypes respond to selection in phylogenetic contexts (Steppan *et al.* 2002; Hansen and Houle 2008). Applying these

Developmental Approaches to Human Evolution, First Edition. Edited by Julia C. Boughner and Campbell Rolian.

tools to the human skull requires thorough exploration of the determinants and patterns of genetic variation and covariation (Cheverud 1982; Ackermann and Cheverud 2004; Hallgrímsson *et al.* 2007b; Hallgrímsson and Lieberman 2008; Hallgrímsson *et al.* 2009). This is because the ability of the skull to respond to selection to produce evolutionary change depends critically on the structure of variation and covariation within the skull. Variation and covariation structure depends both on the variation in developmental processes that is present in populations, and on the tendency of those developmental processes to integrate that variation into patterns of covariation (Hallgrímsson *et al.* 2009).

We have argued in the past that cranial covariation structure is determined to a substantial degree by key developmental processes such as chondrocranial and brain growth and facial prominence outgrowth (Hallgrímsson *et al.* 2007a, 2009). Such processes drive patterns of integrated shape change that are distributed throughout the skull (Hallgrímsson *et al.* 2009; Martínez-Abadías *et al.* 2012a). Processes that determine covariation patterns, however, act at different times during development, altering or obscuring the effects of processes that occurred earlier (Hallgrímsson *et al.* 2009). Observed covariance structures are the product of the overlay of multiple such processes. Importantly, covariance structure can evolve through changes in the relative amount of variance in covariance-generating processes, through changes to the developmental interactions among processes, or the addition or elimination of processes. These are the key elements of the Palimpsest Model of morphological integration (Hallgrímsson *et al.* 2009).

Using the Palimpsest Model, we have used perturbations to craniofacial development in mouse models to investigate the processes that determine covariation structure. With this approach, we have shown that mutations that influence the growth of the chondrocranium produce a common pattern of integrated shape change throughout the mouse skull (Hallgrímsson *et al.* 2006, 2007a, 2009). We have further shown that this same integrated pattern is also a prominent component of covariation structure in the human skull (Martínez-Abadías *et al.* 2012a). Specifically, this study showed that the shape vectors that characterize the integrated effects of variation in chondrocranial growth in mouse mutants also correspond to a significant axis of covariation in humans, explaining >7% of the total shape variance in a large sample of human crania.

The axis of covariation generated by variation in chondrocranial growth is particularly relevant to understanding human evolution because of its relationship to brain size. The spatial packing hypothesis holds that the relative size of the brain and the cranial base impacts the overall shape of the skull. As the brain enlarges relative to the length of the cranial base, the cranial base flexes and the face is pulled posteriorly to lie more directly inferior to the anterior cranial fossa (Biegert 1963). This hypothesis has been tested and found to be consistent with comparative data in primates including hominids (Ross and Ravosa 1993; Lieberman *et al.* 2000; Bastir *et al.* 2010). Hallgrímsson and Lieberman have also tested the hypothesis in mouse mutants with altered growth of the chondrocranium and brain and found that increasing brain size or reducing chondrocranial size produces the predicted changes in the basicranial angle in mice (Hallgrímsson *et al.* 2007a; Hallgrímsson and Lieberman 2008; Lieberman *et al.* 2008).

These same studies have also shown that the widths of the face, cranial base, and neurocranium are highly integrated.

There are strong reasons to believe variation in the growth and size of the chondrocranium plays a central role in determining the overall variation of the shape of the skull. In other words, the causal arrows flow from the chondrocranium to the dermatocranial portions of the neurocranium and face rather than the reverse. First, the chondrocranium, and particularly the basicranium, develops earlier and completes a larger portion of its growth earlier than the face or neurocranium (De Beer 1937; Bastir *et al.* 2006). Secondly, the chondrocranium grows at synchondroses which resemble the growth plates of long bones (De Beer 1937; Scott 1958). These "cranial growth plates" are sites at which intrinsically regulated growth occurs. The fact that mutations can influence chondrocranial growth via effects on these growth centers (Young *et al.* 2006; Matsushita *et al.* 2009; Di Rocco *et al.* 2014; Heuzé *et al.* 2014) suggests strongly that the chondrocranium exhibits intrinsically regulated growth. Further, the demonstration that such mutations also have effects that are not confined to the chondrocranium but rather distributed throughout the skull speaks to the ability of the chondrocranium to drive variation throughout the skull (Hallgrímsson *et al.* 2006, 2007a).

In this study, we build on the central and integrative role of chondrocranial growth in the development of the skull and ask the question of how human craniofacial shape would respond to selection on different aspects of basicranial form. We use evolutionary quantitative genetics and a unique human cranial collection in which relatedness can be reconstructed from church records (Sjøvold 1984; Martínez-Abadías *et al.* 2009a, 2009b, 2012a, 2012b) to estimate the response of craniofacial shape to particular selection gradients. Specifically, we test three selection gradients that mimic the effects of (i) the brachymorph mutation which exhibits reduced chondrocanial growth (Hallgrímsson *et al.* 2006), (ii) a shorter cranial base, and (iii) a more flexed cranial base. We show that selection along these gradients produces globally integrated patterns of response for craniofacial shape that have significant implications for the potential role of variation in chondrocranial growth in human evolution.

Materials and Methods

We analyzed a sample of 390 complete crania from the ossuary of Hallstatt (Austria), which provides a large collection of human skulls with associated genealogical data. Skulls can be individually identified thanks to their decorations, which usually include the names of the individuals. Parish records permit the reconstruction of genealogical relationships, making it possible to estimate directly key quantitative genetic parameters (such as the heritability and the genetic variance-covariance G matrix) for skull shape. The sampled individuals were mainly adults (91%) from both sexes (41% females; 59% males), and a small proportion of the skulls showed slight dysmorphologies (12%). To reconstruct the genealogies of the Hallstatt population we compiled the complete records of births, marriages, and deaths from 1602 to 1900, which included 18,134 individuals. A total of 350 analyzed skulls fall into the

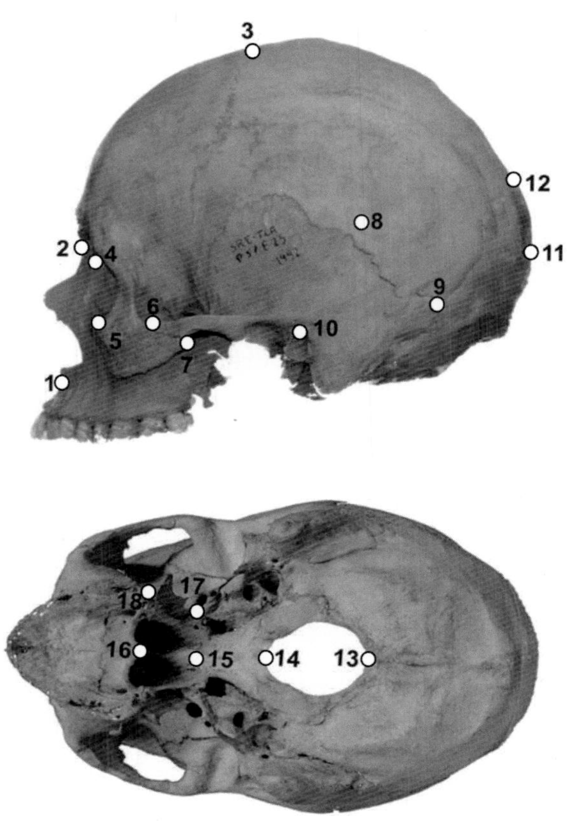

Figure 2.1. Three-dimensional (3D) landmark coordinates representing human left hemiskull. Modified from Martínez-Abadías *et al.* (2012a).

extended and multigenerational genealogies. A more detailed description of this sample can be found elsewhere (Sjøvold 1984; Martínez-Abadías *et al.* 2009a).

Human skull shape was characterized using the three-dimensional coordinates of a set of 18 anatomical landmarks distributed over the left side of the skull that were recorded with a Microscribe G2X digitizer (Solution Technologies, Inc., Oella, MD, USA) (Figure 2.1). Landmark definitions are provided in Table 2.1. This sample is homologous to a dataset including mouse strains of wildtype *C57Bl/6J* and mutant brachymorph mouse model (*Papps2-/Papps2-*, the Jackson Laboratory) that we previously used to assess the phenotypic effect of a genetic mutation altering chondrocranial growth and whose results will be used in the present analysis (Hallgrímsson *et al.* 2006; Martínez-Abadías *et al.* 2012b).

Geometric morphometric techniques were used to capture size and shape variation from the coordinate data representing the human skull. Skull size was estimated as centroid size, the square root of the sum of squared distances of all the landmarks of a skull from their center of gravity (Dryden and Mardia 1998). To extract shape information, we used a Generalized Procrustes Analysis (GPA) superimposition, a procedure that superimposes configurations of landmarks by shifting them to a

Table 2.1. Anatomical definitions of landmarks displayed in Figure 2.1.

Code	Landmark	Definition
1	Subspinale	Deepest point seen in the profile below the anterior nasal spine
2	Nasion	Midline point where the two nasal bones and the frontal intersect
3	Bregma	Ectocranial point where the coronal and sagittal sutures intersect
4	Maxillofrontale	Point where the anterior lacrimal crest meets the fronto-maxillary suture (left)
5	Zygoorbitale	Point where the orbital rim intersects the zygomatico-maxillary suture (left)
6	Superior zygo-temporal	Superior point of the suture between temporal and zygomatic bones (left)
7	Inferior zygo-temporal	Inferior point of the suture between temporal and zygomatic bones (left)
8	Euryon	Most lateral point of the braincase perpendicular to the sagittal plane (left)
9	Asteryon	Point where the lamboidal, parietal, and occipital sutures meet (left)
10	Porion	Most superior point on the margin of the external auditory meatus (left)
11	Opisthocranion	Most posterior point of the skull in the medial sagittal plane
12	Lambda	Midline point of the intersection of the sagittal and lamboidal sutures
13	Opisthion	Midline point on the posterior margin of the foramen magnum
14	Basion	Midline point on the anterior margin of the foramen magnum
15	Hormion	Most posterior midline point on the vomer
16	Posterior nasal spine	Point on vomer-palatine junction
17	Foramen ovale	Most lateral point on the margin of the foramen ovale (left)
18	Alveolar point	Posterior limit of the maxillary alveolar arch at the pterygo-alveolar suture

common position, rotating and scaling them to a standard size until a best fit of corresponding landmarks is achieved (Dryden and Mardia 1998). We performed a principal component (PC) analysis on the resulting Procrustes-coordinates to reduce the dimensionality of the data, which was necessary due to computational limitations in the quantitative genetic analysis. All morphometric analyses were performed using MorphoJ 1.06a (Klingenberg 2011).

From the landmark coordinates we also estimated two traits that can be directly associated with chondrocranial growth: chondrocranial length, measured as the linear distance between the landmarks hormion and opisthocranion; and an estimate of cranial base angle (CBA), measured as the angle between the landmarks nasion, hormion, and basion. Note that we used this CBA measure because no internal landmarks (such as sella) were available for the Hallstatt sample.

The effects of chondrocranial length and cranial base flexion on human cranial shape variation were further explored by computing multivariate regressions of skull shape (represented by the Procrustes shape coordinates) on the predictor variables (chondrocranial length and cranial base flexion). The regressions were pooled within sexes to account for sexual dimorphism. Finally, we estimated the correlation between chondrocranial length and cranial base flexion by linear regression between the two traits and ran a permutation test (10,000 randomization rounds) against the null hypothesis of independence. To determine whether this relationship is conserved among mammals, we separately computed the linear regression between chondrocranial length and cranial base flexion using the wildtype *C57Bl/6J* and the mutant brachymorph (*Papps2-/Papps2-*, the Jackson Laboratory) mice samples. For more details about these mice strains see previous work (Hallgrímsson *et al.* 2006, 2009; Martínez-Abadías *et al.* 2012b).

Quantitative Genetic Analysis

To estimate the phenotypic, genetic, and environmental components of variation of skull shape we combined multivariate methods of geometric morphometrics and quantitative genetics following Klingenberg and Leamy (2001). This approach preserves the multivariate nature of shape while detecting complex patterns of shape change (Klingenberg and Monteiro 2005). Skull shape is thus treated as a whole, preserving the anatomical relationships between the different structures and regions that made up the skull and accounting for the covariation patterns among them.

According to standard approaches to quantitative genetics, the phenotypic variation of a trait ($\mathbf{V_P}$) can be decomposed into its components of genetic ($\mathbf{V_G}$) and environmental ($\mathbf{V_E}$) variation by the expression $\mathbf{V_P} = \mathbf{V_G} + \mathbf{V_E}$ (Falconer and Mackay 1996; Lynch and Walsh 1998). The genetic variation can be further apportioned into its additive ($\mathbf{V_A}$), dominance ($\mathbf{V_D}$) and $\mathbf{V_I}$ (gene interaction or epistasis term). This decomposition relies on the phenotypic resemblance between relatives and it is possible provided there are associated genetic or demographic data. The phenotypic variation is thus obtained from the direct measurement of the trait; the additive genetic variation is estimated as the phenotypic covariation between relatives; and finally, all the variation that cannot be explained by familial relationship is considered as residual environmental variation.

As the input for the multivariate quantitative genetic analysis of skull shape, we used the first 24 shape PCs, which account for 90.9% of the total phenotypic skull shape variation of the Hallstatt sample. To decompose the **P** matrix, the phenotypic variance-covariance matrix for skull shape, into its genetic (**G** matrix) and environmental (**E** matrix) components we used restricted maximum likelihood (REML) methods. These analytical methods are advantageous in contrast to parent-offspring regression or sib analyses because they incorporate multigenerational information from unbalanced datasets (Konigsberg 2000). Moreover, they are not limited by assumptions of non-assortative mating, inbreeding, or selection (Kruuk 2004). REML methods are usually applied under a mixed linear model, the so-called animal model, which jointly accounts for fixed and random effects in order to describe the phenotype of each individual (Lynch and Walsh 1998). The phenotypic variance is broken down into its components of additive genetic value and other random environmental and fixed effects. The components of variance are estimated by an iterative procedure that maximizes the likelihood of observing the actual data (Lynch and Walsh 1998). REML methods maximize only the portion of the likelihood that does not depend on the fixed effects, which are assumed to be known without error (e.g. sex, size, etc.). For this analysis we defined a multivariate model that included the first 24 shape PCs as random dependent variables, skull centroid size and age as covariates, and an individual's sex and deformation status as fixed effects. REML analyses were performed as implemented by the software package VCE6 (Groeneveld *et al.* 2010).

The resulting genetic (**G**), phenotypic (**P**), and environmental (**E**) covariance matrices were imported into MorphoJ 1.06 (Klingenberg 2011) and converted back to the space of the original landmark coordinates from the coordinate system of PC scores (Klingenberg and Leamy 2001). To examine the main patterns of phenotypic and genetic variation, we performed a PCA on each of the **G** and **P** covariance matrices. To quantify the overall relatedness of the **G** and **P** matrices, we computed the matrix correlation between them and evaluated it with a matrix permutation test, as adapted for geometric morphometrics (Klingenberg and McIntyre 1998).

As a complementary analysis, we performed a univariate quantitative genetic analysis to estimate the heritability of chondrocranial length and CBA. The narrow-sense heritability ($\mathbf{h}^2$) of each trait was computed as the proportion of phenotypic variance attributable to additive genetic effects, $\mathbf{h}^2 = \mathbf{V}_A/\mathbf{V}_P$ (Lynch and Walsh 1998). We computed the heritabilities of each trait using several models that accounted for different fixed effects or covariates, such as sex, deformation, and age, using the SOLAR 7.2.5 software package (Almasy and Blangero 1998). SOLAR tests the significance of each covariate and computes the amount of variation explained by the significant ones at the $p<0.10$ level. The final model computed the narrow sense heritabilities retaining the significant covariates only.

Response to Selection

To estimate the response to selection we applied the multivariate breeder's equation (Lande 1979; Lande and Arnold 1983) as modified for geometric morphometric data (Klingenberg and Leamy 2001),

$$\Delta\mu = \mathbf{GP}^{-1}\mathbf{s} = \mathbf{G}\beta$$

where $\boldsymbol{\Delta\mu}$ is the response to selection; **G** is the additive genetic covariance matrix; **P** is the phenotypic covariance matrix; **s** is the selection differential; and $\boldsymbol{\beta}$ is the selection gradient. The selection response $\boldsymbol{\Delta\mu}$ is a vector that reflects the shape change between the phenotypic trait mean after and before the selection episode. The selection differential **s** and the selection gradient $\boldsymbol{\beta}$ specify the magnitude and direction of selection. Together with estimates of **G** and **P**, either of these can be used as selection vectors to estimate the response to selection (Klingenberg and Monteiro 2005, Martínez-Abadías *et al.* 2012a).

From the multivariate breeder's equation (Lande 1979; Lande and Arnold 1983) it is clear that the evolution of a given trait is not only a function of the additive genetic variance and selection on that trait, but also of the genetic covariance between it and other genetically linked traits (Lande 1979; Lande and Arnold 1983). Therefore, the covariation patterns between traits reflected in **G** can be considered as genetic constraints because they will deflect the response vector $\boldsymbol{\Delta\mu}$ from the originally selected direction of change imposed by the selection gradient $\boldsymbol{\beta}$ (Hansen and Houle 2008).

We applied this quantitative genetic model for evolutionary responses to selection to explore how human skull shape would respond to selective events altering chondrocranial growth as well as chondrocranial length and cranial base angle. To do this we defined three selection gradients ($\boldsymbol{\beta}$) simulating the following effects: (1) the effect of a mutation in the phosphoadenosine-phosphosulfate synthetase 2 gene, which in the inbred mice strain brachymorph mouse model (*Papps2-/Papps2-*, the Jackson Laboratory) affects chondrocranial development by reducing its growth (Kurima *et al.* 1998; ul Haque *et al.* 1998); (2) a shortening of the cranial base; and (3) an increase of cranial base flexion. The selection gradients were constructed using a graphical user interface in MorphoJ (Klingenberg 2011), which enables the user to drag the landmark points of the mean shape configuration to specify a shape change. To simulate the effect of the *Papps2* mutation we computed the average effect of the genetic mutation as the difference between the average shape of a reference *C57Bl/6J* wildtype strain and the average shape of the brachymorph mutated strain, as previously done in Martínez-Abadías *et al.* (2012b). To simulate shortening of the cranial base we shifted the landmarks hormion and opisthocranion towards more posterior and anterior positions, respectively. To simulate an increase in cranial base flexion we shifted the hormion towards a more superior position.

For each simulation, the specified landmark shifts representing $\boldsymbol{\beta}$ were projected onto the tangent space to shape space (Dryden and Mardia 1998) to ensure that the selection gradient was in the same space as the variation characterized by the **G** and **P** matrices. This can result in smaller shifts of other landmarks to compensate for changes in overall position, orientation, and size. To make shape changes visible, we amplified the magnitude of selection gradients of the respective shape variable.

Finally, as the predicted selection response $\boldsymbol{\Delta\mu}$ is usually in a different direction from the direction of selection that was entered into the analysis, we decomposed this total response into components of direct and correlated response to selection following Klingenberg and Leamy (2001). The direct response is a scaled version of the selection gradient and these two vectors are in the same direction in shape space; whereas the correlated response is a vector perpendicular to the direct response. All the magnitudes of the responses to selection are measured in Procrustes distance units.

Results

Chondrocranial Dimensions

The multivariate regression of human hemicranial shape on chondrocranial length showed that this trait explains 6.8% of total morphological variation ($p<0.0001$); whereas CBA explains a lower but still significant percentage (2.8%; $p<0.0001$)

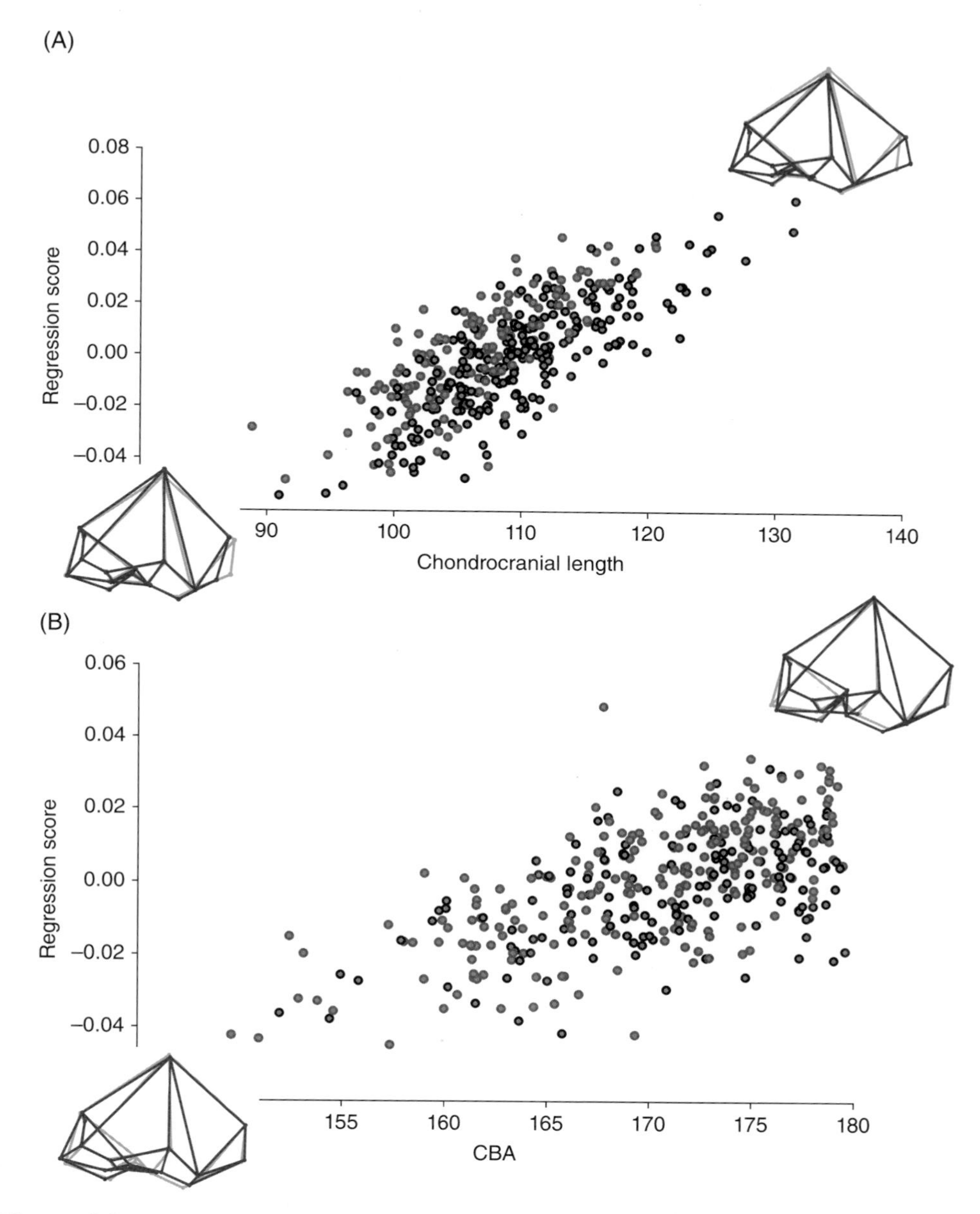

Figure 2.2. Multivariate regression of chondrocranial length (A) and cranial base angle (CBA) (B) on cranial shape in a modern human population. Regressions are pooled within sexes (orange females, blue males).

(Figure 2.2, color plate). The association between chondrocranial length and CBA differs between humans and mice (Figure 2.3). In humans there is a positively low but significant correlation between chondrocranial length and CBA (r^2=0.03; p=0.0005) (Figure 2.3A). In mice there is a moderate and negative significant correlation between chondrocranial length and CBA, especially in the mutant brachymorph

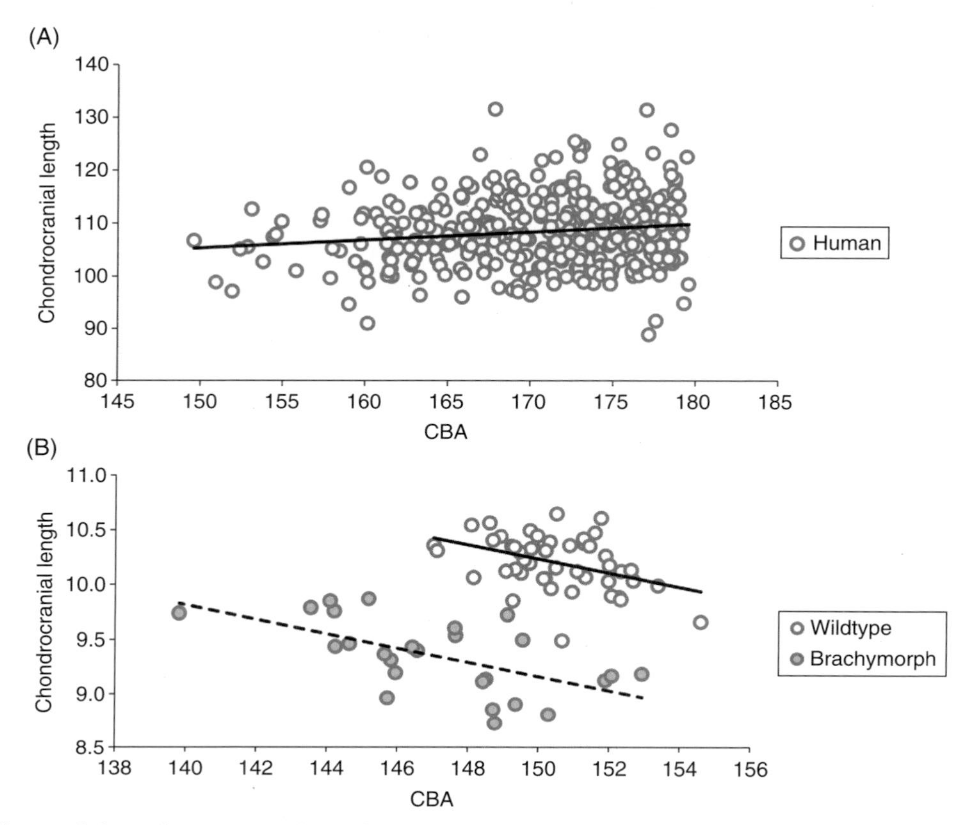

Figure 2.3. Linear regression of cranial base angle (CBA) on chondrocranial length in humans (A) and mice (B).

sample (r^2=0.34; p=0.0023), which explains as much as twice the total morphological variation compared to the *C57Bl/6J* wildtype sample (r^2=0.16; p=0.0033) (Figure 2.3B).

Patterns of Genetic and Phenotypic Variation

PC analysis showed that the genetic variation in the human sample is concentrated over the first 15 dimensions. The first PC accounts for 25.7% of the total variance and the subsequent eigenvalues decline rapidly, with the last seven PCs (from PC16 to PC 23) accounting for less than 0.05 of the total variance in the **G** matrix (Figure 2.4A). For the phenotypic covariance matrix, variation is more evenly distributed across many dimensions of the shape space and the decline of total variance explained by each eigenvector is more gradual (Figure 2.4B).

The shape changes associated with the first PC of the **G** matrix (Figure 2.4A) include a retraction of the lower face and expansion of the cranial vault, as well as flexion of the skull (this description is of a change of the PC1 in the negative direction). The shape changes associated with the first PC of the **P** matrix (Figure 2.4B)

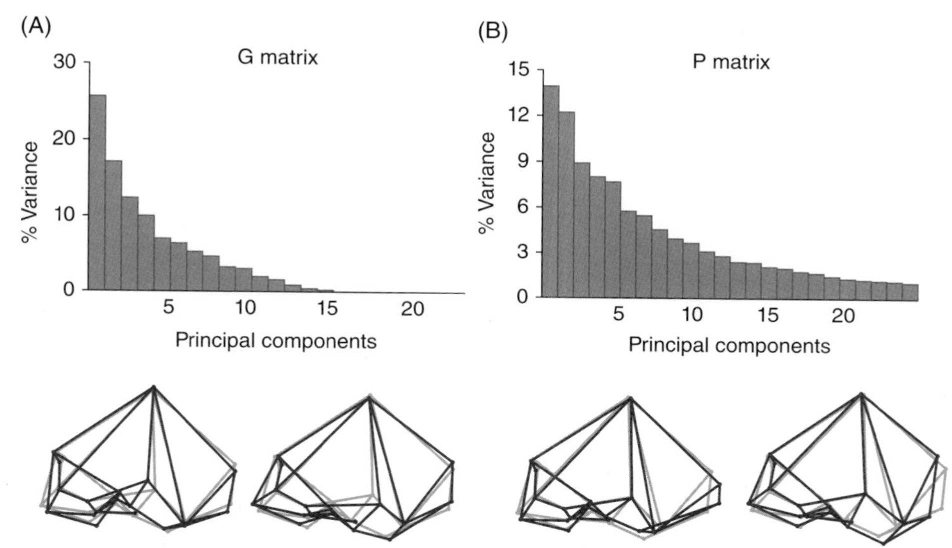

Figure 2.4. Principal component analysis of the genetic (A) and phenotypic (B) covariance matrices. Histograms show eigenvalues of the G and P matrices as percentages of the total variance in the respective covariance matrix. Dark gray wireframes display shape changes associated with the first PC of the G and the P matrices (the light gray wireframes show the overall mean shape configuration). Left: change in the direction with positive sign; right: change in the direction with negative sign.

also include a retraction of the lower face, but the expansion of the cranial vault is constrained to the posterior-occipital region of the cranial vault, and no major shape changes are detected at the hinge region between the face and the cranial base, reflecting changes in the flexion of the skull. Therefore, although the shape changes depicted by the first PCs of the **G** and the **P** matrices are quite similar, there is not a perfect correspondence between the major axes of variation of the genetic and the phenotypic covariance matrices.

The matrix correlation between the **G** and **P** matrices is high: the matrix correlation including diagonal blocks of within-landmark variances and covariances is 0.82 (p<0.0001), whereas it is 0.72 (p<0.0001) without the diagonal blocks and thus including only the covariances among landmarks. This indicates that, overall, the **P** and **G** matrices resemble each other fairly closely.

Univariate Heritabilities

The heritability analyses indicate that there is a moderate amount of additive genetic variation underlying both traits. The heritability of chondrocranial length is 0.33±0.11 and significant (p=0.00048). The heritability of CBA is 0.37±0.13 and significant (p=0.00075). For both traits, sex and age were significant covariates, whereas the status of deformation was not significant and thus was removed from the final models.

Response to Selection

The first simulation deals with the effect of the *Papps2* mutation affecting chondrocranial growth (Figure 2.5, top). The selection gradient simulating the mutation effect encompasses a global skull-shape change simultaneously affecting landmarks from the face, the cranial vault, and the cranial base. The total response to selection also affects the entire skull and involves an anterior movement of the facial complex, a contraction of the cranial vault (the bregma moves backward and inferiorly, the landmarks located on the occipital, such as lambda and opisthocranion, are shifted forward and superiorly, and the euryon, which is located on the parietal marking the greatest width, moves medially), and a decrease of cranial base flexion (hormion, basion, and opistion move inferiorly) (Figure 2.5A). This total response consists of a direct response, which is a reduced version of the selection gradient, and a correlated response affecting all the landmarks throughout the skull. The magnitude of the correlated response is almost three times as large as the direct response, which means that the direction of response has been deflected from the direction of the selection gradient by an angle of 71° (Table 2.2). This indicates that genetic constraints have a substantial effect on the response to selection.

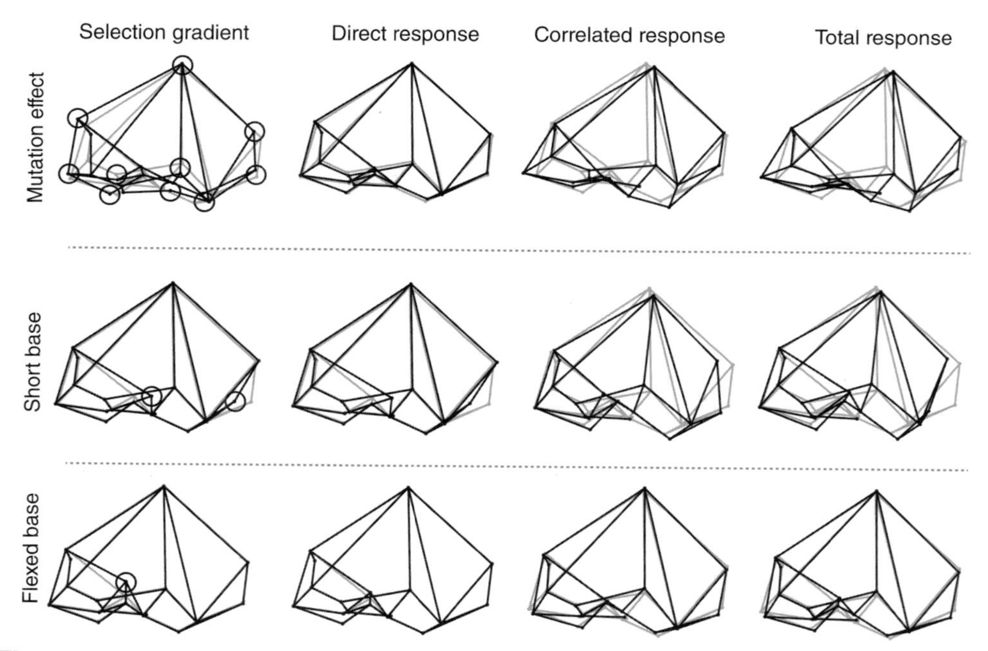

Figure 2.5. Hypothetical selection on skull shape. Effect of the *Papps2* mutation on chondrocranial growth (top). Shortening of the cranial base (middle). Flexion of the cranial base (bottom). For each selection scenario, the changes from the dark gray to the light gray wireframes show from left to right: the selection gradient (scaled to an arbitrary magnitude of shape change), the direct response, the correlated response, and the total response. The light gray wireframes show the overall mean shape configuration. Landmarks used to define the selection gradients are marked with circles.

Table 2.2. Decomposition of the total response to selection into components of direct and correlated response for the three selection scenarios. The angle between the direct response and the total response is an indication of the deflection by genetic constraints. The numbers indicate the magnitude of the respective responses in units of Procrustes distance.

	Response			
	Total	Direct	Correlated	Angle
BRACH effect	0.150	0.049	0.142	71.0
Short base	0.160	0.092	0.130	54.7
Flexed base	0.069	0.020	0.066	72.9

The second simulation represents a hypothetical shortening of the cranial base (Figure 2.5, middle). The selection gradient concerns a localized effect on two landmarks (hormion and opisthocranion) but elicits a pervasive total response to selection. The major effect is a marked reduction of the posterior cranial base associated with a slight reduction of cranial base flexion and protrusion of the face (Figure 2.5, middle). Overall, the shortening of the cranial base induces a more brachymorphic skull shape (shorter and wider skulls). The magnitude of the correlated response exceeds that of the direct response, and the total response to selection is deflected from the direction of the selection gradient by an angle over 54° (Table 2.2).

The third simulation concerns an increase of cranial base flexion by simply shifting the hormion upwards and produces a total response of selection that involves changes throughout the skull and encompasses well-known derived characters of modern humans: increased cranial base flexion, facial retraction, and expansion of the entire cranial vault (Figure 2.5, bottom). The high influence of the correlated response deflects the total response to selection from the vector of the selection gradient by an angle over 72° (Table 2.2). It is interesting to note that the total response to selection to cranial base flexion induces a shape change that is very similar to the shape change associated with the first PC of the **G** matrix (Figure 2.4A).

Discussion

Quantitative genetic tools can be used to define genetic constraints in multivariate phenotypic data when the underlying genetic covariation structure can be estimated and this information can be used to predict evolutionary responses to selection (Lande and Arnold 1983). Although phenotypic covariance matrices do tend to be similar to genetic covariance matrices (Cheverud 1988), as is the case in our data (Martínez-Abadías *et al.* 2009a), it is obviously preferable to estimate genetic covariance structure based on known relatedness. Here, we are able to do this in a large human cranial sample in which we can assign relatedness among individuals using church records. Genetic covariance structure can significantly influence how complex morphologies respond to selection. The total response to selection can be broken down into a component that is in the same direction as the selection gradient and reflects the direct response to the induced selected shape changes, and a component that is perpendicular to the selection gradient, the correlated response, which is due to covariation between traits and reflects indirect shape changes caused by selection (Steppan *et al.* 2002). In this chapter, we have explored the craniofacial shape consequences of selection acting directly on the shape and size of the chondrocranium. Our results show unequivocally that selection for changes in chondrocranial shape produce globally integrated changes throughout the skull. Understanding the nature of these changes has important implications for explaining the pattern of morphological change in the evolutionary history of the hominid skull.

The shape changes that we see in human craniofacial shape in response to changes in chondrocranial size and shape are partially consistent with the spatial packing model and our previous work on mice with perturbed craniofacial development. Simulation of the effect of the *Papps2* mutation and simulation of shortening of the cranial base produce a more spherical, wider, and anteroposteriorly shortened neurocranium, as expected under the spatial packing hypothesis. Surprisingly, in both simulations the cranial base becomes less flexed and the face becomes less retracted as the cranial base becomes shorter (Figure 2.5, top and middle). The spatial packing hypothesis (Hallgrímsson and Lieberman 2008; Lieberman *et al.* 2008; Lieberman 2011) predicts the increase in neurocranial globularity and width but also a decrease in the cranial base angle and a retracted face, which is the opposite of what we observe here. By contrast, the selection for a more flexed cranial base does produce the expected pattern of change. Flexion of the cranial base is associated with retrusion of the face and a more globular neurocranium (Figure 2.5, bottom).

Overall, our results point to the integrated nature of the human skull, showing that for any simulation scenario, the total response to selection is pervasive and affects all the regions of the skull (Figure 2.5). However, only one of the simulations (i.e., the increase in cranial base flexion) fits the predictions of the spatial packing hypothesis. When we simulate the effects of the *Papps2* mutation or shortening of the chondrocranial length, we find that the total response to selection involves a more brachymorph neurocranium, but not a more flexed cranial base and a more retracted face. The apparent deviations from the predictions of the spatial packing hypothesis are potentially interesting. One possible explanation could be low variance and/or covariance in the features predicted (facial prognathism and cranial base flexion). This is unlikely given the documented heritable variation in these features in this sample (Martínez-Abadías *et al.* 2009a) and the reported high degree of covariation between the face and the cranial base in this (Martínez-Abadías *et al.* 2009a) and other human samples (Hallgrímsson *et al.* 2007a; Lieberman *et al.* 2008; Bastir and Rosas 2009; Martínez-Abadías *et al.* 2012a). Another possibility is that our results imply that it is more likely that the evolutionary changes of the human skull involved changes in the cranial base flexion rather than in chondrocranial length. This should be taken with caution, however, because the shape changes that we observe in the Hallstatt human sample are the final result of an evolutionary process and there might not be enough variation left in these traits to disentangle the actual selection and evolutionary forces that shaped the human skull in the past. Finally, it is possible that the relationship between brain size, cranial base length, and cranial base angle is more complex than the current conceptualization of the spatial packing model implies. Indeed, we found that the relationship between chondrocranial length and cranial base flexion is significantly different in humans and mice, even in the *Papps2* mutant strain (Figure 2.3). This result underscores that the relationship between cranial base dimensions (length and flexion) is not conserved among mammals. The most interesting implication of the results of this study is the extent to which the changes in chondrocranial form produce globally integrated changes throughout the skull. In particular, the finding that the correlated response to these changes is much larger than the direct response speaks to the integrated and, therefore, constrained nature of variation in craniofacial shape. This is true of the chondrocranium in part because of its central location, early development, and intrinsically determined growth. The same may also be true, however, of other major groups of processes that influence the shape of the skull. The growth of the brain would be one such process. Understanding how selection for increased brain size during the course of human evolution has produced integrated changes throughout the skull is a major issue in paleoanthropology that could be explored using the same methods used here. This issue is important because it will help determine what aspects of craniofacial shape in hominids need special adaptive explanations and what aspects are simply correlated responses to changes in brain size because of the developmentally determined integration structure of the human skull (Lieberman 2011).

Chondrocranial growth is clearly a central determinant of variation in craniofacial form. To what extent might selection have acted on chondrocranial size or shape during the course of human evolution? This is a difficult question to answer, obviously. While it is fairly clear that selection has acted on brain size, albeit indirectly via its behavioral correlates, or on tooth size and shape as well as related aspects of jaw functional morphology (Ackermann and Cheverud 2004), it is perhaps less clear that the growth or size of the chondrocranium was a direct target of selection. It is possible that cranial base length, angle, or shape could be selected on secondarily to changes in brain size. Another possibility is that chondrocranial shape would itself respond indirectly to selection on facial shape and brain size. A more plausible scenario, though, concerns selection on the cranial base as an adaptation to bipedalism. The foramen magnum is shifted anteriorly in bipedal hominids, presumably to better balance the head on the vertebral column. The resulting shape transformation of the basicranium is complex (Nevell and Wood 2008), involving differential changes in growth at the basicranial synchondrosis and changes in the angle of the petrous temporal among others. The results of our study and related work on mice underscore the fact that the developmental changes that would have driven these changes in cranial base would have been accompanied by distributed changes in the overall shape of the skull.

Disentangling the developmental interactions that underlie the generation of craniofacial morphology is essential to understand the evolutionary history of the human craniofacial complex. Understanding how the skull is integrated can help predict craniofacial morphology in incomplete specimens in the fossil record. Certain anatomical configurations (such as a highly flexed cranial base and a small brain) may be developmentally unlikely if not impossible under normal, non-pathological conditions in human populations. More importantly, understanding the developmentally based structure of craniofacial variation can help disentangle which anatomical changes are by-products of others and which require unique evolutionary explanations. For example, the results presented here highlight the extent to which changes in chondrocranial shape drive integrated variation throughout the skull. Understanding how variation in the chondrocranium drives integrated variation in the skull can help disentangle which aspects of shape are simply side effects of changes in chondrocranial shape. Arriving at such an understanding of craniofacial variation will require an integrated approach that relies not only on the fossil record but also on studies of natural variation in humans and our relatives as well as both natural and experimentally induced variation in model organisms.

Acknowledgments

We thank Drs J. Boughner and C. Rolian for inviting us to participate at the 83rd American Association of Physical Anthropologists's Wiley Symposium "Evolutionary Developmental Anthropology: An Evo-Devo Approach to Understanding Primate and Human Evolution" and to contribute to this book. We are also indebted to authorities and museum curators of the following institutions: Parish and Musealverein of Hallstatt (Hallstatt, Austria), Institut für Anatomie (Innsbruck, Austria), and Naturhistorisches Museum Wien and Österreichisches Museum für Volkskunde (Vienna, Austria) for allowing us to measure their valuable cranial collections and for their help during fieldwork. We gratefully acknowledge support from the National Institutes of Health (1R01DE021708, R01DE018500, and R01DE018500-02S1) and the National Science and Engineering Research Council (#238992-12) to BH as well as Wenner Gren Foundation for Anthropological Research (7149) and Beca de Postgrado Caja Madrid to NM-A.

References

Ackermann, R. R., and Cheverud, J. M. 2004. Detecting genetic drift versus selection in human evolution. *Proceedings of the National Academy of Sciences, USA* 101(52):17946–17951.

Almasy, L., and Blangero, J. 1998. Multipoint quantitative-trait linkage analysis in general pedigrees. *American Journal of Human Genetics* 62(5):1198–1211.

Bastir, M., and Rosas, A. 2009. Mosaic evolution of the basicranium in *Homo* and its relation to modular development. *Evolutionary Biology* 36(1):57–70.

Bastir, M., Rosas, A., and O'Higgins, P. 2006. Craniofacial levels and the morphological maturation of the human skull. *Journal of Anatomy* 209(5):637–654.

Bastir, M., Rosas, A., Stringer, C. *et al.* 2010. Effects of brain and facial size on basicranial form in human and primate evolution. *Journal of Human Evolution* 58(5):424–431.

Biegert, J. 1963. The evaluation of characteristics of the skull, hands and feet for primate taxonomy. *In*: S. L. Washburn (ed.) *Classification and human evolution*. Chicago, Aldine. pp.116–145.

Cheverud, J. M. 1982. Phenotypic, genetic, and environmental integration in the cranium. *Evolution* 36(3):499–516.

Cheverud, J. M. 1988. A comparison of genetic and phenotypic correlations. *Evolution* 42(5):958–968.

De Beer, G. 1937. The development of the vertebrate skull. Oxford: Clarendon Press. pp.xxiii, 552.

Di Rocco, F., Biosse Duplan, M., Heuzé, Y. *et al.* 2014. FGFR3 mutation causes abnormal membranous ossification in achondroplasia. *Human Molecular Genetics* 23(11):2914–2925.

Dryden, I. L., and Mardia, K. V. 1998. *Statistical shape analysis*. Chichester: John Wiley & Sons, Ltd.

Falconer, D. S., and Mackay, T. F. C. 1996. *Introduction to quantitative genetics*. Harlow: Longman.

Groeneveld, E., Kovač, M., and Mielenz, N. 2010. *VCE 6. Users guide and reference manual*, Version 6(2).

Hallgrímsson, B., Brown, J. J., Ford-Hutchinson, A. F. *et al.* 2006. The brachymorph mouse and the developmental-genetic basis for canalization and morphological integration. *Evolution & Development* 8(1):61–73.

Hallgrímsson, B., Jamniczky, H., Young, N. M. *et al.* 2009. Deciphering the Palimpsest: Studying the relationship between morphological integration and phenotypic covariation. *Evolutionary Biology* 36(4):355–376.

Hallgrímsson, B., and Lieberman, D. E. 2008. Mouse models and the evolutionary developmental biology of the skull. *Integrative and Comparative Biology* 48(3):373–384.

Hallgrímsson, B., Lieberman, D. E., Liu, W. *et al.* 2007a. Epigenetic interactions and the structure of phenotypic variation in the cranium. *Evolution & Development* 9(1):76–91.

Hallgrímsson, B., Lieberman, D. E., Young, N. M. *et al.* 2007b. Evolution of covariance in the mammalian skull. *Novartis Foundation Symposium* 284:164–185; discussion 185–190.

Hansen, T. F., and Houle, D. 2008. Measuring and comparing evolvability and constraint in multivariate characters. *Journal of Evolutionary Biology* 21:1201–1219.

Heuzé, Y., Martínez-Abadías, N., Stella, J. M. *et al.* 2014. Quantification of facial skeletal shape variation in fibroblast growth factor receptor-related craniosynostosis syndromes. *Birth Defects Research Part A: Clinical and Molecular Teratology* 100(4):250–259.

Klingenberg, C. P. 2011. MorphoJ: an integrated software package for geometric morphometrics. *Molecular Ecology Resources* 11:353–357.

Klingenberg, C. P., and Leamy, L. J. 2001. Quantitative genetics of geometric shape in the mouse mandible. *Evolution* 55(11):2342–2352.

Klingenberg, C. P., and McIntyre, G. S. 1998. Geometric morphometrics of developmental instability: Analyzing patterns of fluctuating asymmetry with Procrustes methods. *Evolution* 52(5):1363–1375.

Klingenberg, C. P., and Monteiro, L. R. 2005. Distances and directions in multidimensional shape spaces: Implications for morphometric applications. *Systematic Biology* 54(4):678–688.

Konigsberg, L. W. 2000. Quantitative variation and genetics. *In*: S. Stinson, B. Bogin, and D. O'Rourke (eds) *Human biology: An evolutionary and biocultural perspective*. Oxford: Wiley-Blackwell. pp.135–162.

Kruuk, L. E. 2004. Estimating genetic parameters in natural populations using the "animal model." *Philosophical Transactions of the Royal Society B: Biological Sciences* 359(1446):873–890.

Kurima, K., Warman, M. L., Krishnan, S. *et al.* 1998. A member of a family of sulfate-activating enzymes causes murine brachymorphism. *Proceedings of the National Academy of Sciences, USA* 95(15):8681–8685.

Lande, R. 1979. Quantitative genetic analysis of multivariate evolution, applied to brain:body size allometry. *Evolution* 33(1):402–416.

Lande, R., and Arnold, S. J. 1983. The measurement of selection on correlated characters. *Evolution* 37(6):1210–1226.

Lieberman, D. E. 2011. *The evolution of the human head.* Cambridge, MA: Cambridge University Press.

Lieberman, D. E., Hallgrímsson, B., Liu, W. *et al.* 2008. Spatial packing, cranial base angulation, and craniofacial shape variation in the mammalian skull: Testing a new model using mice. *Journal of Anatomy* 212(6):720–735.

Lieberman, D. E., Ross, C. F., and Ravosa, M. J. 2000. The primate cranial base: Ontogeny, function, and integration. *American Journal of Physical Anthropology* Suppl. (31):117–169.

Lynch, M., and Walsh, B. 1998. *Genetics and analysis of quantitative traits.* Sunderland, MA: Sinauer. pp.xvi, 980.

Martínez-Abadías, N., Esparza, M., Sjøvold, T. *et al.* 2009a. Heritability of human cranial dimensions: Comparing the evolvability of different cranial regions. *Journal of Anatomy* 214(1):19–35.

Martínez-Abadías, N., Esparza, M., Sjovold, T. *et al.* 2012a. Pervasive genetic integration directs the evolution of human skull shape. *Evolution* 66(4):1010–1023.

Martínez-Abadías, N., Mitteroecker, P., Parsons, T. E. *et al.* 2012b. The developmental basis of quantitative craniofacial variation in humans and mice. *Evolutionary Biology* 39(4):554–567.

Martínez-Abadías, N., Paschetta, C., de Azevedo, S. *et al.* 2009b. Developmental and genetic constraints on neurocranial globularity: Insights from analyses of deformed skulls and quantitative genetics. *Evolutionary Biology* 36(1):37–56.

Matsushita, T., Wilcox, W. R., Chan, Y. Y. *et al.* 2009. FGFR3 promotes synchondrosis closure and fusion of ossification centers through the MAPK pathway. *Human Molecular Genetics* 18(2):227–240.

Nevell, L., and Wood, B. 2008. Cranial base evolution within the hominin clade. *Journal of Anatomy* 212(4):455–468.

Ross, C. F., and Ravosa, M. J. 1993. Basicranial flexion, relative brain size, and facial kyphosis in nonhuman primates. *American Journal of Physical Anthropology* 91(3):305–324.

Scott, J. H. 1958. The cranial base. *American Journal of Physical Anthropology* 16(3):319–348.

Sjøvold, T. 1984. A report on the heritability of some cranial measurements and non-metric traits. *In*: G. N. Van Vark and W. W. Howells (eds) *Multivariate statistical methods in physical anthropology*. Dordrecht: Reidel Publishing Company. pp.223–246.

Steppan, S. J., Phillips, P. C., and Houle, D. 2002. Comparative quantitative genetics: Evolution of the G matrix. *Trends in Ecology & Evolution* 17(7):320–327.

ul Haque, M. F., King, L. M., Krakow, D. *et al.* 1998. Mutations in orthologous genes in human spondyloepimetaphyseal dysplasia and the brachymorphic mouse. *Nature Genetics* 20(2):157–162.

Young, B., Minugh-Purvis, N., Shimo, T. *et al.* 2006. Indian and sonic hedgehogs regulate synchondrosis growth plate and cranial base development and function. *Developmental Biology* 299(1):272–282.

Chapter 3

The Tooth of the Matter: The Evo-Devo of Coordinated Phenotypic Change

Julia C. Boughner
Department of Anatomy and Cell Biology, College of Medicine, University of Saskatchewan, Saskatoon, SK, Canada

Introduction

There are several notable trends in the course of human evolution, among them habitual bipedality, bigger brains, and greater manual dexterity. Another trend seen across fossil hominins is the evolutionary shift towards smaller jaws and teeth, especially the premolars and molars (e.g., Dart 1948; Robinson 1954; Wood and Aiello 1998; Wood 2009, and summarized in Clement and Hillson 2013 (Figure 3.1, color plate)). As almost a century of work has shown, the teeth of different primate species and geographic populations develop at different times and grow to different sizes (Swindler 1976, and as discussed in Scott and Lockwood 2004). Often these species also have distinct mandible shapes and sizes (e.g., Simons 1976; Lucas 1981; Boughner 2011; Cofran 2014), which raises the question of the degree to which inter-population diversity among tooth development schedules is due to differences in jaw size and spatial relationships between the developing jaw and its dentition (Dean and Beynon 1991; Dean and Wood 1981).

This line of reasoning is based on the assumption that jaw tissues instruct dental tissues where and when to form. The mandible (i.e., dentary bone) and the lower permanent molars are particularly useful for probing the level of developmental autonomy between jaws and teeth. This is because while space in the jawbone is held by the deciduous dentition for permanent incisors, canine, and premolars, space has to be created *de novo* for the permanent molars to develop and emerge into (Nanci 2012). Thus one reasonable starting place is to query if space availability in the jaw dictates when permanent molars start to form (Dean and Wood 1981). A subsequent question is whether different space availability due to variation in jaw size and shape singularly generates taxonomic variation in timing of molar initiation and eruption.

Developmental Approaches to Human Evolution, First Edition. Edited by
Julia C. Boughner and Campbell Rolian.

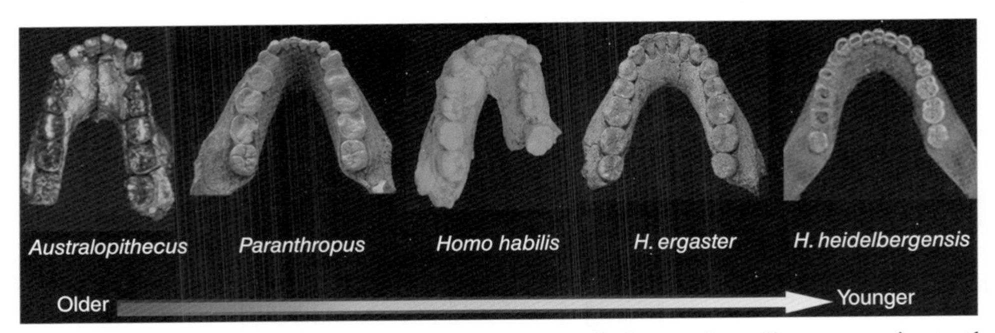

Figure 3.1. The trend in hominins towards more gracile jaws and smaller post-canine teeth, particularly from *Homo ergaster* (and *H. erectus*, not shown) onwards. Many factors, including ecological, physiological, and socio-cultural, sculpted human craniodental morphology. Explaining at the genetic level how change in tooth size and jaw robusticity occur may help to structure plausible explanations of how or why this trend happened. Images used with permission from the Smithsonian Institute and *Science* (295:15 Feb) 2002.

In primates as in other toothed vertebrates, teeth and jaws are functionally and spatially integrated, and yet how this relationship is established and maintained over developmental and evolutionary time is poorly understood. As will be explored in-depth, if anything, contemporary evolutionary and developmental genetic studies of odontogenesis suggest that the mandible is influenced by, or responds to, the developing teeth (Weiss *et al.* 1998; Boughner and Dean 2004; Caumul and Polly 2005). Further, this review will show that the current state of knowledge suggests that the two systems – teeth and jaws – develop independently of each other despite evolving together.

The reasons for the significant size-reduction in the hominin jaw apparatus (defined here as the upper and lower jaw skeletons and dentitions) are unknown and intriguing; not least because eating and vocalizing are vitally important behaviors to all human, hominin, and other primate groups (and, indeed, most mammals). So why the macroevolutionary change (small teeth, more gracile jaws) in the hominin lineage specifically? In particular, this trend towards gracile jaws and teeth is salient to understanding modern human evolution in the context of cultural adaptations that include cutting, pulverizing, cooking, and otherwise processing our food (Caumul and Polly 2005; von Cramon-Taubadel 2011); particularly as reduced molar size and increased food processing were concurrent with the rise of *Homo erectus* (Wood 2009; Organ *et al.* 2011). A culture of food preparation alleviates the need for the jaw apparatus to physically break down high-energy yet tough foods, such as meats and hard nuts. Of course, one must consider both sides of this issue (e.g., Eng *et al.* 2013): was the evolution of smaller teeth and jaws the impetus to process hard/tough foods with fire and tools; or did a cultural practice of food processing allow energy to be diverted elsewhere (i.e., brain growth) from building big teeth and jaws? If robust jawbones and large teeth alongside massive chewing muscles are no longer essential to meet minimum caloric intakes, then selection to grow and maintain a massive jaw apparatus is arguably relaxed. Thus, among hominins, better problem-solving abilities matched with improved tool-making skills may have led in part to the evolution of smaller jaws and teeth. But while this particular theory may explain the coordinated evolutionary decrease of tooth and jaw sizes in hominins, it does not account for the coordinated evolution of teeth and jaws across a range of different phenotypes (e.g., from extremes of prognathism to orthognathism; and different dental formulae) in non-human primates. Also, a dramatic example of the opposite direction of a size-reduction trend are the derived hyper-robust jaws and mega-molars of Paranthropines in which jaw and tooth development and evolution were still sufficiently coordinated with each other to function properly. Different and multiple factors surely drive the macroevolution of the primate jaw apparatus (e.g., Ross and Iriarte-Diaz 2014).

Whatever the impetus, hominid teeth and jaws evolved together in generally synchronized ways that enabled their owners to eat well enough to survive and reproduce successfully. Presumably the processes acting on the jaw apparati of our fossil ancestors and cousins are the same ones operating today in humans and our nearest livings relations, great apes, if not other primate taxa (Stock *et al.* 1997; Kavanagh *et al.* 2007; Jernvall and Thesleff 2012). Are differences within and among hominin taxa in tooth developmental schedule and mandibular proportion coincidental or causative of each other? From hyper-robust to hyper-gracile phenotypic extremes, the question driving this review chapter is: *How did hominin teeth and jaws evolve to function together properly regardless of variation in size and shape?* This chapter, which is by no means a comprehensive review, focuses on the developmental biology of teeth and jaws in the context of primate and human oral evolution.

When natural selection acts on a phenotype, what is also being selected for is how that phenotype develops; the genes and development are, in effect, just along for the ride. Thus developmental-genetic information about how teeth and jaws grow gives useful insights into how these structures evolve by identifying genetic and epigenetic sources of phenotypic variation, flexibility, and constraint (Stock *et al.* 1997; McCollum and Sharpe 2001; Kavanagh *et al.* 2007; Jussila and Thesleff 2012). Determining the sources of ontogenetic variation within the dentition – a system resilient to perturbations (Garn *et al.* 1959; Lewis and Garn 1960) – is key to understanding what regulates dental development so tightly, and arguably critical to grasping the mechanisms by which the dentition evolves. The thesis that will be explored here is of autonomy between dentition and jawbone: during the earlier stages of their development, tooth and jaw tissues are blind to each other's existence, and are instead coordinated in time by a third, external factor such as the body's central clock located in the suprachiasmatic nucleus of the brain (Boughner and Hallgrímsson 2008).

To test this hypothesis of autonomy, it is necessary to investigate how teeth develop (Figure 3.2) versus how jaws develop (Figure 3.3, color plate) to define exactly what the role of one system may play for the other. These insights began to accrue in earnest about a hundred years ago with pre-genetic developmental studies that tried to explain when (i.e., initiation time), what (i.e., tooth class), and where teeth form in the jaw based on insights from mouse and lizard model organisms.

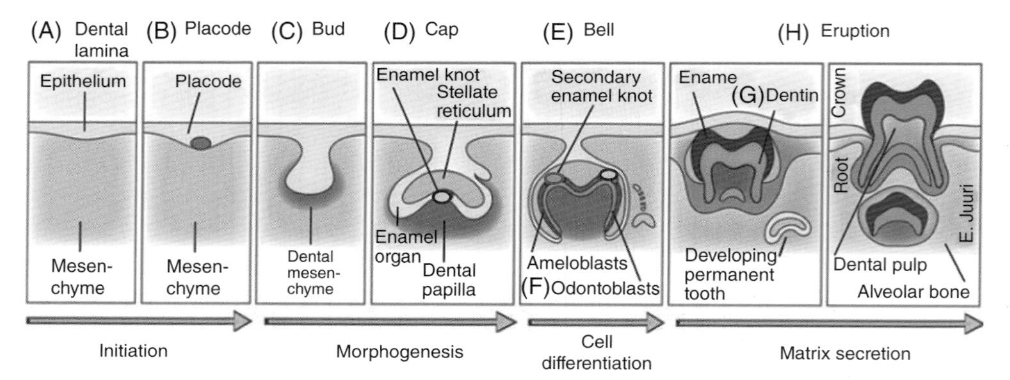

Figure 3.2. Classic stages of odontogenesis. Initiation is defined by the visible thickening of the dental lamina (A) and formation of a tooth placode (B), morphogenesis through bud (C), cap (D), and bell (E) stages, when enamel knots develop and orchestrate cusp formation. Next, enamel- and dentine-secreting cells differentiate (F) and begin to lay down the mineralized tissues of the tooth crown and roots (G) even as the tooth begins to erupt through the jawbone (H). Adapted with permission from Jussila and Thesleff (2012).

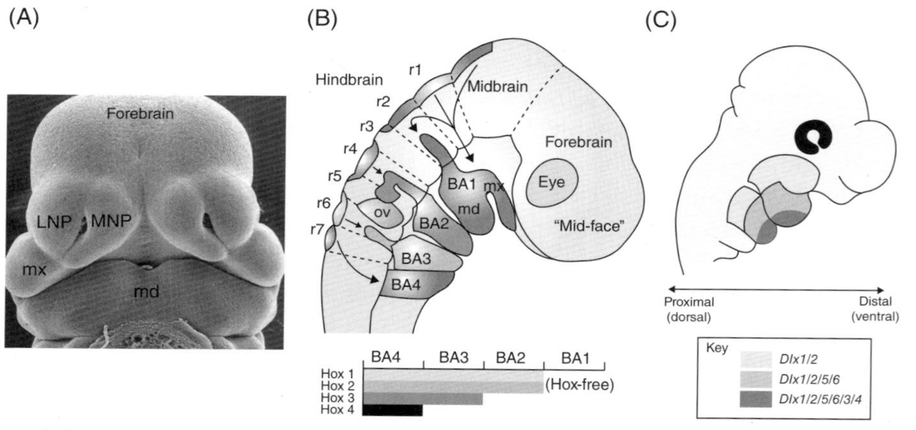

Figure 3.3. Classic stages of jaw morphogenesis in a mouse embryo model (embryonic day 10–11). The maxillary and mandibular prominences form (A) via cranial neural crest cell migration and proliferation (B). The identity of each prominence is patterned by genes (e.g., nested expression of *Dlx1-6*) (C) to eventually form distinct upper and lower jaw skeletons. Adapted with permission from (A) Lippincott Williams & Wilkins (2007); (B) *Nature Reviews Neuroscience* 8 (2007); (C) Minoux and Rijli, *Development* (2010). Legend: BA, brachial arch; LNP, lateral nasal prominence; md, mandibular prominence; MNP, medial nasal prominence; mx, maxillary prominence; r, rhombomere.

Tooth Developmental Biology: Genetic Determinants of Tooth Class, Position, and Form

In the late 19th and early 20th centuries, important work was published on segmental development in vertebrates. Huxley and de Beer's (1934) pioneering research on limb development, and Goodrich (1913) and others' comparative studies on regional differentiation in vertebrate bodies markedly influenced zoologist and paleontologist Percy Butler's work on the evolution of heterodonty – different tooth types, or classes, in the same dentition – in mammals (Butler 1937, 1939). From the works above, Butler understood that types and locations of vertebrae varied among orders, for example, mammals and birds. He suggested that the vertebral column formed from pluripotential cells capable of being any type of vertebrae, for example, cervical or thoracic (Butler 1939), depending on which developmental path these vertebral cells were directed down. Butler then applied this model of vertebral morphogenesis to the dental lamina, proposing that all tooth primordia were capable of becoming any tooth type. As is now known to be true, Butler correctly believed that all the teeth of an individual's dentition formed from the same source tissues, via the same developmental mechanisms, and were supported by the same physiological systems (Thesleff and Sharpe 1997; Chai *et al.* 2000; Jernvall and Thesleff 2000). Based on this generally homogenous developmental environment, Butler surmised that tooth differentiation must be directed by local factors. From Schour's (1934) neurovascular experiments on developing teeth, and prior to contemporary understanding of developmental genetics, Butler constructed his "Field Theory" around the idea that identical tooth primordia developed within the influence of a morphogenetic field, active in the underlying mesenchyme of the dentary or maxilla/premaxilla bones. Expressed in different concentrations, these morphogens activated specific potentials (e.g., tooth classes) in undifferentiated embryonic cells according to their positions in the field (Figure 3.4A, color plate). Both degree and location of the morphogenetic field determined tooth identity (Butler 1939).

As is now common knowledge, although poorly understood in Butler's time, vertebrate teeth form via highly conserved processes that begin with cross-signaling

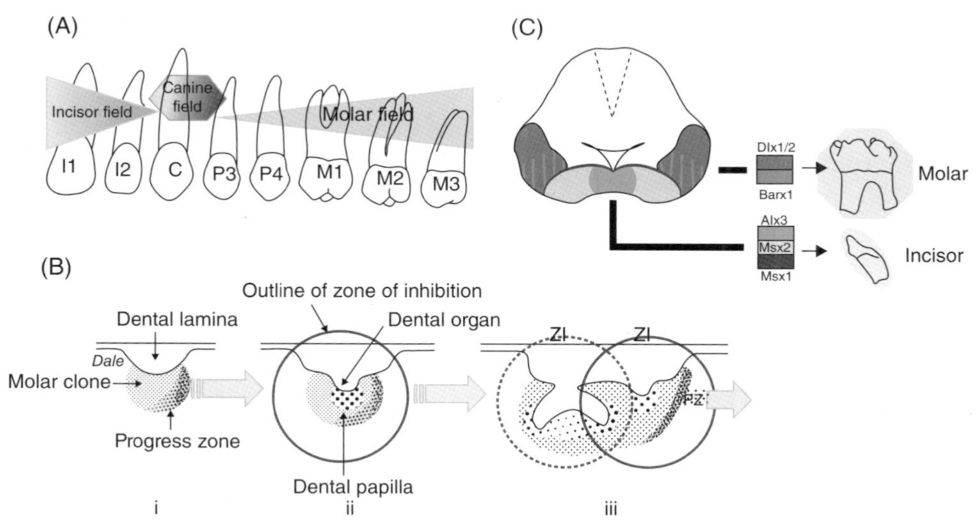

Figure 3.4. Progression of the models crafted to explain how tooth identity as well as initiation time are determined, first via the Field Theory (A), the Clone Model (B), and more recently the Molecular Field Model (C). Based on Ten Cate (1994) and Cobourne and Sharpe (2003), published in Boughner and Hallgrímsson 2008.

between the dental epithelium and the underlying dental mesenchyme (Figure 3.2A–B). This genetic "cross-talk" is detectable via modern assays for gene expression; however, it is also visible as a thickening of the dental lamina in the precise location of the presumptive tooth (Figure 3.2A–B). As a quick primer, several factors work in concert to produce the size and shape of a given tooth including its occlusal (biting/chewing) surface (Figure 3.2C–H). These factors include: gene regulatory networks (groups of interacting genes that regulate gene expression and its protein products, often in a positive/negative feedback loop); signaling centers (structures from which mRNA and proteins arise, and organize tissue outgrowth and morphogenetic fields, for example); and cellular dynamics (i.e., cell proliferation and apoptosis that structures the topography of the tooth crown). All these factors and processes will be discussed in-depth later in the chapter.

Importantly, Butler published his work at a time when studies of dental development were stalled and perhaps a little confused for the lack of a strong theoretical framework upon which to build and test new ideas. As the Field Theory emerged, morphogenetic fields were the most reasonable resolution to many developmental questions. Regardless, Butler appreciated that the Field Theory left several fundamental questions unanswered: for example, it poorly accounted for differences in absolute tooth number between taxa; the source of the morphogenetic substance remained a mystery; and the signal(s) that defined and then sustained a given morphogenetic field for the total period of permanent tooth development (in excess of 18 years in humans) was not explained. Dahlberg (1945) expanded on the Field Theory with his studies of human dental development, which highlighted to him the subtle but notable morphological variation within a single tooth class (e.g., I1/I2; M1/M2/M3). To account for these intra-class differences, Dahlberg (1945) proposed that morphogen concentrations decreased, cline-like, across a field and within a region to generate variation in tooth size and shape within each class. This, Dahlberg (1945) concluded, also explained morphological similarities between adjacent classes, that is, the molariform shape of most posterior premolars.

Building on Glasstone's (1938, 1967) tissue culture work, which indicated that tooth morphology was regulated intrinsically by the tooth organ, Lumsden (1979) filled more gaps in scientific knowledge of how tooth identity was determined by experimentally showing that molar teeth formed properly and completely when presumptive molar tissue (i.e., molar dental lamina) was cultured in a non-oral environment. Within four weeks of growth in the anterior eye chambers of adult mice, embryonic molar tissues younger than 12 days formed first and second permanent molars; and dental lamina aged between 12 and 16 days formed two or, often, three adult molar teeth. The molars developed in normal sequence (M1, M2, M3) into normal crown shapes (Lumsden 1979). Lumsden's work strongly suggested that the influence of a morphogenetic field was unnecessary to both the normal appearance and differentiation of (molar) teeth. This powerful developmental study using mouse, the most prevalent mammalian model system, offered strong evidence that tooth identity was at least to some degree programmed intrinsically, and not by external factors. This finding is supported by reports of human ovarian and cranial teratomas, natural experiments where teeth are morphologically recognizable despite forming outside of the mouth (Beaty and Ahn 2014; Dean and Munro 2005), and by recent experimental cultures of rodent teeth where all three molars developed recognizably *in vitro* (Kavanagh *et al.* 2007). In summary, these works add credence to the working hypothesis that the dentition is developmentally autonomous from the jaw.

In parallel to mouse studies, other steps that moved this research area forward came via experimental work done in a reptilian model system. In the early-to-mid-20th century, very little was understood about the ontogeny of reptilian tooth replacement. Rather than addressing tooth class and form, Osborn focused on understanding what determined tooth position in polyphyodont dentitions (i.e., continuously replaced teeth). He proposed (Osborn 1971) that once a tooth germ began to develop, it emitted an inhibitory molecule that formed a spherical "zone of inhibition" around the young tooth (Figure 3.4B, color plate). As the tooth developed, the sphere decreased; successive tooth germs formed from competent dental tissues positioned at the posterior limit of (and thus outside) the sphere. Osborn (1978) suggested that this "zone of inhibition" was maintained by each developing tooth primordium (i.e., clone) around itself to prohibit the activation of subsequent primordia in that zone. The posterior progression of the clone alongside the zone of inhibition dictated the number of primordia that developed in a tooth class, and hence, the number of teeth in the jaw (Osborn 1978). Each tooth class developed from a single clone of proliferating cells. Osborn's "Clone Model" rested upon this theory of the *intrinsic* control of tooth formation and differentiation. Similar to Lumsden, Glasstone ad others' conclusions, rather than suggesting that an extrinsic morphogenetic field differentiated regions of the dental lamina as in the Field Theory, Osborn (1978) proposed that the fate of tooth germs was predetermined in the dental lamina.

Whether Osborn intended it, his work was the first to address spatial relationships between developing jaws and teeth. A major implication of the Clone Model was that there was an acute developmental relationship between time of tooth initiation and the length of jawbone that could house young teeth. The total length of jaw through which the clone might travel restricted the posterior extension of the proliferating clone and the number of primordia it could form (Osborn 1978). This dictated the rate at which the clone escaped the influence of the zone of inhibition, and, hence, the times at which successive tooth germs began to develop, or "initiate." This was an important idea that helped explain the synchronous growth of two different systems, the jaws and the teeth. Indeed, the idea of inhibitory zones, and even "activation zones," has re-emerged in the last decade of developmental-genetic work (Chen *et al.* 1996; Jernvall and Thesleff 2000; Kavanagh *et al.* 2007), as have the Field Theory and Clone Model (Fraser *et al.* 2006; Mitsiadis and Smith 2006; Ribeiro *et al.* 2013). Indeed, a good case has been made for the synthesis of the Clone Model and Field Theory – that is, nested zones of specification (Fields) alongside activation and inhibition (Clones) – with contemporary knowledge of genetic patterning as a framework for understanding odontogenesis (Townsend *et al.* 2009).

In the years after Osborn published his Clone Model, embryological studies continued to build on what was known about the formation of presumptive dental tissues, and what has been discovered since that time is the bedrock of what is accepted today about how a tooth forms (e.g., Kollar and Baird 1969). Kollar and

Lumsden (1979) defined three main phases of tooth development – initiation, morphogenesis, and cell differentiation (Figure 3.2). Teeth develop from two main tissue types: epithelium and mesenchyme. The mesenchyme originates from a special cell population called cranial neural crest cells (CNCC). Early in embryonic development these CNCC leave the developing spinal cord (neural crest) and migrate into the primordial buds of facial tissue (Figure 3.3, color plate) (Chai *et al.* 1998). For oral (versus pheryngeal) dentitions, early interactions between neural crest-derived mesenchyme and oral epithelium are critical to make the dental lamina competent to form teeth and thus trigger initiation (Thesleff *et al.* 1995; Thesleff and Sharpe 1997; Jernvall and Thesleff 2000). Localised cell proliferation along the dental lamina forms a series of epithelial swellings that, with continued cell division, invaginate the underlying mesenchyme (Ten Cate 1994) (Figure 3.2A–B). The locations of these invaginations correspond to the positions of the future deciduous teeth. Next, morphogenesis includes the development of the tooth germ as far as the bell stage (Figure 3.2E), when the infolding of germ tissues forms cusp patterns, and the tooth takes shape. In the final phase, cell differentiation, interactions between epithelium and mesenchyme lead to the differentiation of ameloblasts and odontoblasts that then rhythmically secrete over time the hard tissues (enamel and dentin, respectively) of the crown, with enamel on the surface, dentin below, and roots made of dentin (Ten Cate 1994) (Figure 3.2D–H).

In the past 20 or so years, scientific understanding of not only the stages but also the developmental genetics of odontogenesis has progressed in leaps and bounds. Butler's Field Theory served as inspiration for the Molecular Field Theory (a.k.a. the odontogenic homeobox code model) (Sharpe 1995) (Figure 3.4C, color plate), where substantial progress was made in terms of laying down a framework based on the particular roles of discrete genes (e.g. Thomas *et al.* 1997; Plikus *et al.* 2005) (versus non-specific morphogens) that laid the foundation on which to investigate the gene regulatory networks regulating particular stages of odontogenesis as well as tooth formation in general across vertebrates (Fraser *et al.* 2009; Fraser and Smith 2011). Contemporary molecular studies have further clarified not only the sources and identities of the tissues involved in the formation of the dental lamina and tooth germs but also specific genes and indeed gene regulatory networks (Kaski *et al.* 1996; Thesleff 2006; Jussila and Thesleff 2012; Kim *et al.* 2012). In some developmental processes, certain genes appear to be functionally redundant in the presence of other members of the same gene family (Thesleff and Sharpe 1997), potentially buffering odontogenesis from perturbations. During tooth formation, some genes are expressed only in a specific tissue (i.e., dental mesenchyme), at specific stages of development, and in particular structures such as the primary and secondary enamel knots (Figure 3.2D–E). These "knots" are signaling centers – master organizers of cusp morphogenesis – that arise when mesenchyme condenses locally along the margin of the dental papilla and the enamel organ/stellate reticulum (Figure 3.2D–E). Enamel knots are now known to be essential to establish tooth cusp position, number, and shape (Thesleff and Sharpe 1997; Thesleff *et al.* 2001), and to be the structures via which cusp number and morphology evolve (Jernvall and Thesleff 2012). Extensive and interactive scientific online libraries detailing not only genes but signaling/regulatory networks now serve as platforms for future discoveries (http://compbio.med.harvard.edu/ToothCODE/ O'Connell *et al.* 2012; http://bite-it.helsinki.fi/ Kaski *et al.* 1996).

Osborn's Clone Theory was revitalized via the Inhibitory Cascade Model (e.g., Kavanagh *et al.* 2007; Renvoisé *et al.* 2009) where distance between successive molar germs influences the timing of initiation (but absolute space in the jawbone for the molars does not have an effect). Further to the molecular control of odontogenesis and crown size, *in silico* experiments have shown that only a few variables, if not a single parameter (i.e., cell number at a given tooth cusp), changed under fixed conditions of cell proliferation and cell death can simulate a range of different cusp and crown morphologies seen across living mammalian and reptilian taxa, and even within a given species (Salazar-Ciudad and Jernvall 2010). This model has been supported by statistical analyses of micro-CT scans of fossil hominin teeth, showing that a combination of early genetic crown patterning and later enamel deposition sculpts the species-specific topography of the occlusal surface and crown morphology (Skinner *et al.* 2009, 2010).

Developmental-genetic studies have also significantly improved collective understanding of the timing of odontogenesis. Teeth also develop in a tightly time-regulated fashion as evidenced by the development of periodic growth lines (i.e., short-period daily cross-striations; longer-period striae of Retzius and, at the tooth surface, perikymata) (e.g., Boyde 1989; Bromage and Dean 1985; Dean 1987, and see the chapter by Lacruz in this volume). In primates, the timing of tooth initiation as well as emergence typically varies among individuals by months, and sometimes years (e.g., Nanda 1960; Smith *et al.* 1994, 2007; Liversidge 2008). Conversely, once enamel and dentin begin to be secreted, the pace of hard tissue deposition is rhythmically constant and largely immune from perturbation (Garn *et al.* 1959; Lewis and Garn 1960). There is now good molecular evidence that ameloblasts and odontoblasts march in time to a local "peripheral clock" acting autonomously within the tissues of the developing tooth (Lacruz *et al.* 2012 and discussed by Lacruz in this volume). Crown morphology varies based on, for example, the number of ameloblasts (the ameloblast front) and their rate of activity (e.g., Boyde 1964; Lacruz and Bromage 2006). As the timing of hard tissue deposition processes is both regular and specific to a given primate taxon, it is possible to draw inferences about the life history of an individual and a species (Dean 1987; Dean 2006). Primate, if not also mammalian, tooth and jaw development and growth – much like that of other body parts – appear to be orchestrated in time by longer-period rhythms and global factors, such as the central clock of the suprachiasmatic nucleus (Boughner and Hallgrímsson 2008; Bromage *et al.* 2012).

In sum, all of these developmental-genetic and morphogenetic processes are apparently intrinsic to the dental epithelium and mesenchyme and its derivate structures, and occur independently of instruction or stimulus from the jaw tissues (as per the formation of teeth in teratomas, or that tooth morphology appears to be determined intrinsically). Based on this current understanding of the tooth organ and odontogenesis, one could argue that the developing teeth need the jaw as little more than the medium within which to form and to then remain anchored, emphasizing the functional rather than developmental integration of the jaw apparatus. One caveat to this theory is whether timing cues travel between tooth and jaw tissues to coordinate the pace of their development, a question for which new data and insights are needed.

Jaw Morphogenesis: Developmental Genetics of Jaw Identity and Its Differentiation from Tooth

To properly unpack the level of developmental autonomy between tooth and jaw tissues one must understand how not only the dentition but also the jaw forms by itself. Teeth arise directly from within the embryonic jaw primordia, thus from the start tooth and jaw tissues develop in spatially and functionally intimate ways. As noted above, whether they actively and directly cross-talk with each other to synchronize the timing of their development is unclear. All the structures in the lower jaw, including the teeth, develop from the first branchial (a.k.a. pharyngeal) arch, a bar of soft tissue that makes up the lower face of the embryo and goes on to form the mandibular prominence proper (Mina 2001; Cerny *et al.* 2004) (Figure 3.3A, color plate). In contrast, all the structures of the upper jaw develop from several paired mid-facial prominences (Cerny *et al.* 2004; Lee *et al.* 2004). The exact number of prominences varies depending on the vertebrate class, but in primates these prominences are maxillary, medial nasal, and lateral nasal. As discussed above, the lower and upper facial prominences are populated and patterned by a special population of cells – the cranial neural crest – that migrates into the face/head from the neural tube (Figure 3.3B, color plate). These now jaw-competent buds of tissue must merge and fuse together at the appropriate time to form a healthy jaw (Hallgrímsson *et al.* 2004; Thomason *et al.* 2008; Parsons *et al.* 2011; Medio *et al.* 2012). For instance, misexpression of the genes that control cell proliferation and cell death in the upper facial prominences can lead to facial cleft birth defects (e.g., Thomason *et al.* 2008; Kasberg *et al.* 2013 among many others). Although mandibular clefts are for unknown reasons rare to non-existent, early fusion of left and right mandibular prominences is vital to normal jaw development (Fujita *et al.* 2013).

In terms of evolutionary jaw origins, gain-of-function experiments among other genetic assays have revealed that the loss of homeobox, or *Hox*, genes seems to have been requisite for the first jaws to develop (Hunt *et al.* 1991a, 1991b; Grammatopoulos *et al.* 2000; Cohn 2002; Couly *et al.* 2002; Kuratani 2004; Kuraku and Meyer 2009) (Figure 3.3B, color plate). About 40 years ago, the discovery of homeotic genes, a relatively small number of highly conserved molecules, gave unprecedented insight into the developmental patterning of animal bodies. Notable among homeotic genes are *Hox* genes which regulate the number, identity, pattern, and formation of body fields such as appendages and organs, as well as the formation of cell types and the specification of primary body axes (Lewis 1978; Mark *et al.* 1997; Carroll *et al.* 2001). These *bauplan* genes are so conserved that they recur and are redeployed across virtually all vertebrates and invertebrates (for a review see Pascual-Anaya *et al.* 2013. While homeobox-containing genes (versus *Hox* genes *sensu stricto*) are vital to tooth and jaw morphogenesis (recently reviewed by Doshi and Patil 2012), interestingly the absence of *Hox* gene expression does not seem requisite for development of the teeth (James *et al.* 2002) as it is for the jaws. This stark difference in developmental programs alongside evidence that the earliest jawed vertebrates lacked teeth (Fraser *et al.* 2009; Finarelli and Coates 2011) and that the first sets of teeth were not integrated with the jaw skeleton (Rücklin *et al.* 2012; Donoghue and Rücklin 2014), bolsters the idea that teeth and jaws have discrete evolutionary and developmental origins and thus a significant amount of independence from each other's influence.

The vertebrate jaw is defined by its hinged joint connecting and allowing the deliberate movement of the lower jaw against the upper jaw (Cerny *et al.* 2010). The upper jaw contributes to the mid-facial skeleton and is thus intimately linked in development, form, and function to sensory organs, respiratory structures, and the brain (Hu and Marcucio 2009; Sugahara *et al.* 2011; Chong *et al.* 2012; Richtsmeier and Flaherty 2013). Conversely, the lower jaw skeleton (i.e., the dentary bone in primates and other mammals) is connected to the rest of the skull only at the jaw joint. Thus, compared to the upper jaw, the lower jaw develops under less influence from the rest of the head, even if both jaws are functionally integrated with each other. In this respect, compared to the upper jaw, the lower jaw is a simpler experimental model with which to tease apart the developmental influence of the jaw on the dentition and vice versa.

In the past decade several essential genes in patterning jaw identity have been identified and placed into evolutionary context. Many of these genes appear to have originated prior to when gnathostome (jawed) vertebrates branched off from agnathan (jawless) animals, notably hagfishes and lampreys (Kuraku *et al.* 2010; Gillis *et al.* 2013). As discussed earlier, the absence of *Hox* gene expression in the first branchial arch seems prerequisite for jaw formation (Hunt *et al.* 1991a). Building on this finding, the nested and precise spatio-temporal expression of endothelin (*Edn*) and distalless (*Dlx*) genes (Beverdam *et al.* 2002; Depew *et al.* 2002, 2005; Ozeki *et al.* 2004; Vieux-Rochas *et al.* 2010), alongside other genes/gene families, such as *Gsc*, *Hand*, *Fgf*, and *Bmp* (e.g., Trumpp *et al.* 1999; Tucker *et al.* 1999) in this signaling pathway, is essential to instruct the mandibular arch to become a lower (versus an upper) jaw (Figure 3.3C, color plate). With their different roles and phenotypes manifested via different developmental programs (Jeong *et al.* 2008; Minoux and Rijli 2010), the mammalian upper and lower jaws typically look distinct from each other even while the shapes of upper and lower dental arcades match. Similar to odontogenesis, gene regulatory pathways and signaling cascades not only determine jaw identity (Ruest *et al.* 2004; Sun *et al.* 2012) but also fine-tune jaw morphology via localized patterning within each embryonic jaw primordium (Trumpp *et al.* 1999; Tucker *et al.* 1999; Aggarwal *et al.* 2010; Gitton *et al.* 2010; Vieux-Rochas *et al.* 2010). Deciphering the gene regulatory networks that pattern the upper jaw skeleton is more challenging if for no other reason than several local developmental events (i.e. formation of sensory organs such as the nose) occur simultaneously.

From these insights arise many questions, chief among them whether species-specific phenotypes are established *in utero* and/or derive in whole or in part via postnatal growth. Micro-computed tomography imaging and 3D morphometric analyses of the facial prominences of brachiocephalic (Crf4) and wildtype (C57BL/6J) mouse embryos, neonates, and adults found that strain-specific phenotypes are visible at least as early as embryonic day (E) 10 (Boughner *et al.* 2008). Thus a template for the adult phenotype appears to be established before birth. After birth, phenotypic differences seem further entrenched via allometric growth patterns; although at least within the same genus (e.g., *Pan*) the mandible, for example, may develop along the same growth trajectory (Boughner and Dean 2008). Between at least *Homo* and *Pan*, different postnatal growth trajectories appear to contribute substantially to adult phenotype only before the deciduous teeth emerge (Singh 2013). Conversely, a recent comparison of postnatal growth trajectories between recent humans and robust australopithecines found genus-level differences up until the emergence of M_2 (Cofran 2014). Together these findings underscore that it is useful if not imperative to study prenatal and postnatal ontogenies to understand the developmental and, by corollary, evolutionary origins of a particular jaw phenotype.

A Brief History of Wondering How Teeth "Know" When and How Fast to Form in the Jawbone

While the genetic processes controlling jaw identity and form, and tooth identity, position, and number, are becoming very well characterized (Thesleff 2003), there remains a black box of information regarding how tooth and jaw tissues change in tandem as primates grow and evolve. Here we return to the question of whether space in the mandible influences the timing of tooth initiation, and the working hypothesis that teeth and jaws develop largely "blind" to each other. Boughner and Dean (2008; 2004) used 3D morphometric and radiographic analyses of chimpanzee (*Pan troglodytes*), bonobo (*P. paniscus*), and baboon (*Papio anubis*) skeletal mandibles to quantify species variation in jaw form and space against timing of permanent molar initiation. In *Papio*, there is significant delay of months between the successive initiation of adjacent permanent molars (M1, M2, M3) such that these teeth develop in sequence but with time between (Swindler and Meekins 1991; Dirks *et al.* 2002). This delay might reflect a lack of space – and the need to grow more of it – in the jawbone behind the deciduous molars/premolars. Conversely, in *Pan*, M1, M2 and M3 initiate in quick succession (Anemone *et al.* 1991; Reid *et al.* 1998), suggesting that adequate space for all three molars is already available. However, space in excess of the minimum needed to fit the permanent molars was seen in both *Pan* and *Papio*. Thus, in these species, space was not a rate-limiting step to molar initiation (Boughner and Dean 2004) and taxonomic variation in the timing of odontogenesis does not seem to be a by-product of jaw form. Subsequent work has supported the hypothesis that space in the jaw does not constrain times of molar onset, implying that schedule of dental development is independent of mandible form (Boughner 2011; Boughner and Dean 2008).

A growing body of evidence continues to indicate that the dentition and jaw skeleton are two developmentally discrete systems that have evolved in complement, and under very strong pressure to become and remain functionally integrated (Gómez-Robles and Polly 2012; Paradis *et al.* 2013). Currently, there is ongoing debate about the degree to which the earliest teeth and jaws were evolutionarily, developmentally, and functionally integrated with each other (Soukup *et al.* 2008; Fraser *et al.* 2009, 2010; Finarelli and Coates 2011; Fraser and Smith 2011; Rücklin *et al.* 2011, 2012). The take-home message at this time is that while the earliest teeth and jaws were not initially well integrated, teeth and jaws later developed and evolved in tightly coordinated ways.

The mouse is the longest-standing and best-established experimental model system to study mammalian odontogenesis (e.g., Thesleff *et al.* 1987; Chen *et al.* 1996; Peterková *et al.* 1996; Thesleff 2006; Kim *et al.* 2012), as well as mandible morphogenesis and growth (e.g. Cheverud *et al.* 1991; Leamy 1993; Depew *et al.* 2002; Klingenberg *et al.* 2004; Ruest *et al.* 2004; Siahsarvie *et al.* 2012), even as it translates to primates (Willmore *et al.* 2009; Paradis *et al.* 2013). Better understood is how the jawbone is resorbed to facilitate tooth eruption (Wise *et al.* 2002; Wise and King 2008; Wise 2009), not mineralization. However, early studies in mutant mouse strains noted that teeth continue to develop, albeit potentially malformed, despite abnormal invasion of bone into the tooth organ (Gruneberg 1937; Diamond 1944). In recent years, contemporary genetic studies using engineered knock-out mice have shown that factors such as parathyroid hormone-related protein (PTHrP) (Mekaapiruk *et al.* 2002) and an Opg/Rank/Rankl feedback loop (Ohazama *et al.* 2004; Alfaqeeh *et al.* 2013) are critical to control osteoclast activity in alveolar bone to continue to make space for the growing tooth. More recently, with the advent of high-resolution 3D imaging capabilities, using mouse models it has been possible to reconstruct in 3D the changing spatial organization of tooth organs, including the dynamic balance of alveolar bone deposition and resorption as the tooth form and then begins to erupt (Hovorakova *et al.* 2005; Chlastakova *et al.* 2011). An important next step is direct visualization in 3D embryonic teeth and the surrounding jaw, in order to statistically assess the spatial organization of these structures relative to one another (Raj *et al.* 2014); for example, whether molar size and/or timing of onset is intrinsically controlled only (Kavanagh *et al.* 2007) or also influenced by adjacent non-dental tissues.

If signals from the jaw tissues and/or, space available in the jaw direct onset time of tooth formation then it is valuable to examine the earliest embryonic stages of odontogenesis, when space would be expected to have significant influence. In particular, the *p63* gene global knock-out mouse mutant (*p63*$^{-/-}$) (Mills *et al.* 1999; Yang *et al.* 1999) has already been used successfully to study the earliest onset of odontogenesis (Laurikkala *et al.* 2006; Rufini *et al.* 2006). Boughner and colleagues are now using this toothless mouse model with an essentially normal mandible (i.e., dentary bone) to understand the developmental genetics of healthy jaw formation in the complete absence of the dentition. This *p63*$^{-/-}$ mutant is rare among mouse models of disrupted odontogenesis because virtually all other mutants also have malformed jaws. This is because key genes (e.g., *Msx1*, *Shh*, *Bmp4*, *Fgf8*) that have been globally, or non-specifically, knocked-down or knocked-out in the embryonic face are essential for both healthy jaw and tooth development. Many of these genes are deployed in different parts (e.g., head, limbs) of the developing body; thus, even if a gene is vital to tooth and jaw formation, this does not necessarily mean that these tissues are regulated by the same gene networks. Paradis and colleagues (2013) combined micro-CT imaging with 3D geometric morphometrics to compare mandible morphology in mice with normal tooth and jaw development (*p63*$^{+/+}$, *p63*$^{+/-}$) against mice (*p63*$^{-/-}$) with normal jaw development despite the early and complete failure of the dentition to form. As this *p63*-null mutation is lethal, mice do not live past birth because they cannot eat or drink properly; however, at least by E18, normal mandible shape was fully recognizable. We saw only two differences in the mandibles of these toothless mice. The first shape change was lack of alveolar bone, which as a derivative of the dental papilla (Hall, 2015), would be expected to be absent in edentate mice. The second shape difference we attributed to a 24–48-hour (1–2 embryonic stages) developmental delay (Figure 3.5, color plate) possibly due to the general increase in developmental noise (Willmore and Hallgrímsson 2005) caused by the global loss of *p63* expression in the embryo. While this tumor-suppressor transcription factor seems extraneous to dentary bone formation, *p63* is vital to odontogenesis not least because it regulates a suite of other genes that act downstream to maintain the dental epithelium (Laurikkala *et al.* 2006), enabling its early and essential interactions with the dental mesenchyme. Conversely, the dentary bone arises largely from mesoderm (Figure 3.3B, color plate) accounting for why the lower jaw skeleton was little affected by the complete loss of *p63* expression (Paradis *et al.* 2013).

Next, we investigated the effect of "toothlessness" on the upper jaw skeleton, composed of multiple bones, notably in mouse (and primate) the premaxillae and

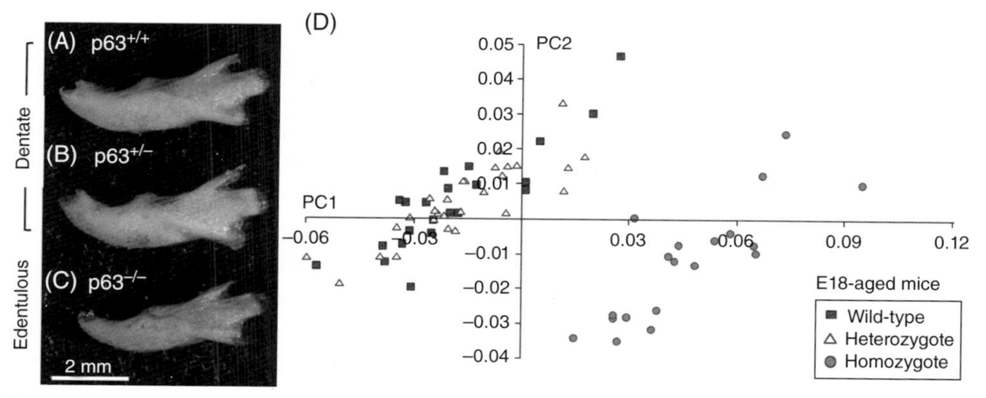

Figure 3.5. Compared to dentate (A [*p63*$^{+/+}$], B [*p63*$^{+/-}$]) mice, mandible (i.e. dentary bone) morphogenesis occurs virtually unperturbed in toothless (C [*p63*$^{-/-}$]) mice by embryonic day (E) 18 (birth occurs around E21). The toothless mandibles lack alveolar bone, which either never forms or is resorbed soon after forming. After a principal component analysis (PCA), the mandible shapes of the toothless *p63*$^{-/-}$ mice appear to match the mandible shapes of dentate mice that are 1–2 days advanced in their development (PC1): thus an E18-aged *p63*$^{-/-}$ mandible looks like a normal E16–17-aged mandible. Otherwise shape variance is comparable (PC2) to that of toothed mice littermates. Adapted from Paradis *et al.* (2013).

maxillae which also house the upper dentition. Contrary to what we saw in the "toothless" *p63*$^{-/-}$ mandible where morphogenesis was virtually unperturbed, we noted that the premaxillae and maxillae were malformed in this mouse mutant. It is not yet clear whether the malformed upper jaw phenotype is due to the absence of teeth or some other symptom of the loss of *p63* activity from the epithelium of the face, including disrupted signaling pathways normally active in that part of the head. Based on these *p63*-centric developmental studies, if the upper jaws and teeth are developmentally interdependent, then this seems to be at later stages such as tooth eruption and emergence (as discussed above), as supported by a large body of evidence ranging from classic studies by Bjork *et al.* (1956) and Richardson (1987) to a more recent review by Wise (2009). Our results suggest an intriguing scenario of varying degrees of ontogenetic separation between jaws and teeth, and developmental-genetic differences between the upper and lower jaws and dentitions. We continue to explore the *p63* gene regulatory network to glean more information about the signaling pathways driving mandible morphogenesis in the absence of odontogenesis to help identify molecular mechanisms of evolutionary change.

Further, We propose that there is rigid selection for cohesive developmental change between teeth and jaws (Gómez-Robles and Polly 2012) such that they fit together properly, orchestrated by the central body clock alongside peripheral clocks (Boughner and Hallgrímsson 2008) to locally control the timing of tooth morphogenesis (Lacruz *et al.* 2012) if not also jaw morphogenesis. While evidence has yet to be produced of peripheral clocks acting in the developing jaw, recent work suggests that genes are expressed in a circadian manner in the bones of the mouse skull (Zvonic *et al.* 2007) and jaw (Gafni *et al.* 2009). The jaw apparatus (i.e., teeth and bone(s)) appears to be a system in which at least a part of the developmental networks is different yet the whole is integrated. In this model, tooth and jaw tissues develop via parallel processes, including distinct core gene regulatory networks, that are strongly integrated by function (Boughner 2014).

What mechanisms would allow coordinated evolution despite independent gene regulatory networks (GRNs)? The elaborate details (reviewed in Ettensohn 2013) are far from being worked out for even a single GRN, let alone two or more GRNs acting and evolving in concert. Studies to date suggest, among other theories, that similar or same suites of genes redeployed in relatively simple, modular ways (e.g. Salazar-Ciudad *et al.* 2000) tend towards a certain type of robustness (i.e., degeneracy, Whitacre and Bender 2010) that may enable evolvability and thus complex phenotypic change without being lethal to the organism. Specific to the jaw apparatus, coordinated change might occur at the level of the GRN guiding CNCC formation and migration (Sauka-Spengler and Bronner-Fraser 2008) – as noted above, the CNCC contribute to both the presumptive teeth and jaws. Later jaw-tooth coordination might happen via different mechanisms, such as physical interactions at the borders of these tissues (i.e., the interface of the tooth follicle/tooth proper and alveolar/cortical jawbone) that trigger a feedback loop of gene expression and cellular movements (Mammoto *et al.* 2011; O'Connell *et al.* 2012). With the expectation that the specific developmental processes will be characterized in time, in theory, it is selection for this tight functional integration that drove change in GRNs, for instance, to facilitate the evolution of the great range of dental-gnathic phenotypes seen across primates, rodents, and quite probably most vertebrates.

Implications for Human and Primate Evo-Devo Studies

Altogether, what does the current state of knowledge suggest about human evolution, particularly regarding our bigger brains and smaller mouths? First, spatial packing of a bigger brain in modern humans has driven the rotation of the face and jaws beneath the cranium and the corresponding flexure of the cranial base (Hallgrímsson *et al.* 2006; Lieberman *et al.* 2008) irrespective (to a certain point at least) of diet and jaw function. Cultural food preparation including cutting and cooking almost certainly played roles in our ability to adapt our diets to fuel our big brains (Eng *et al.* 2013). Related to observed and modeled changes in hominin facial form as accounted for by brain evolution, a recent geometric morphometric study showed that the jaws and teeth of *Homo* respond to selection pressures (intensified or relaxed) in integrated ways dictated largely but not entirely by function (i.e., mastication) (Gómez-Robles and Polly 2012). The finding that jaws and teeth are integrated partly by function raises interesting questions about the role that function played in developmental-evolutionary origins of *Paranthropus* megadont teeth and hyper-robust jawbones (Clement and Hillson 2013; Cofran 2014). Queries about the extent to which function (i.e., versus phylogeny and/or developmental constraints) drives morphology are particularly on-point at this time because recent experimental work in mouse showed that a soft food diet (i.e., function) relaxed developmental constraint and decreased integration strength across modules of the remodeled mandible (Anderson *et al.* 2014). Thus selection for function (i.e., eating certain foods) seems to be an important force integrating jaw and tooth morphology; but other factors affect integration, too. To test if the results in mouse hold true for primates – especially in terms of probing, to paraphrase Anderson and colleagues (2014), if new patterns of covariation might open in primates new avenues of selection (i.e., the distinct mandible robusticity of *Homo floresiensis*, Daegling *et al.* 2014) – it would be useful to compare direction and strength of integration not only among paranthropines and australopithecines but also across modern human populations and non-human primate taxa that eat habitually softer versus harder diets.

Second, because the dentition and jaw skeleton function in different even if complementary ways, developmental autonomy between these two systems allows different directions of phenotypic change and integration. Relevant to Mid-Pleistocene *Homo*, this variation includes the retromolar space, where mandibles grow larger on the anterior-posterior plane than the length of the corpus and thus the amount of space occupied by all teeth, perhaps for functional reasons. However, the hypothesis that masticatory and/or paramasticatory function (e.g., heavy loading of the incisors and canines) in Neanderthals drove the anterior migration of the dentition/creation of a retromolar space has not been supported by recent studies (reviewed in Clement *et al.* 2012). Conversely, recent work suggests that a retromolar space is partly a product of allometry (Harvati *et al.* 2011) and genetic drift (Holton and

Franciscus 2008). Here reappears the theme of function accounting for some, but not all, patterns of integration in the primate jaw apparatus. Also, the dramatic effect that the loss of a single gene (i.e., *p63*) has on the dentition (i.e., tooth agenesis) while leaving at least the lower jaw relatively free from perturbation suggests that the dentition is afforded a high level of developmental and thus evolutionary flexibility that does not put the jaw skeleton, or the rest of the organism, at risk. At least in theory, this lack of pleiotropy allows for dramatic local change in the dentition, which is so varied and distinct among primates that teeth are regularly and sometimes exclusively used to identify a genus or species (e.g., Kramarz *et al.* 2012).

Lastly, the timing of odontogenesis does not appear to be dictated by jaw form or signaling, supporting the idea that different ages of tooth mineralization truly reflect different tempos of childhood development and life history. The processes coordinating development of the jaw apparatus are likely to be conserved across Primates and thus useful for understanding the evolutionary origins of a wide variety of extant and fossil tooth and jaw forms. This work includes modeling constraints on phenotype to probe the interplay of culture and biology on teeth and jaws in the genus *Homo*.

Conclusions

The foundational question remains, how do distinct yet spatially and functionally integrated tissue types develop and consequently evolve into a broad spectrum of viable phenotypes? The fossil record shows an amazing amount of morphological diversity, from hyper-robust to highly gracile teeth and jaws that nonetheless develop in tandem to fit and function properly together as an animal grows. The current data suggest integration in the absence of pleiotropy, at least in early patterning genes such as *p63* that act in distinct regulatory networks driving either jaw or tooth morphogenesis. Other gene families such as *Msx*, *Bmp*, and *Fgf* that are active in jaw and dental tissues may help orchestrate the coevolution of these two systems in ways that are as yet unknown. Future work is needed to address questions of functional, evolutionary and developmental integration between jaws and teeth, and between upper and lower jaws, using developmental genetic techniques alongside *in silico* gene network modeling, and newer systematic approaches, to make sense of the enormous complexity of "black box" morphogenetic mechanisms that translate genotype into phenotype. For physical anthropologists, this information is needed to decipher the processes that have enabled the considerable morphological diversity seen across primates living and extinct.

Acknowledgements

The research presented here was supported by an NSERC Discovery Grant #402148 and NSERC Research Tools & Instruments Grant #458799-2014 to JCB (co-applicant), as well as funds from the College of Medicine/Division of Biomedical Science, University of Saskatchewan.

References

Aggarwal, V. S., Carpenter, C., Freyer, L. *et al.* 2010. Mesodermal Tbx1 is required for patterning the proximal mandible in mice. *Developmental Biology* 344(2):669–681.

Alfaqeeh, S. A., Gaete, M., and Tucker, A.S. 2013. Interactions of the tooth and bone during development. *Journal of Dental Research* 92(12):1129–1135.

Anderson, P., Renaud, S., and Rayfield, E. 2014. Adaptive plasticity in the mouse mandible. *BMC Evolutionary Biology* 14(1):85.

Anemone, R. L., Watts, E. S. and Swindler, D. R. 1991. Dental development of known-age chimpanzees, *Pan troglodytes* (primates, pongidae). *American Journal of Physical Anthropology* 86(2):229–241.

Beaty, N. B., and Ahn, E. 2014. Adamantinomatous craniopharyngioma containing teeth. *New England Journal of Medicine* 370(9):860–860.

Beverdam, A., Merlo, G. R., Paleari, L. *et al.* 2002. Jaw transformation with gain of symmetry after *Dlx5*/*Dlx6* inactivation: Mirror of the past? *Genesis* 34(4):221–227.

Bjork, A., Jensen, E., and Palling, M. 1956. Mandibular growth and third molar impaction. *Acta Odontologica Scandinavica* 14:231–272.

Boughner, J. C. 2011. Making space for permanent molars in growing baboon (*Papio anubis*) and great ape (*Pan paniscus* and *P. troglodytes*) mandibles: Possible ontogenetic strategies and solutions. *Anatomy Research International* 2011.

Boughner, J. C. 2014. Hominin craniodental evolution and autonomy of mandible and tooth morphogenesis. *American Journal of Physical Anthropology* 153(S58):216.

Boughner, J. C., and Dean, M. C. 2004. Does space in the jaw influence the timing of molar crown initiation? A model using baboons (*Papio anubis*) and great apes (*Pan troglodytes, Pan paniscus*). *Journal of Human Evolution* 46(3):255–277.

Boughner, J. C., and Dean, M. C. 2008. Mandibular shape, ontogeny and dental development in bonobos (*Pan paniscus*) and chimpanzees (*Pan troglodytes*). *Evolutionary Biology* 35(4):296.

Boughner, J. C., and Hallgrímsson, B. 2008. Biological spacetime and the temporal integration of functional modules: A case study of dento-gnathic developmental timing. *Developmental Dynamics* 237(1):1–17.

Boughner, J. C., Wat, S., Diewert, V.M. *et al.* 2008. Short-faced mice and developmental interactions between the brain and the face. *Journal of Anatomy* 213(6):646–662.

Boyde, A. 1964. *The structure and development of mammalian enamel.* University of London.

Boyde, A. 1989. Enamel. *In*: B. K. B. Berkovizt, A. Boyde, R. M. Frank *et al.* (eds) *Teeth: handbook of microscopic anatomy*. Berlin: Springer-Verlag. pp.409–473.

Bromage, T. G. and Dean, M. C. 1985. Re-evaluation of the age at death of immature fossil hominids. *Nature* 317(6037):525–527.

Bromage, T. G., Hogg, R. T., Lacruz, R. S., and Hou, C. 2012. Primate enamel evinces long period biological timing and regulation of life history. *Journal of Theoretical Biology* 305:131–144.

Butler, P. M. 1937. Studies of the mammalian dentition. *I. The teeth of Centetes ecaudatus and its allies. Proceedings of the Zoological Society of London* B107(1):103–132.

Butler, P. M. 1939. Studies of the mammalian dentition, differentiation of the post-canine dentition. *Proceedings of The Zoological Society of London* B109(1):1–36.

Carroll, S. B., Grenier, J. K., and Weatherbee, S. D. 2001. *From DNA to diversity: Molecular genetics and the evolution of animal design*. Oxford: Blackwell Scientific.

Caumul, R. and Polly, P. D. 2005. Phylogenetic and environmental components of morphological variation: Skull, mandible, and molar shape in marmots (*Marmota*, Rodentia). *Evolution* 59(11):2460–2472.

Cerny, R., Cattell, M., Sauka-Spengler, T. *et al.* 2010. Evidence for the prepattern/cooption model of vertebrate jaw evolution. *Proceedings of the National Academy of Sciences, USA* 107(40):17262–17267.

Cerny, R., Lwigale, P., Ericsson, R. *et al.* 2004. Developmental origins and evolution of jaws: New interpretation of "maxillary" and "mandibular." *Developmental Biology* 276(1): 225–236.

Chai, Y., Bringas, P., Shuler, C. *et al.* 1998. A mouse mandibular culture model permits the study of neural crest cell migration and tooth development. *International Journal of Developmental Biology* 42:87–94.

Chai, Y., Jiang, X., Ito, Y. *et al.* 2000. Fate of the mammalian cranial neural crest during tooth and mandibular morphogenesis. *Development* 127(8):1671–1679.

Chen, Y., Bei, M., Woo, I. *et al.* 1996. *Msx1* controls inductive signaling in mammalian tooth morphogenesis. *Development* 122(10):3035–3044.

Cheverud, J. M., Hartman, S. E., Richtsmeier, J. T., and Atchley, W. R. 1991. A quantitative genetic analysis of localized morphology in mandibles of inbred mice using finite element scaling. *Journal of Craniofacial Genetics and Developmental Biology* 11:122–137.

Chlastakova, I., Lungova, V., Wells, K. *et al.* 2011. Morphogenesis and bone integration of the mouse mandibular third molar. *European Journal of Oral Sciences* 119(4):265–274.

Chong, H. J., Young, N. M., Hu, D. *et al.* 2012. Signaling by *Shh* rescues facial defects following blockade in the brain. *Developmental Dynamics* 241(2):247–256.

Clement, A. F., and Hillson, S. W. 2013. "Do larger molars and robust jaws in early hominins represent dietary adaptation?" A new study in tooth wear. *Archaeology International* 16:59–71.

Clement, A. F., Hillson, S. W., and Aiello, L. C. 2012. Tooth wear, Neanderthal facial morphology and the anterior dental loading hypothesis. *Journal of Human Evolution* 62(3):367–376.

Cofran, Z. 2014. Mandibular development in *Australopithecus robustus*. *American Journal of Physical Anthropology* 154(3):436–446.

Cohn, M. J. 2002. Evolutionary biology: Lamprey Hox genes and the origin of jaws. *Nature* 416(6879):386–387.

Couly, G., Creuzet, S., Bennaceur, S. *et al.* 2002. Interactions between *Hox*-negative cephalic neural crest cells and the foregut endoderm in patterning the facial skeleton in the vertebrate head. *Development* 129(4):1061–1073.

Daegling, D. J., Patel, B. A., and Jungers, W. L. 2014. Geometric properties and comparative biomechanics of *Homo floresiensis* mandibles. *Journal of Human Evolution* 68:36–46.

Dahlberg, A. 1945. The changing dentition of man. *Journal of the American Dental Association* 32:676–690.

Dart, R. 1948. The infancy of *Australopithecus*. *Robert Broom Commemorative Volume*. Cape Town: Royal Society of South Africa. pp.143–152.

Dean, M. C. 1987. Growth layers and incremental markings in hard tissues: A review of the literature and some preliminary observations about enamel structure in *Paranthropus boisei*. *Journal of Human Evolution* 16(2):157.

Dean, M. C. 2006. Tooth microstructure tracks the pace of human life-history evolution. *Proceedings of the Royal Society of London, Series B: Biological Sciences* 273(1603):2799–2808.

Dean, M. C., and Beynon, A. D. 1991. Tooth crown heights, tooth wear, sexual dimorphism and jaw growth in hominoids. *Zeitschrift für Morphologie und Anthropologie* 78(3): 425–440.

Dean, M. C., and Munro, C. F. 2005. Enamel growth and thickness in human teeth from ovarian teratomas (dermoid cysts). *In*: E. Żądzińska (ed.) *13th International Symposium on Dental Morphology*. Lodz: Poland Wydawnictwo Uniwersytetu Lodzkiego. pp.371–382.

Dean, M. C., and Wood, B. A. 1981. Developing pongid dentition and its use for ageing individual crania in comparative cross-sectional growth studies. *Folia Primatologica (Basel)* 36(1–2):111–127.

Depew, M. J., Lufkin, T., and Rubenstein, J. L. 2002. Specification of jaw subdivisions by *Dlx* genes. *Science* 298(5592):381–385.

Depew, M. J., Simpson, C. A., Morasso, M., and Rubenstein, J. L. R. 2005. Reassessing the *Dlx* code: The genetic regulation of branchial arch skeletal pattern and development. *Journal of Anatomy* 207(5):501–561.

Diamond, M. 1944. The patterns of growth and development of the human teeth and jaws. *Journal of Dental Research* 23:273–303.

Dirks, W., Reid, D. J., Jolly, C. J. *et al.* 2002. Out of the mouths of baboons: Stress, life history, and dental development in the Awash National Park hybrid zone, Ethiopia. *American Journal of Physical Anthropology* 118(3):239.

Donoghue, P. C. J. and Rücklin, M. 2014. The ins and outs of the evolutionary origin of teeth. *Evolution & Development* doi:10.1111/ede.12099.

Doshi, R. R., and Patil, A. S. 2012. A role of genes in craniofacial growth. *Institute of Integrative Omics and Applied Biotechnology Journal* 3(2):19–36.

Eng, C. M., Lieberman, D. E., Zink, K. D., and Peters, M. A. 2013. Bite force and occlusal stress production in hominin evolution. *American Journal of Physical Anthropology* 151(4):544–557.

Ettensohn, C. A. 2013. Encoding anatomy: Developmental gene regulatory networks and morphogenesis. *Genesis* 51(6):383–409.

Finarelli, J. A., and Coates, M. I. 2011. First tooth-set outside the jaws in a vertebrate. *Proceedings of the Royal Society of London, Series B: Biological Sciences* 279:775–779.

Fraser, G. J., Cerny, R., Soukup, V. *et al.* 2010. The odontode explosion: The origin of tooth-like structures in vertebrates. *BioEssays* 32:808–817.

Fraser, G. J., Graham, A., and Smith, M. M. 2006. Developmental and evolutionary origins of the vertebrate dentition: Molecular controls for spatio-temporal organisation of tooth sites in osteichthyans. *Journal of Experimental Zoology Part B: Molecular and Developmental Evolution* 306B(3):183–203.

Fraser, G. J., Hulsey, C. D., Bloomquist, R. F. *et al.* 2009. An ancient gene network is co-opted for teeth on old and new jaws. *PLoS Biol* 7(2):e1000031.

Fraser, G. J., and Smith, M. M. 2011. Evolution of developmental pattern for vertebrate dentitions: An oro-pharyngeal specific mechanism. *Journal of Experimental Zoology Part B Molecular and Developmental Evolution* 316:99–112.

Fujita, K., Taya, Y., Shimazu, Y. *et al.* 2013. Molecular signaling at the fusion stage of the mouse mandibular arch: Involvement of insulin-like growth factor family. *International Journal of Developmental Biology* 57(5):399–406.

Gafni, Y., Ptitsyn, A. A., Zilberman, Y. *et al.* 2009. Circadian rhythm of osteocalcin in the maxillomandibular complex. *Journal of Dental Research* 88(1):45–50.

Garn, S. M., Lewis, A. B., and Polacheck, D. L. 1959. Variability of tooth formation. *Journal of Dental Research* 38(1):135–148.

Gillis, J. A., Modrell, M. S., and Baker, C. V. H. 2013. Developmental evidence for serial homology of the vertebrate jaw and gill arch skeleton. *Nature Communications* 4:1436.

Gitton, Y., Heude, É., Vieux-Rochas, M. *et al.* 2010. Evolving maps in craniofacial development. *Seminars in Cell & Developmental Biology* 21(3):301.

Glasstone, S. 1938. A comparative study of the development in vivo and in vitro of rat and rabbit molars. *Proceedings of the Royal Society of London B* 126:315–330.

Glasstone, S. 1967. Morphodifferentiation of teeth in embryonic mandibular segments in tissue culture. *Journal of Dental Research* 46:611–614.

Gómez-Robles, A., and Polly, P. D. 2012. Morphological integration in the hominin dentition: Evolutionary, developmental, and functional factors. *Evolution* 66(4):1024–1043.

Goodrich, E. S. 1913. Metameric segmentation and homology. *Quarterly Journal of Microscopic Science* 59:227–2248.

Grammatopoulos, G. A., Bell, E., Toole, L. *et al.* 2000. Homeotic transformation of branchial arch identity after Hoxa2 overexpression. *Development* 127(24):5355–5365.

Gruneberg, H. 1937. The relations of endogenous and exogenous factors in bone and tooth development. *Journal of Anatomy* 71(Pt 2):236–244.1.

Hall, B.K. 2015. Bone and cartilage: Developmental and evolutionary skeletal biology. Elsevier: London.

Hallgrímsson, B., Brown, J. J. Y., Ford-Hutchinson, A. F. *et al.* 2006. The brachymorph mouse and the developmental-genetic basis for canalization and morphological integration. *Evolution & Development* 8(1):61–73.

Hallgrímsson, B., Dorval, C. J., Zelditch, M. L., and German, R. Z. 2004. Craniofacial variability and morphological integration in mice susceptible to cleft lip and palate. *Journal of Anatomy* 205(6):501–517.

Harvati, K., Singh, N., and López, E. 2011. A three-dimensional look at the Neanderthal mandible. *In*: S. Condemi and G.-C. Weniger (eds) *Continuity and discontinuity in the peopling of Europe*. Springer: Netherlands. pp.179–192.

Holton, N. E., and Franciscus, R. G. 2008. The paradox of a wide nasal aperture in cold-adapted Neandertals: A causal assessment. *Journal of Human Evolution* 55(6):942–951.

Hovorakova, M., Lesot, H., Peterka, M., and Peterkova, R. 2005. The developmental relationship between the deciduous dentition and the oral vestibule in human embryos. *Anatomy and Embryology* 209(4):303–313.

Hu, D., and Marcucio, R. S. 2009. A *Shh*-responsive signaling center in the forebrain regulates craniofacial morphogenesis via the facial ectoderm. *Development* 136(1):107–116.

Hunt, P., Gulisano, M., Cook, M. *et al.* 1991a. A distinct *Hox* code for the branchial region of the vertebrate head. *Nature* 353(6347):861–864.

Hunt, P., Whiting, J., Nonchev, S. *et al.* 1991b. The branchial *Hox* code and its implications for gene regulation, patterning of the nervous system and head evolution. *Development Suppl.* 2:63–77.

Huxley, J., and de Beer, G. R. 1934. *The elements of experimental embryology*. Cambridge: Cambridge University Press.

James, C. T., Ohazama, A., Tucker, A. S., and Sharpe, P. T. 2002. Tooth development is independent of a *Hox* patterning programme. *Developmental Dynamics* 225(3):332–335.

Jeong, J., Li, X., McEvilly, R. J. *et al.* 2008. *Dlx* genes pattern mammalian jaw primordium by regulating both lower jaw-specific and upper jaw-specific genetic programs. *Development* 135(17):2905–2916.

Jernvall, J., and Thesleff, I. 2000. Reiterative signaling and patterning during mammalian tooth morphogenesis. *Mechanisms of Development* 92(1):19–29.

Jernvall, J., and Thesleff, I. 2012. Tooth shape formation and tooth renewal: Evolving with the same signals. *Development* 139(19):3487–3497.

Jussila, M., and Thesleff, I. 2012. Signaling networks regulating tooth organogenesis and regeneration, and the specification of dental mesenchymal and epithelial cell lineages. *Cold Spring Harbor Perspectives in Biology* 4(4).

Kasberg, A. D., Brunskill, E. W., and Steven Potter, S. 2013. SP8 regulates signaling centers during craniofacial development. *Developmental Biology* 381(2):312–323.

Kaski, M., Nieminen, P., Sahlberg, C. *et al.* 1996. *Gene expression in tooth (www database)*. University of Helsinki.

Kavanagh, K. D., Evans, A. R., and Jernvall, J. 2007. Predicting evolutionary patterns of mammalian teeth from development. *Nature* 449(7161):427–432.

Kim, K. M., Lim, J., Choi, Y.-A. *et al.* 2012. Gene expression profiling of oral epithelium during tooth development. *Archives of Oral Biology* 57(8):1100–1107.

Klingenberg, C. P., Leamy, L. J., and Cheverud, J. M. 2004. Integration and modularity of quantitative trait locus effects on geometric shape in the mouse mandible. *Genetics* 166(4):1909–1921.

Kollar, E. J., and Baird, G. R. 1969. The influence of the dental papilla on the development of tooth shape in embryonic mouse tooth germs. *Journal of Embryology and Experimental Morphology* 21(1):131–148.

Kollar, E. J., and Lumsden, A. G. 1979. Tooth morphogenesis: The role of the innervation during induction and pattern formation. *Journal de Biologie Buccale* 7(1):49–60.

Kramarz, A. G., Tejedor, M. F., Forasiepi, A. M., and Garrido, A. C. 2012. New early Miocene primate fossils from northern Patagonia, Argentina. *Journal of Human Evolution* 62(1):186–189.

Kuraku, S., and Meyer, A. 2009. The evolution and maintenance of *Hox* gene clusters in vertebrates and the teleost-specific genome duplication. *International Journal of Developmental Biology* 53(5–6):765–773.

Kuraku, S., Takio, Y., Sugahara, F. *et al.* 2010. Evolution of oropharyngeal patterning mechanisms involving *Dlx* and endothelins in vertebrates. *Developmental Biology* 341(1): 315–323.

Kuratani, S. 2004. Evolution of the vertebrate jaw: Comparative embryology and molecular developmental biology reveal the factors behind evolutionary novelty. *Journal of Anatomy* 205(5):335–347.

Lacruz, R. S., and Bromage, T. G. 2006. Appositional enamel growth in molars of South African fossil hominids. *Journal of Anatomy* 209(1):13.

Lacruz, R. S., Hacia, J. G., Bromage, T. G. *et al.* 2012. The circadian clock modulates enamel development. *Journal of Biological Rhythms* 27(3):237–245.

Laurikkala, J., Mikkola, M. L., James, M. *et al.* 2006. *p63* regulates multiple signalling pathways required for ectodermal organogenesis and differentiation. *Development* 133(8): 1553–1563.

Leamy, L. J. 1993. Morphological integration of fluctuating asymmetry in the mouse mandible. *Genetica* 89:139–153.

Lee, S. H., Bedard, O., Buchtova, M. *et al.* 2004. A new origin for the maxillary jaw. *Developmental Biology* 276(1):207–224.

Lewis, A. B., and Garn, S. M. 1960. The relationship between tooth formation and other maturational factors. *The Angle Orthodontist* 30(2):70–77.

Lewis, E. B. 1978. A gene complex controlling segmentation in *Drosophila*. *Nature* 276(5688):565–570.

Lieberman, D. E., Hallgrímsson, B., Liu, W. *et al.* 2008. Spatial packing, cranial base angulation, and craniofacial shape variation in the mammalian skull: Testing a new model using mice. *Journal of Anatomy* 212(6):720–735.

Liversidge, H. M. 2008. Timing of human mandibular third molar formation. *Annals of Human Biology* 35(3):294–321.

Lucas, P. W. 1981. An analysis of canine size and jaw shape in some Old and New World non-human primates. *Journal of Zoology* 195(4):437–448.

Lumsden, A. 1979. Pattern formation in the molar dentition of the mouse. *Journal de Biologie Buccale* 7:77–103.

Mammoto, T., Mammoto, A., Torisawa, Y.-s. *et al.* 2011. Mechanochemical control of mesenchymal condensation and embryonic tooth organ formation. *Developmental Cell* 21(4):758–769.

Mark, M., Rijli, F. M., and Chambon, P. 1997. Homeobox genes in embryogenesis and pathogenesis. *Pediatric Research* 42(4):421–429.

McCollum, M., and Sharpe, P. T. 2001. Evolution and development of teeth. *Journal of Anatomy* 199(Pt 1–2):153–159.

Medio, M., Yeh, E., Popelut, A. *et al.* 2012. Wnt/ß-catenin signaling and *Msx1* promote outgrowth of the maxillary prominences. *Frontiers in Physiology* 3.

Mekaapiruk, K., Suda, N., Hammond, V. E. *et al.* 2002. The influence of parathyroid hormone-related protein (PTHrP) on tooth-germ development and osteoclastogenesis in alveolar bone of PTHrP-knock out and wild-type mice in vitro. *Archives of Oral Biology* 47(9):665.

Mills, A. A., Zheng, B., Wang, X. J. *et al.* 1999. *p63* is a *p53* homologue required for limb and epidermal morphogenesis. *Nature* 398(6729):708–713.

Mina, M. 2001. Regulation of mandibular growth and morphogenesis. *Critical Reviews in Oral Biology & Medicine* 12(4):276–300.

Minoux, M., and Rijli, F. M. 2010. Molecular mechanisms of cranial neural crest cell migration and patterning in craniofacial development. *Development* 137(16):2605–2621.

Mitsiadis, T. A., and Smith, M. M. 2006. How do genes make teeth to order through development? *Journal of Experimental Zoology Part B: Molecular and Developmental Evolution* 306B(3):177–182.

Nanci, A. 2012. *Ten Cate's oral histology: Development, structure and function*. UK: Elsevier. p.400.

Nanda, R. S. 1960. Eruption of human teeth. *American Journal of Orthodontics* 46:363–378.

O'Connell, D., Ho, J., Mammoto, T. *et al.* 2012. A Wnt-Bmp feedback circuit controls intertissue signaling dynamics in tooth organogenesis. *Science Signaling* 5:ra4.

Ohazama, A., Courtney J. M., and Sharpe P. T. 2004. Opg, Rank, and Rankl in tooth development: Co-ordination of odontogenesis and osteogenesis. *Journal of Dental Research* 83(3):241–244.

Organ, C., Nunn, C. L., Machanda, Z., and Wrangham, R. W. 2011. Phylogenetic rate shifts in feeding time during the evolution of *Homo*. *Proceedings of the National Academy of Sciences, USA* 108(35):14555–14559.

Osborn, J. 1971. The ontogeny of tooth succession in *Lacerta vivipara* Jacquin (1787). *Proceedings of the Royal Society of London, Series B: Biological Sciences* 179:261–289.

Osborn, J. 1978. Morphogenetic gradients: Fields vs. clones. *In*: P. Butler and K. Joysey (eds) *Development, function and evolution of teeth*. New York: Academic Press.

Ozeki, H., Kurihara, Y., Tonami, K. *et al.* 2004. Endothelin-1 regulates the dorsoventral branchial arch patterning in mice. *Mechanisms of Development* 121(4):387.

Paradis, M. R., Raj, M. T., and Boughner, J. C. 2013. Jaw growth in the absence of teeth: The developmental morphology of edentulous mandibles using the *p63* mouse mutant. *Evolution & Development* 15(4):268–279.

Parsons, T. E., Schmidt, E. J., Boughner, J. C. *et al.* 2011. Epigenetic integration of the developing brain and face. *Developmental Dynamics* 240(10):2233–2244.

Pascual-Anaya, J., D'Aniello, S., Kuratani, S., and Garcia-Fernandez, J. 2013. Evolution of *Hox* gene clusters in deuterostomes. *BMC Developmental Biology* 13(1):26.

Peterková, R., Lesot, H., Vonesch, J. L. *et al.* 1996. Mouse molar morphogenesis revisited by three-dimensional reconstruction. *I. Analysis of initial stages of the first upper molar development revealed two transient buds. International Journal of Developmental Biology* 40:1009–1016.

Plikus, M. V., Zeichner-David, M., Mayer, J.-A. *et al.* 2005. Morphoregulation of teeth: Modulating the number, size, shape and differentiation by tuning Bmp activity. *Evolution & Development* 7(5):440–457.

Raj, M. T., Prusinkiewicz, M., Cooper, D. M. L. *et al.* 2014. Technique: Imaging earliest tooth development in 3D using a silver-based tissue contrast agent. *The Anatomical Record* 297(2):222–233.

Reid, D. J., Schwartz, G. T., Dean, C., and Chandrasekera, M. S. 1998. A histological reconstruction of dental development in the common chimpanzee, *Pan troglodytes*. *Journal of Human Evolution* 35(4–5):427–448.

Renvoisé, E., Evans, A. R., Jebrane, A. *et al.* 2009. Evolution of mammal tooth patterns: New insights from a developmental prediction model. *Evolution* 63(5):1327–1340.

Ribeiro, M. M., de Andrade, S. C., de Souza, A. P., and Line, S. R. P. 2013. The role of modularity in the evolution of primate postcanine dental formula: Integrating jaw space with patterns of dentition. *The Anatomical Record* 296(4):622–629.

Richardson, M. E. 1987. Lower third molar space. *Angle Orthodontist* 57:155.

Richtsmeier, J., and Flaherty, K. 2013. Hand in glove: Brain and skull in development and dysmorphogenesis. *Acta Neuropathologica* 125(4):469–489.

Robinson, J. T. 1954. Prehominid dentition and hominid evolution. *Evolution* 8(4):324–334.

Ross, C. F., and Iriarte-Diaz, J. 2014. What does feeding system morphology tell us about feeding? *Evolutionary Anthropology* 23(3):105–120.

Rücklin, M., Donoghue, P. C. J., Johanson, Z. *et al.* 2012. Development of teeth and jaws in the earliest jawed vertebrates. *Nature* 491(7426):748–751.

Rücklin, M., Giles, S., Janvier, P., and Donoghue, P. C. J. 2011. Teeth before jaws? Comparative analysis of the structure and development of the external and internal scales in the extinct jawless vertebrate *Loganellia scotica*. *Evolution & Development* 13(6):523–532.

Ruest, L.-B., Xiang, X., Lim, K.-C. *et al.* 2004. Endothelin-A receptor-dependent and -independent signaling pathways in establishing mandibular identity. *Development* 131(18):4413–4423.

Rufini, A., Weil, M., McKeon, F. *et al.* 2006. *p63* protein is essential for the embryonic development of vibrissae and teeth. *Biochemical and Biophysical Research Communications* 340(3):737–741.

Salazar-Ciudad, I., Garcia-Fernandez, J., and Sole, R. V. 2000. Gene networks capable of pattern formation: From induction to reaction-diffusion. *Journal of Theoretical Biology* 205(4):587–603.

Salazar-Ciudad, I., and Jernvall, J. 2010. A computational model of teeth and the developmental origins of morphological variation. *Nature* 464(7288):583.

Sauka-Spengler, T., and Bronner-Fraser, M. 2008. A gene regulatory network orchestrates neural crest formation. *Nature Reviews Molecular Cell Biology* 9(7):557–568.

Schour, I. 1934. Effect of tooth injury on other teeth. I. The effect of a fracture confined to one or two incisors and their investing tissues on the other incisors in the rat. *Physiological Zoology* 7:304–329.

Scott, J. E., and Lockwood, C. A. 2004. Patterns of tooth crown size and shape variation in great apes and humans and species recognition in the hominid fossil record. *American Journal of Physical Anthropology* 125(4):303–319.

Sharpe, P. 1995. Homeobox genes and orofacial development. *Connective Tissue Research* 32(1–4):17–25.

Siahsarvie, R., Auffray, J.-C., Darvish, J. *et al.* 2012. Patterns of morphological evolution in the mandible of the house mouse *Mus musculus* (Rodentia: Muridae). *Biological Journal of the Linnean Society* 105(3):635–647.

Simons, E. L. 1976. The nature of the transition in the dental mechanism from pongids to hominids. *Journal of Human Evolution* 5(5):511–528.

Singh, N. 2013. *Ontogenetic study of allometric variation in Homo and Pan mandibles. The Anatomical Record.*

Skinner, M. M., Evans, A., Smith, T. *et al.* 2010. Brief communication: Contributions of enamel-dentine junction shape and enamel deposition to primate molar crown complexity. *American Journal of Physical Anthropology* 142(1):157–163.

Skinner, M. M., Gunz, P., Wood, B. A. *et al.* 2009. Discrimination of extant *Pan* species and subspecies using the enamel–dentine junction morphology of lower molars. *American Journal of Physical Anthropology* 140(2):234–243.

Smith, B. H., Crummet, T. L., and Brandt, K. L. 1994. Ages of eruption of primate teeth: A compendium for aging individuals and comparing life histories. *Yearbook of Physical Anthropology* 37(S19):177–231.

Smith, T. M., Reid, D. J., Dean, M. C. *et al.* 2007. Molar development in common chimpanzees (*Pan troglodytes*). *Journal of Human Evolution* 52(2):201–216.

Soukup, V., Epperlein, H.-H., Horacek, I., and Cerny, R. 2008. Dual epithelial origin of vertebrate oral teeth. *Nature* 455(7214):795–798.

Stock, D. W., Weiss, K. M., and Zhao, Z. 1997. Patterning of the mammalian dentition in development and evolution. *BioEssays* 19:481–490.

Sugahara, F., Aota, S., Kuraku, S. *et al.* 2011. Involvement of Hedgehog and FGF signalling in the lamprey telencephalon: Evolution of regionalization and dorsoventral patterning of the vertebrate forebrain. *Development* 138(6):1217–1226.

Sun, Y., Teng, I., Huo, R. *et al.* 2012. Asymmetric requirement of surface epithelial β-catenin during the upper and lower jaw development. *Developmental Dynamics* 241(4):663–674.

Swindler, D. R. 1976. *Dentition of living primates*. London: Academic Press.

Swindler, D. R., and Meekins, D. 1991. Dental development of the permanent mandibular teeth in the baboon, *Papio cynocephalus*. *American Journal of Human Biology* 3(6):571.

Ten Cate, A. R. 1994. *Oral histology: Development, structure, and function*. London: Mosby.

Thesleff, I. 2003. Developmental biology and building a tooth. *Quintessence International* 34(8):613–620.

Thesleff, I. 2006. The genetic basis of tooth development and dental defects. *American Journal of Medical Genetics Part A* 140A(23):2530–2535.

Thesleff, I., Keranen, S., and Jernvall, J. 2001. Enamel knots as signaling centers linking tooth morphogenesis and odontoblast differentiation. *Advances in Dental Research* 15(1):14–18.

Thesleff, I., Mackie, E., Vainio, S., and Chiquet-Ehrismann, R. 1987. Changes in the distribution of tenascin during tooth development. *Development* 101(2):289–296.

Thesleff, I., and Sharpe, P. 1997. Signalling networks regulating dental development. *Mechanisms of Development* 67(2):111–123.

Thesleff, I., Vaahtokari, A., Kettunen, P., and Aberg, T. 1995. Epithelial-mesenchymal signaling during tooth development. *Connective Tissue Research* 32(1–4):9–15.

Thomas, B. L., Tucker, A. S., Qui, M. *et al.* 1997. Role of *Dlx-1* and *Dlx-2* genes in patterning of the murine dentition. *Development* 124(23):4811–4818.

Thomason, H. A., Dixon, M. J., and Dixon, J. 2008. Facial clefting in Tp63 deficient mice results from altered *Bmp4*, *Fgf8* and *Shh* signaling. *Developmental Biology* 132(1):273–282.

Townsend, G., Harris, E. F., Lesot, H. *et al.* 2009. Morphogenetic fields within the human dentition: A new, clinically relevant synthesis of an old concept. *Archives of Oral Biology* 54:S34.

Trumpp, A., Depew, M. J., Rubenstein, J. L. R *et al.* 1999. Cre-mediated gene inactivation demonstrates that *Fgf8* is required for cell survival and patterning of the first branchial arch. *Genes and Development* 13(23):3136–3148.

Tucker, A. S., Yamada, G., Grigoriou, M. *et al.* 1999. Fgf-8 determines rostral-caudal polarity in the first branchial arch. *Development* 126(1):51.

Vieux-Rochas, M., Mantero, S., Heude, E. *et al.* 2010. Spatio-temporal dynamics of gene expression of the Edn1-*Dlx5/6* pathway during development of the lower jaw. *Genesis* 48(6):262.

von Cramon-Taubadel, N. 2011. Global human mandibular variation reflects differences in agricultural and hunter-gatherer subsistence strategies. *Proceedings of the National Academy of Sciences, USA* 108(49):19546–19551.

Weiss, K. M., Stock, D. W., and Zhao, Z. 1998. Dynamic interactions and the evolutionary genetics of dental patterning. *Critical Reviews in Oral Biology & Medicine* 9(4):369–398.

Whitacre, J., and Bender, A. 2010. Degeneracy: A design principle for achieving robustness and evolvability. *Journal of Theoretical Biology* 263(1):143–153.

Willmore, K., and Hallgrímsson, B. 2005. Within individual variation: Developmental noise versus developmental stability. *In*: B. Hallgrímsson and H. B. K. (eds) *Variation: A central concept in biology*. London: Elsevier. pp.191–218.

Willmore, K., Roseman, C. C., Rogers, J. *et al.* 2009. Comparison of mandibular phenotypic and genetic integration between baboon and mouse. *Evolutionary Biology* 36(1):19–36.

Wise, G. E. 2009. Cellular and molecular basis of tooth eruption. *Orthodontics & Craniofacial Research* 12(2):67.

Wise, G. E, Frazier-Bowers, S., and D'Souza, R. N. 2002. Cellular, molecular, and genetic determinants of tooth eruption. *Critical Reviews in Oral Biology & Medicine* 13(4):323–335.

Wise, G. E., and King, G. J. 2008. Mechanisms of tooth eruption and orthodontic tooth movement. *Journal of Dental Research* 87(5):414–434.

Wood, B., and Aiello, L. C. 1998. Taxonomic and functional implications of mandibular scaling in early hominins. *American Journal of Physical Anthropology* 105(4):523–538.

Wood, B. 2009. Where does the genus *Homo* begin, and how would we know? *In*: F. E. Grine *et al.* (eds) *The first humans: Origin and early evolution of the genus* Homo. Dordrecht, Netherlands: Springer Science+Business Media B.V. pp.17–28.

Yang, A., Schweitzer, R., Sun, D. *et al.* 1999. *p63* is essential for regenerative proliferation in limb, craniofacial and epithelial development. *Nature* 398(6729):714.

Zvonic, S., Ptitsyn, A. A., Kilroy, G. *et al.* 2007. Circadian oscillation of gene expression in murine calvarial bone. *Journal of Bone and Mineral Research* 22(3):357–365.

Chapter 4
Genetic Regulation of Amelogenesis and Implications for Hominin Ancestors

Rodrigo S. Lacruz
Department of Basic Science and Craniofacial Biology, New York University College of Dentistry, New York, NY, USA

Introduction

The recognition that the evolutionary history of species has sometimes been interpreted based solely on the analysis of dental remains is testimony to the wide yet species-specific range of variation in tooth shape and the durability of tooth enamel. In its matured or mineralized form, dental enamel can be regarded as a living fossil preserving less than 2% of organic material. The durability of enamel is all the more impressive when one considers its constant exposure to a pathogen-rich and highly acidic environment in the oral cavity, and its constant use in chewing relatively hard objects. During amelogenesis (the process of enamel formation), enamel is a soft gel-like matrix consisting of a small number of enamel-specific extracellular proteins. These proteins act as a scaffold for the growth of the enamel crystals. The main function of teeth is to break down food particles. Thus, evolutionary adaptations to different diets have resulted in an impressive range of variation in the mammalian tooth shapes and underlying enamel microstructures that have presumably been optimized through evolutionary time to increase tooth performance. Besides these critical functional aspects of mammalian teeth, it is possible to identify microstructural characteristics of mineralized enamel that are associated with daily or circadian growth. These microstructural features, known as cross striations, have been used as a key source for interpreting the life history of our hominin ancestors (Dean *et al.* 2001; Lacruz *et al.* 2008; Dean 2010).

Perhaps one of the most unexpected evolutionary innovations identified in some early hominins (e.g., *Australopithecus*, *Paranthropus*) was the relatively large size of the molar teeth compared to those of *Homo sapiens* (McHenry 2002). A contributing factor to tooth size is the thickness of the enamel covering the tooth. Early African hominins of the genera *Australopithecus* and *Paranthropus* showed greater molar occlusal enamel thickness relative to *H. sapiens* (Olejniczak *et al.* 2008). The development of enamel thickness is related to three processes: (i) the number of cells involved, (ii) the lifespan of the cells, (iii) and the amount of enamel proteins secreted by the cells on a daily basis. The daily growth of enamel is the main focus of this

Developmental Approaches to Human Evolution, First Edition. Edited by
Julia C. Boughner and Campbell Rolian.

chapter. Recent research into the molecular basis of cross striations suggests that their development is associated with the activity of the circadian clock. The circadian clock is the body's central timekeeper, located in the suprachiasmatic nucleus (SCN) of the brain and responsible for regulating activity patterns over a roughly 24-hour period. Secondary to this "master clock," peripheral clocks are evident in many organs, such as the liver, heart, kidney, and as shown here, also in enamel. Based on this research, a hypothesis is proposed here to explain a possible physiological function of a peripheral clock in enamel. Finally, this information is used to interpret how modifications of the activity of the enamel clock influenced hominin evolution.

Tooth and Enamel Development

The development of teeth involves complex processes of heterotypic and homotypic cell interactions followed by differentiation of tissue-specific secretory cells (Thesleff and Hurmerinta 1981; Jernvall and Thesleff 2012). Tooth morphogenesis is similar to the development of other organs in that tissue interactions are mediated by reciprocal signals from ectodermally-derived epithelium and neural crest mesenchyme (Thomas and Sharpe 1998). Ameloblasts derived from the epithelium secrete enamel matrix while odontoblasts derived from the mesenchyme secrete dentine matrix. Tooth morphogenesis and enamel development form a developmental continuum. Tooth morphogenesis includes the bud, cap, and bell stages, whereas enamel development can be described largely as a two-stage process of secretion followed by maturation (Nanci 2008). The secretory stage is initiated after epithelial cells of the enamel organ transform into ameloblasts via ultrastructural, size, and polarity changes. These cells then become highly polarized and fully functional secretory cells (Figure 4.1, color plate). These cells are tightly bound by junctional complexes and form cell cohorts as they move away from the dentine. Their movement follows a predetermined pathway that may vary from species to species and also from tooth type to tooth type.

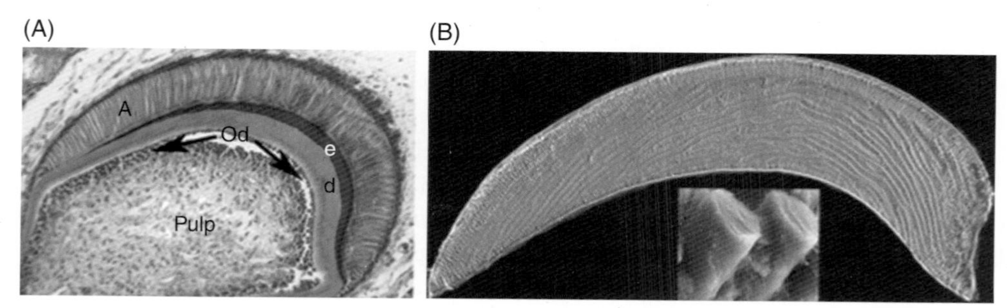

Figure 4.1. (A) Coronal section of a mouse incisor at postnatal day 2 showing the secretory ameloblast layer (A), enamel (e), odontoblast layer (Od), dentine (d), and pulp chamber. (B) Scanning electron micrograph of a coronal section of a mouse incisor at approximately 3 weeks of age showing the architecture of the mineralized enamel. Inset in B shows detail of enamel prisms (or rods) which contain the enamel crystallites.

Ameloblasts synthesize and secrete a number of enamel-specific proteins during the secretory stage. The most abundant of these is amelogenin, which comprises about 90% of the secreted ameloblast products. In the extracellular domain, amelogenin self-assembles into spheres at specific sites of mineral growth, thus limiting growth direction (Paine *et al.* 2001). Other extracellular proteins contributing to forming enamel are ameloblastin and enamelin (Bartlett *et al.* 2006). Enamel crystals formed during the secretory stage naturally require ions and minerals to initiate nucleation and thus some ion transport is needed at this stage (Lacruz *et al.* 2010a). The thin enamel crystals observed in the secretory stage are surrounded by matrix and fluids that are subsequently removed in the maturation stage (Smith and Nanci 1995). In rodents for instance, maturation starts once the enamel has reached its full thickness. The ameloblast cells then reorganize their ultrastructural characteristics while also reducing their height (Smith 1998). Maturation is characterized by an increase in ion transport functions and breakdown and resorption of protein matrix leading to the mineralization of the tissue (Smith 1998; Lacruz *et al.* 2012a). Enamel takes almost twice as long to be mineralized as to be secreted. Once mineralization has been completed, under a scanning electron microscope it can be seen that the crystals are arranged as bundles, with each bundle commonly called an enamel prism or rod (Boyde 1989). It appears that physiological changes occurring during the secretory stage may manifest as recognizable markings (growth lines) in the mineralized tooth. Analysis of these markings reveals a repetitive daily pattern (Boyde 1989; Bromage 1991; Smith 2006; Antoine *et al.* 2009). Most notably, microstructures known as cross striations found in the enamel of most mammals have been directly associated with daily incremental enamel growth. That is, a cross striation is formed in the enamel about every 24–27 hours. This evidence is based on experimental studies using vital labels (Bromage 1991; Smith 2006) and by counting the cross striations in individuals for whom the age at death was known (Boyde 1989; Antoine *et al.* 2009). By counting periodic growth lines in the teeth and matching these counts with the progress of tooth eruption and emergence, the pace and period of a hominin's maturation or life history could be reconstructed compared to its closest living relatives (i.e., *Pan troglodytes*, *P. paniscus*). One shortcoming of these studies is that they were performed using histological sections of mineralized (fully formed) enamel and hence, lacking proximal mechanistic data, could not account for temporal variation observed in enamel formation.

The Molecular Circadian Clock in Enamel

In recent years, an increasing interest has built towards identifying a molecular mechanism that may explain daily growth variation in enamel. The circadian clock is one likely candidate.

The core of the mammalian circadian clock system is controlled by a group of neurons located in the SCN of the hypothalamus (Reppert and Weaver 2002). The SCN largely operates via light inputs received in retinal ganglion cells and transmitted through the hypothalamic tract to the SCN using glutamate as the neurotransmitter (Reppert and Weaver 2002; Schibler 2005). Although light is the main signal (zeitgeber) and the most widely studied, several non-photic stimuli are also known, including ambient temperature, food availability, and physical activity. At the molecular level, the central clock consists of two interlocked positive-negative transcriptional feedback loops (Stratmann and Schibler 2006). The positive branch of the system is controlled by Brain and Muscle ARNT (Arylhydrocarbon Receptor Nuclear Translocator)-Like Protein (BMAL) and Circadian Locomotor Output Cycles Kaput (CLOCK) (Hirayama and Sassone-Corsi 2005). CLOCK (and its paralog NPAS2) and BMAL induce the expression of a number of genes by binding to critically important E-box elements for circadian activity in the promoter regions of downstream genes (Reppert and Weaver 2002; Okamura 2004; Ripperger *et al.* 2006). The CLOCK-BMAL complex activates the repressors PER (PER 1–3) and CRY (Cryptochrome) (CRY1 and CRY2) and also activates the expression of the orphan nuclear receptor REV-ERB alpha, which represses *Bmal* transcription (Preitner *et al.* 2002). CRY and PER are made in the cytoplasm and dimerize and upon reaching certain levels, at which time they can enter the nucleus and bind to CLOCK and BMAL through protein-protein interaction mechanisms (Shearman *et al.* 2000; Hirayama and Sassone-Corsi 2005). This binding represses the activity of BMAL and CLOCK. Post-translational regulation adds complexity to the system by maintaining a delay between activation and repression (Gallego and Virshup 2007) but in general the molecular organization of the circadian cycle takes ~24hrs (Mohawk *et al.* 2012). Most peripheral organs (heart, liver, kidney, etc.) express local circadian oscillations but require input from the SCN, which acts as the master regulator of these peripheral clocks (Mohawk *et al.* 2012). It appears that every cell in the body contains a circadian clock and that about 3–10% of the mRNA in a given tissue shows circadian rhythms (Panda *et al.* 2002; Hughes *et al.* 2009). An important characteristic of peripheral clocks is that genes under circadian regulation are non-overlapping in the different tissues, such that cellular physiology is temporally controlled in a tissue-dependent manner (Dibner *et al.* 2010; Mohawk *et al.* 2012).

The existence of an enamel clock has become apparent in recent years and several reports have contributed significantly to a better understanding of the mechanism of activity of this clock. Studies conducted by our group and others reported the expression of clock proteins in enamel cells (Zheng *et al.* 2011; Lacruz *et al.* 2012b; Zheng *et al.* 2013) as well as anti-phase daily oscillations of mRNA expression of *Per2* and *Bmal1* in enamel cell cultures (Lacruz *et al.* 2012b; Zheng *et al.* 2013). These latter studies are important because anti-phase activity of the antagonists *Bmal1* and *Per2* is a hallmark of circadian clock activity. It has also been proposed that clock genes may affect the expression of enamel genes (Athanassiou-Papaefthymiou *et al.* 2011; Lacruz *et al.* 2012b; Zheng *et al.* 2013). Examples include the regulation of *Amelx* and *Klk4*. Both *Amelx* and *Klk4* gene expression are up-regulated in the ameloblast-like HAT-7 cell line following *Bmal1* overexpression, suggesting direct positive induction of these enamel genes by *Bmal1* (Zheng *et al.* 2013). For *Amelx*, which lacks an E-box in the promoter regions proximal to the transcription start site, we proposed that activation of clock genes likely occurs by an indirect route (Lacruz *et al.* 2012b). NFYa is known to be a potent positive inducer of *Amelx* and has a highly conserved E-box element in its promoter (Lacruz *et al.* 2012b). We suggest that this E-box may act as a binding site for the clock proteins to induce *Amelx* expression. Further evidence that *Amelx* expression is regulated by clock genes comes from the identification of daily oscillations of *Amelx* and *Nfya* mRNA in enamel cell cultures using serum synchronized ameloblast-like LS8 cells (Lacruz *et al.* 2012b). Moreover, mRNA isolated from molar enamel organs dissected from mouse pups every 4 hours on postnatal days 2 through 4 showed that both *Nfya* and *Amelx* oscillated *in vivo* (Lacruz *et al.* 2012b). These data combined strongly suggest that the circadian clock modulates enamel development. This raises the question: what is the physiological function of this "enamel clock"?

Possible Function of the "Enamel Clock": Is It Linked to Cross Striations?

Physiological and metabolic processes are largely coordinated by the circadian clock through transcription (Feng and Lazar 2012). The physiological relevance of maintaining a rhythm in peripheral tissues can be exemplified by the regulation of glucose secretion in the liver. The liver clock mediates glucose homeostasis by regulating its export from the liver when needed to counterbalance glucose abundance derived from normal daily feeding cycles (Lamia *et al.* 2008). Some clues to the physiological relevance of the enamel clock have been recently reported (Lacruz *et al.* 2012b). It is well known that bicarbonate production and transport is critical for normal enamel development (Smith 1998; Lacruz 2010a, 2010b). Bicarbonate functions in enamel to buffer acidic pH stress arising from the release of hydrogen ions (H^+) during the nucleation of crystals (Smith 1998). We have recently reported that the mRNA expression of genes involved in bicarbonate production (*Car2*) and transport (*Slc4a4*) increase during the night, a time when *Amelx* expression decreases (Figure 4.2) (Lacruz *et al.* 2012b). Given that

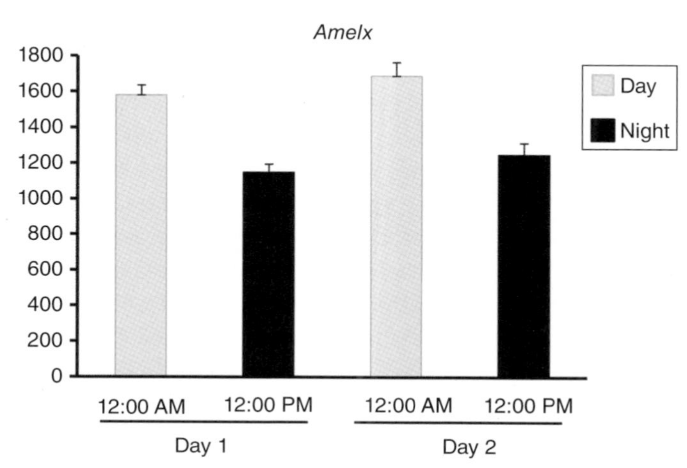

Figure 4.2. Real time PCR of amelogenin (*Amelx*) mRNA expression in mouse molar homogenates dissected at 12:00AM and 12:00PM from animals of the same litter at postnatal day 2 during a consecutive period of 48 hours. Primers as well as details of dissection and RNA isolation were described by Lacruz *et al.* (2012b). *Amelx* mRNA expression decreases significantly during the night period (ANOVA, $p<0.05$). *Amelx* values are relative to β-actin.

Amelx makes about 90% of the secretory products of ameloblasts, the increase of *Car2* and *Slc4a4* mRNA during low periods of *Amelx* mRNA production suggests that the circadian clock mediates some ameloblast activities by elevating specific cell functions above basal levels of normal activity on a daily/nightly basis (Lacruz *et al.* 2012b). Why *Amelx* increases during the day in a nocturnal animal remains to be investigated. It is also unknown whether the same pattern of day-night segregation regulates the enamel clock in diurnal animals.

Early studies by Okada (Okada 1943) on the formation of daily lines in dentine associated with daily changes in acid-base balance provide valuable information for understanding the link between the circadian clock and the formation of cross striations. These studies described how daily changes in acid-base balance in the blood plasma were reflected in the formation of daily lines in dentine. This prompted us to speculate that fluctuations in extracellular pH may affect the mineralization of enamel and the formation of cross striations. Local acidic conditions in the enamel zone can be brought about by an increase in CO_2 and/or in nucleation activity, the latter resulting in the release of H^+ ions. Carbonic anhydrase 6 (*Car6*) is expressed in the extracellular compartment of the ameloblasts within the enamel zone (Smith *et al.* 2006). Given that *Car6* is also involved in CO_2 production, it is plausible that *Car6* may be under circadian regulation in enamel and that this may lead to a net daily decrease of pH in the enamel zone. To test this hypothesis, the expression profile of *Car6* was analyzed in mouse molar homogenates over 48 hours and it was found that *Car6* mRNA levels oscillate with peaks of activity during the night (Figure 4.3). Moreover, the discharge of H^+ ions associated with crystal nucleation described above is likely buffered by the increase in bicarbonate transport. As previously

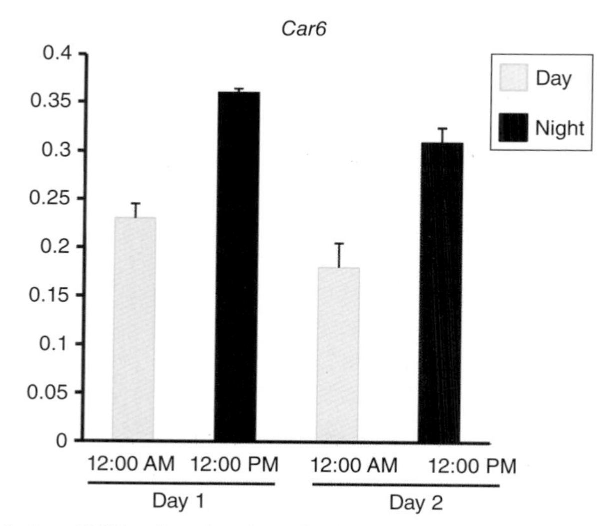

Figure 4.3. Real time PCR of carbonic anhydrase 6 (*Car6*) mRNA expression in mouse molar homogenates dissected at 12:00AM and 12:00PM from animals of the same litter at postnatal day 2 during a consecutive period of 48 hours. Primers as well as details of dissection and RNA isolation were described by Lacruz *et al.* (2012b). *Car6* mRNA expression increases significantly during the night period (ANOVA, $p<0.05$). *Car6* values are relative to β-actin.

described (Lacruz *et al.* 2012b), bicarbonate transporter activity increases during the night also. Thus the limited data available suggest that acid-base balance is likely disrupted daily in the enamel zone, affecting mineralization. Although this evidence will require formal testing, the hypothesis proposed here is that the interplay between amelogenin production during the day, which has been suggested to mediate pH levels around neutral values (Smith *et al.* 1996), coupled with an increase in carbonic anhydrase activity and bicarbonate transport as well as other cell functions during the night, leads to an overall net mineral gain in enamel occurring at different times during the day. It is further hypothesized that these periods of mineral gain occur at different times than the periods of matrix secretion, such that these functions alternate their activity peaks in a circadian manner. Such a mechanism could lead to the development of cross striations in enamel, as depicted in Figure 4.4. A proximal correlate to this scenario is found in bone development. Rodent calvarial bone examined over 6 days at 1-hour intervals using a *Per1*-luciferase model identified bursts of mineral deposition followed by little deposition with a circadian periodicity (McElderry *et al.* 2013). Further evidence in support of this hypothesis derives from studies describing differences in mineral concentration at the sites of cross striations.

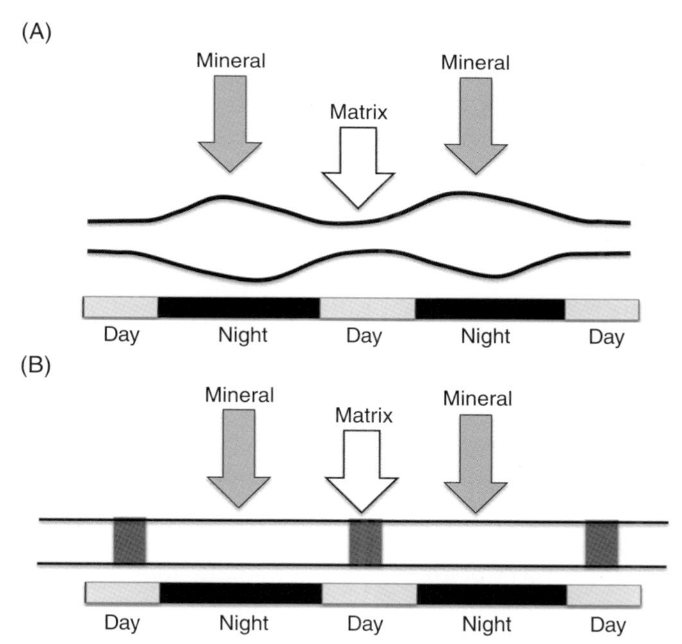

Figure 4.4. Schematic representation of the development of cross striations. (A) An example of cross striations observed with an electron microscope, appearing as alternating constrictions and varicosities. Mineral deposition is hypothesized to dominate in the varicosity portions along the enamel rod whereas matrix deposition is high during the periods of constrictions. (B) Cross striations seen in transmitted light microscopy appear as dark lines along prisms. It is hypothesized that the mineralization phase occurs during the bright areas along the prisms and that matrix is secreted during the dark line (cross striation) zones. Mineral deposition peaks during the night whereas matrix deposition increases during the day.

In the mammalian species examined, these sites showed reduced mineral content compared to adjacent areas along the enamel rods (Boyde 1979; Driessen *et al.* 1984), providing chemical data in support of the notion that there is differential capacity for mineral absorption/transport throughout the day.

Two important points should be highlighted here. First, both matrix secretion and mineralization take place continuously through the day and night regardless of where activity peaks occur. The shift in *Amelx* production during the night to about half of the daytime level allows the increase in mineral transport and changes in acid-base balance. The second issue is that even at its relatively lower expression level during the night, *Amelx* mRNA is more than 100-fold more abundant than that of *Car6* or the bicarbonate transporters reported previously (Lacruz *et al.* 2012b).

Indirect evidence also suggests that the transport of other minerals required for enamel development may be regulated by the circadian clock. A likely example is calcium. In SCN neuronal activity, Ca^{2+} influx is essential for the transmission of photic input (Hamada *et al.* 1999). Ca^{2+} levels also modulate the activity of the peripheral clock of the liver (Báez-Ruiz and Diaz-Muñoz 2011), doing so through Ca^{2+} pumps similar to those that we have recently identified in the enamel organ (Lacruz *et al.* 2012a). Given the relevance of Ca^{2+} in the formation of enamel (Smith 1998; Hubbard 2000), it would be unsurprising to identify that Ca^{2+} influx in enamel is regulated by the circadian clock. Daily shifts in ameloblast activity similar to those described here may be the normal physiological determinant for the formation of daily cross striations.

Daily Growth of Enamel and Its Relevance in Hominin Evolution

The most comprehensive studies to date on the daily growth of African Plio-Pleistocene hominin enamel are those reported in Dean *et al.* (2001) as well as those described by my colleagues and I (Lacruz and Bromage 2006; Lacruz *et al.* 2008). Dean *et al.* (2001) made a groundbreaking histological study showing differences in daily growth in *Australopithecus* and early *Homo* relative to *H. sapiens*, with the latter showing slower daily rates. Subsequently we reported values for 33 teeth representing nine different taxa (Lacruz *et al.* 2008). The bulk of our studies were conducted using a portable confocal scanning optical microscope (PCSOM) specifically designed to non-destructively study mineralized tissues of early hominins (Bromage *et al.* 2009). Results from these works confirmed that *Australopithecus*, *Paranthropus*, and early *Homo* developed their molar enamel by daily growth rates higher than those of modern humans (Dean *el al.* 2001; Lacruz and Bromage 2006; Lacruz *et al.* 2008; Dean 2010). These data confirmed a previous report by Beynon and Wood 1987), who analyzed a limited sample of molar teeth with the scanning electron microscope. Our results also confirmed that the molars of both *Paranthropus* species (*P. boisei* and *P. robustus*) had higher daily secretion rates (DSR) in the outer enamel than those observed in any other taxa, averaging >7.0μm a day. This is 25–30% more daily enamel growth than in similar regions of the tooth in *H. sapiens*. Although the species of the *Paranthropus* genus share faster daily growth rate averages than any other hominin, particularly in

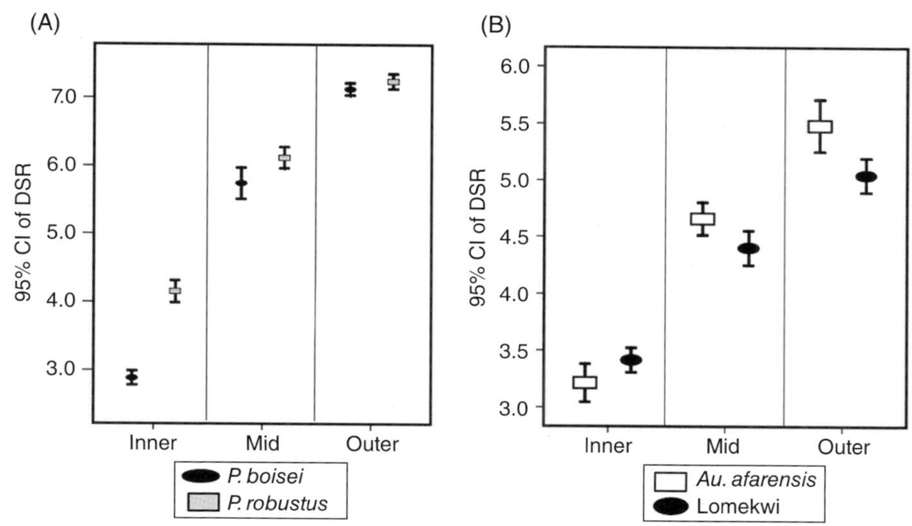

Figure 4.5. Daily secretion rates (DSR) in hominin taxa. (A) Error bars at 95% confidence interval comparing DSR values of *P. boisei* and *P. robustus*. (B) Error bars at 95% confidence interval comparing DSR values of *Au. afarensis* and the Lomekwi/Kataboi material. The *Paranthropus* and *Au. afarensis* samples were reported in Lacruz *et al.* (2008). The Lomekwi and Kataboi sample includes KNM-WT 8556, 38358, 38362, and 16006. Measurements of cross-striation spacing in all teeth were taken following the methodology described by Lacruz *et al.* (2008).

the outer enamel (Lacruz *et al.* 2008), some apparent differences can be observed between them. Figure 4.5A shows error bars at the 95% confidence interval of cross striation measurements, also known as DSR, for the inner, middle, and outer cuspal enamel of the two species of the genus *Paranthropus*. Our reported differences in DSR between *P. boisei* and *P. robustus* were subsequently used to support the hypothesis that the East and South African *Paranthropus* species are not sister taxa, as their molar enamel does not share similar developmental mechanisms (Wood and Harrison 2011). It is interesting to note the high value placed by these authors on variations in DSRs of cells to interpret phylogenetic relationships between early hominins. Additional insights into possible variation in daily growth of enamel may be provided by investigating cross striations in penecontemporaneous hominin demes from nearby localities. For example, following methodology described by Lacruz *et al.* (2008) for the *Paranthropus* data, Figure 4.5B shows error bars at the 95% confidence interval for the inner, middle, and outer cuspal enamel DSR in molars of *Australopithecus afarensis* and a sample of molars from the Kataboi and Lomekwi sites in Kenya. This latter material has been associated with, albeit not assigned to, *Kenaynthropus platyops* and has been dated to ca. 3.5 my (Leakey *et al.* 2001), thus well within the *Au. afarensis* range of 3.7–3.0 my (Kimbel *et al.* 2004). The observed overall daily growth rates of the Kataboi and Lomekwi molar samples follow a different growth trajectory than that of the *Au. afarensis* sample in that the initial rates are higher in the former but then drop relative to the rates observed in *Au. afarensis* in similar areas of the enamel crowns. These growth differences are statistically significant across regions (Mann-Whitney;

$p<0.05$). Studies on modern humans show fairly consistent daily growth rates in molars when cross striations are measured in similar areas of the crown using similar schemes (Beynon *et al.* 1991; Lacruz and Bromage 2006). For extant chimpanzees (*Pan troglodytes*), daily growth values of molar enamel also appear to be consistent (Smith *et al.* 2007). The data presented here thus appear to support growth differences in the molars between *Au. afarensis* and the Kataboi-Lomekwi sample.

In addition to insights into fossil hominin taxonomy and speciation, cross striations can also be uniquely and powerfully informative about the growth, development, and life history of early hominins. Cumulative counts of individual cross striations can provide accurate estimates of crown formation time (Boyde 1989; Antoine *et al.* 2009). Likewise, cross striation averages together with linear enamel thickness values can also provide good estimates of crown formation (Dean *et al.* 2001; Dean 2010; Lacruz and Ramirez Rozzi 2010; Lacruz *et al.* 2012a). Data from these studies and those reported by Beynon and Wood (1987) as well as Ramirez Rozzi (1993, 1998) strongly suggest that despite the relatively larger molar teeth of early hominins, their molar crowns took less time to form than those of *H. sapiens*. The inference made from these results is that early hominin crowns formed in a shorter period of time because they needed to be functional at an earlier age (Dean 2006; Lacruz *et al.* 2008). These data on fast crown development are supported by faster rates of root development in hominins, so it appears that the growth of both crown and root were under selective pressure to have teeth erupt early (Dean 2010; Dean and Cole 2013). These constraints suggest overall faster growth and development of hominins relative to anatomically modern humans.

Discussion

A key characteristic of tooth formation is that once enamel forms, it does not have the capacity to remodel or repair itself, unlike bone. Therefore any evolutionary modifications affecting tooth phenotype, and hence function, occur during its ontogeny (Jernvall and Thesleff 2012). This is indeed one of the research goals in evolutionary developmental biology, to understand how ontogenetic programs are modified and how these modifications produce new phenotypes (Hall 2000, 2003).

In the context of the hominin fossil record, megadontia appears as an evolutionary innovation most evident in the genus *Paranthropus* (McHenry 2002). Constraints acting on ontogenetic programs are also part of the set of pursuits of evo-devo (Hall 2000). For *Paranthropus*, which as far as we can interpret had a shorter growth and development period relative to modern humans (Bromage and Dean 1985; Ramirez Rozzi 1993; Lacruz *et al.* 2008; Dean 2010), it was important to produce a phenotype with very thick molar enamel in a relatively short period of time. Selection acting on tooth development favored two main mechanisms of growth in *Paranthropus*: (i) high daily rates of enamel secretion (Figure 4.5B), and (ii) the recruitment of a large number of ameloblast cells in the cohorts that form the enamel (Beynon and Wood 1987; Ramirez Rozzi 1993; Lacruz and Bromage 2006). More broadly, the relatively larger molars of other hominin taxa, which in most cases also had shorter growth and development periods than humans, were able to produce their thicker enamel via increased rates of daily enamel growth (Beynon and Wood 1987; Dean *et al.* 2001; Lacruz *et al.* 2008). These data also suggest that reductions in enamel daily production rates through the later part of the Pleistocene became possible when enamel thickness was reduced and/or when growth periods were extended, relaxing the constraints on the time allotted to form teeth (Dean *et al.* 2001; Dean 2006; Lacruz *et al.* 2008).

Focusing on the effects of the circadian clock on enamel formation, the data thus far obtained provide strong evidence to suggest that variations in daily enamel growth observed in histological sections are the result of differences in the modulating effect of the circadian clock across primate species. If the enamel clock model proposed here is correct, such modifications are associated with variation in rates of amelogenin production and mineralization, affecting the development of cross striations. As a consequence of this variation, other cell functions are also affected. Thus daily changes in an amelogenin-based growth plan determine the capacity of the ameloblasts to engage in other functions such as bicarbonate or calcium transport and acid-base balance. The amelogenin gene is highly conserved showing identical protein and DNA sequences for humans and chimpanzees (Lacruz *et al.* 2011). Given this high level of homology we can assume that the amelogenin gene was also quite similar in extinct hominins. *Paranthropus* species had the highest daily rates of enamel growth among hominins and likely required a greater daily quantity of mineral transport, accompanied by an increase in other activities such as cell movement, proteolysis, and endocytosis to facilitate mineralization. Similarly, this molecular machinery may have operated at a different level in the *Au. afarensis* and the Kataboi/Lomekwi samples, influencing ontogenetic aspects of daily enamel growth.

Conclusion

The circadian clock mediates the development of the tooth enamel, just as it regulates many other organs including the liver, heart, and kidney. The tooth clock thus mediates the segregation of ameloblast function with an amelogenin dependency over a 24-hour period, meaning that the high volumes of amelogenin required during the secretory stage limit the ameloblast's capacity to maintain high levels of activity of genes participating in other functions, such as mineral build-up. The hominin fossil record provides some understanding of the modifications of circadian clock function associated with different enamel phenotypes. The clearest expression of such modifications to enamel ontogeny is observed in *Paranthropus* molars, which present higher daily growth rates in the outer enamel than any other taxa, contributing to the development of thick enamel molars in this genus relative to other hominins. Other possible examples of changes in molar ontogeny may be those described here for the *Au. afarensis* and the Kataboi/Lomekwi samples, although the phenotype resulting from such a modification is as yet unclear. Based on current understanding of the biological functions of the circadian clock in modulating enamel formation, it is hypothesized here that differences in daily growth of hominin enamel may be related to net changes in the activity of the enamel clock across species.

Acknowledgements

I would like to thank Julia Boughner and Campbell Rolian for their invitation to contribute to this book and especially to Dr Boughner for her editorial suggestions. The work presented here has benefited hugely from discussions over many years with Christopher Dean, Charles Smith, Tim Bromage, Alan Boyde, and Fernando Ramirez Rozzi on various aspects of enamel development. I am thankful particularly to Christopher Dean and Charles Smith for their comments and suggestions that have influenced this contribution and improved the final outcome. I am also indebted to Malcolm Snead and Michael Paine for discussions and help in pursuing this story in their labs at USC. This work was funded by NIH/NIDCR K99/R00 grant (DE 022799) and by the Leakey Foundation.

References

Antoine, D., Hillson, S., and Dean, M. C. 2009. The developmental clock of dental enamel: A test for the periodicity of prisms cross striations in modern humans and evaluation of the most likely sources of error in histological studies of this kind. *Journal of Anatomy* 214(1):45–55.

Athanassiou-Papaefthymiou, M., Kim, D., Harbron, L. *et al.* 2011. Molecular and circadian controls of ameloblasts. *European Journal of Oral Sciences* 119(Suppl. 1):35–40.

Báez-Ruiz, A., and Díaz-Muñoz. M. 2011. Chronic inhibition of endoplasmic reticulum calcium-release channels and calcium-ATPase lengthens the period of hepatic clock gene Per1. *Journal of Circadian Rhythms* 9(6):1–10.

Bartlett, J. D., Ganss, B., Goldberg, M. *et al.* 2006. Protein-protein interactions of the developing enamel matrix. *Current Topics in Developmental Biology* 74:57–115.

Beynon, A. D., and Wood, B. A. 1987. Patterns and rates of enamel growth on the molar teeth of early hominids. *Nature* 326:493–496.

Beynon, A. D., Dean, M. C., and Reid, D. J. 1991. On thick and thin enamel in hominoids. *American Journal of Physical Anthropology* 86:295–310.

Boyde, A. 1979. Carbonate concentration, crystal centres, core dissolution, caries, cross striations, circadian rhythms and compositional contrast in SEM. *Journal of Dental Research* 58b:981–983.

Boyde, A. 1989. "Enamel." *In*: B. K. B. Berkovitz (ed.) *Handbook of microscopic anatomy* vol. 6. Berlin: Springer-Verlag. pp.309–473.

Bromage, T. G. 1991. Enamel incremental periodicity in the pig-tailed macaque: A polychrome fluorescent labelling study of dental hard tissues. *American Journal of Physical Anthropology* 86(2):205–214.

Bromage, T. G. and Dean, M. C. 1985. Re-evaluation of the age at death of immature fossil hominids. *Nature* 317:525–527.

Bromage, T. G., Goldman, H. M., McFarlin, S. C. *et al.* 2009. Confocal scanning optical microscopy of a 3-million-year-old *Australopithecus afarensis* femur. *Scanning* 31(1):1–10.

Dean, M. C. 2006. Tooth microstructure tracks the pace of human life-history evolution. *Proceedings of the Royal Society B: Biological Sciences* 273:2799–2808.

Dean, M. C. 2010. Retrieving chronological age from dental remains of early fossil hominins to reconstruct human growth in the past. *Philosophical Transactions of the Royal Society B: Biological Sciences* 365:3397–3410.

Dean, M. C., and Cole, T. J. 2013. Human life history evolution explains dissociation between the timing of tooth eruption and peak rates of root growth. *PLoS ONE* 8:e54534.

Dean, M. C., Leakey, M.G., Reid, D. *et al.* 2001. Growth processes in teeth distinguish modern humans from *Homo erectus* and earlier hominins. *Nature* 414:628–631.

Dibner, C., Schibler, U., and Albrecht, U. 2010. The mammalian circadian timing system: Organization and coordination of central and peripheral clocks. *Annual Reviews in Physiology* 72:517–549.

Driessens, F. C. M., Heijligers, H. J. M., Borggreven, J. M. P. H., and Woltgens, J. H. M. 1984. Variations in the mineral composition of human enamel on the levels of cross striations and striae of Retzius. *Caries Research* 18(3):237–241.

Feng, D., and Lazar, M. A. 2012. Clocks, metabolism, and the epigenome. *Molecular Cell* 47(2):158–167.

Gallego, M., and Virshup, D. M. 2007. Post-translational modifications regulate the ticking of the circadian clock. *Nature Reviews Molecular Cell Biology* 8(2):139–148.

Hall, B. K. 2000. Evo-devo or devo-evo – does it matter? *Evolution and Development* 2(4):177–178.

Hall, B. K. 2003. Evo-devo: Evolutionary developmental mechanisms. *International Journal of Developmental Biology* 47(7–8):491–495.

Hamada, T., Liou, S.-Y., Fukushima, T. *et al.* 1999. The role of inositol triphosphate-induced Ca^{2+} release from IP_3-receptor in the rat suprachiasmatic nucleus on circadian entrainment mechanism. *Neuroscience Letters* 263(2–3):125–128.

Hirayama, J., and Sassone-Corsi, P. 2005. Structural and functional features of transcription factors controlling the circadian clock. *Current Opinion in Genetics and Development* 15(5):548–556.

Hubbard, M. J. 2000. Calcium transport across the dental enamel epithelium. *Critical Reviews in Oral Biology and Medicine* 11(4):437–466.

Hughes, M. E., DiTacchio, L., Hayes, K. R. *et al.* 2009. Harmonics of circadian gene transcription in mammals. *PLoS Genetics* 5:e1000442.

Jernvall, J., and Thesleff, I. 2012. Tooth shape formation and tooth renewal: Evolving with the same signals. *Development* 139(19):3487–3497.

Kimbel, W. H., Rak, Y., and Johanson, D. C. 2004. *The skull of Australopithecus afarensis.* Oxford: Oxford University Press.

Lacruz, R. S., and Bromage, T. G. 2006. Appositional enamel growth in molars of South African fossil hominids. *Journal of Anatomy* 209(1):13–20.

Lacruz, R. S., and Ramirez Rozzi, F. V. 2010. Molar crown development in *Australopithecus afarensis*. *Journal of Human Evolution* 58(2):210–206.

Lacruz, R. S., Dean, M. C., Ramirez Rozzi, F. V., and Bromage, T. G. 2008. Megadontia, patterns of enamel secretion, and striae periodicity in Plio-Pleistocene fossil hominins. *Journal of Anatomy* 213(2):148–158.

Lacruz, R. S., Hacia, J. G., Bromage, T. G. *et al.* 2012a. The circadian clock modulates enamel development. *Journal of Biological Rhythms* 27(3):237–245.

Lacruz, R. S., Lakshminarayanan, R., Bromley, K. M. *et al.* 2011. Structural analysis of a repetitive protein sequence motif in strepsirrhine primate amelogenin. *PLoS ONE* 6(3):e18028.

Lacruz, R. S., Nanci, A., Kurtz, I. *et al.* 2010a. Regulation of pH during amelogenesis. *Calcified Tissue International* 86(2):91–103.

Lacruz, R. S., Nanci, A., White, S. N. *et al.* 2010b. The sodium bicarbonate cotransporter (NBCe1) is essential for the normal development of mouse dentition. *Journal of Biological Chemistry* 285(32):24432–24438.

Lacruz, R. S., Smith, C. E., Bringas, Jr., P. *et al.* 2012b. Identification of novel candidate genes involved in mineralization of dental enamel by genome-wide transcript profiling. *Journal of Cellular Physiology* 227(5):2264–2275.

Lamia, K. A., Storch, K.-F., and Weitz, C. J. 2008. Physiological significance of a peripheral tissue circadian clock. *Proceedings of the National Academy of Sciences of the USA*. 105(39):15172–15177.

Leakey, M. G., Spoor, F., Brown, F. H. *et al.* 2001. New hominin genus from eastern Africa shows diverse middle Pliocene lineages. *Nature* 410:433–440.

McElderry, J. D., Zhao, G., Khmaladze, A. *et al.* 2013. Tracking circadian rhythms of bone mineral deposition in murine calvarial organ cultures. *Journal of Bone and Mineral Research* 28:1846–1854.

McHenry, H. M. 2002. Introduction to the fossil record of human ancestry. *In*: W. C. Hertwig (ed.) *The primate fossil record*. Cambridge: Cambridge University Press. pp.401–405.

Mohawk, J. A., Green, C. B., and Takahashi, J. S. 2012. Central and peripheral circadian clocks in mammals. *Annual Review of Neuroscience* 35:445–462.

Nanci, A. 2008. *Ten Cate's oral histology: Development, structure, and function*, 7th edn. St Louis: Mosby Elsevier.

Okada, M. 1943. Hard tissues of animal body. Highly interesting details of Nippon studies in periodic patterns of hard tissues are described. *Shanghai Evening Post. Medical Edition of September 1943*, pp.15–31.

Okamura, H. 2004. Clock genes in cell clocks: Roles, actions, and mysteries. *Journal of Biological Rhythms* 19(5):388–399.

Olejniczak, A. J., Smith, T. M., Skinner, M. M. *et al.* 2008. Three-dimensional molar enamel distribution and thickness in *Australopithecus* and *Paranthropus*. *Biology Letters* 4(4):406–410.

Paine, M. L., White, S. N., Luo, W. *et al.* 2001. Regulated gene expression dictates enamel structure and tooth function. *Matrix Biology* 20(5–6):273–292.

Panda, S., Antoch, M. P., Miller, B. H. *et al.* 2002. Coordinated transcription of key pathways in the mouse by the circadian clock. *Cell* 109(3):307–320.

Preitner, N., Damiola, F., Lopez-Molina, L. *et al.* 2002. The orphan nuclear receptor REV-ERBα controls circadian transcription within the positive limb of the mammalian circadian oscillator. *Cell* 110(2):251–260.

Ramirez Rozzi, F. V. 1993. Tooth development in East Africa *Paranthropus*. *Journal of Human Evolution* 24(6):429–454.

Ramirez Rozzi, F. V. 1998. Can enamel microstructure be used to establish the presence of different species of Plio-Pleistocene hominids from Omo, Ethiopia? *Journal of Human Evolution* 35(4):543–576.

Reppert, S. M., and Weaver, D. R. 2002. Coordination of circadian timing in mammals. *Nature* 418:935–941.

Ripperger, J. A., and Schibler, U. 2006. Rhythmic CLOCK-BMAL1 binding to multiple E- box motifs drives circadian Dbp transcription and chromatin transitions. *Nature Genetics* 38(3):369–374.

Schibler, U. 2005. The daily rhythms of genes, cells and organs. *EMBO Reports* 6:S1, S9–S13.

Shearman, L. P., Sriram, S., Weaver, D. R. *et al.* 2000. Interacting molecular loops in the mammalian circadian clock. *Science* 288:1013–1019.

Smith, C. E. 1998. Cellular and chemical events during enamel maturation. *Critical Reviews in Oral Biology and Medicine* 9(2):128–161.

Smith, C. E., Issid, M., Margolis, H. C., and Moreno, E. C. 1996. Developmental changes in the pH of enamel fluid and its effects on matrix resident proteinases. *Advances in Dental Research* 10(2):159–169.

Smith, C. E., and Nanci, A. 1995. Overview of changes in enamel organ cells associated with major events in amelogenesis. *International Journal of Developmental Biology* 39(1):153–161.

Smith, C. E., Nanci, A., and Moffatt, P. 2006. Evidence by signal peptide trap technology for the expression of carbonic anhydrase 6 in rat incisor enamel organs. *European Journal of Oral Sciences* 114(s1):147–153.

Smith, T. M. 2006. Experimental determination of the periodicity of incremental features in enamel. *Journal of Anatomy* 208(1):99–103.

Smith, T. M., Reid, D. J., Dean, M. C. *et al.* 2007. Molar development in common chimpanzees (*Pan troglodytes*). *Journal of Human Evolution* 52(2):201–216.

Stratmann, M., and Schibler, U. 2006. Properties, entrainment, and physiological functions of mammalian peripheral oscillators. *Journal of Biological Rhythms* 21(6):494–506.

Thesleff, I., and Hurmerinta, K. 1981. Tissue interactions in tooth development. *Differentiation* 18(2):75–88.

Thomas, B. L., and Sharpe, P. T. 1998. Patterning of the murine dentition by homeobox genes. *European Journal of Oral Sciences* 106(s1):48–54.

Wood, B. A., and Harrison, T. 2011. The evolutionary context of the first hominins. *Nature* 470:347–352.

Zheng, L., Papagerakis, S., Schnell, S. D. *et al.* 2011. Expression of clock proteins in developing tooth. *Gene Expression Patterns* 11(3–4):202–206.

Zheng, L., Seon, Y. J., Mourão, M. A. *et al.* 2013. Circadian rhythms regulate amelogenesis. *Bone* 55(1):158–165.

Chapter 5
Evo-Devo Sheds Light on Mechanisms of Human Evolution: Limb Proportions and Penile Spines

Philip L. Reno
Department of Anthropology, The Pennsylvania State University, University Park, PA, USA

Introduction

Comparative embryology played an important role in anthropological investigations of the first half of the 20th century (Schultz 1925, 1926), but the frequency of these studies waned due to methodological challenges and ethical considerations of working on long-lived, slow reproducing, and often endangered primates. Due to these limitations it has been difficult to link genetic, developmental, and morphological evolution in primates. Evolutionary developmental biology (evo-devo) has emerged from relatively recent inclusion of developmental genetics and embryology into the Modern Synthesis (Gilbert *et al.* 1996; Raff 2000). This has resulted from technological advances in molecular genetics that enable the roles of individual genes to be tracked and tested during embryological development and led to paradigmatic shifts in the understanding of the depth to which many developmental genes and processes are highly conserved (Carroll 2008). Evolutionary developmental anthropology can take advantage of a broad range of phenotypic, genetic, and embryological analyses to identify the genomic and developmental mechanisms underlying the intraspecific variation and evolutionary diversity of human and primate anatomy (Lovejoy *et al.* 1999, 2003; Carroll 2003).

It has long been understood that most anatomical traits are not simply the product of individual genes but emerge from the action of complex and integrated "developmental systems" (Dobzhansky 1956; Buchanan *et al.* 2009). Recent advances in developmental genetics have begun to determine the links between genotype, embryogenesis, and phenotype. Like the phenotypes they encode, genomes and developmental processes share similar properties including recognizable homology, functional diversity, modularity, and specificity (Carroll 2008). While phenotypic homology has been fundamental to evolutionary analyses since their inception, the high conservation of genome sequences, protein functions, gene regulatory networks, and developmental

Developmental Approaches to Human Evolution, First Edition. Edited by Julia C. Boughner and Campbell Rolian.

processes was unexpected (Mayr 1963). The classic example is the similarity in sequence and chromosomal positioning of Homeotic Cluster (HomC) genes that are key to organizing the anterior-posterior axis of the fruit fly (*Drosophila*) and *Hox* genes, which have similar roles patterning the body axes of vertebrates (McGinnis *et al.* 1984; Favier and Dolle 1997). There are even examples of developmental genes and pathways that share deeper homology than the structures they specify. For example, the genes *eyeless* and *Pax6* initiate the development of eyes in insects and vertebrates (Halder *et al.* 1995). Protein function is so conserved in this case that *Pax6* can induce compound eye formation when misexpressed in *Drosophila* (Halder *et al.* 1995). This pervasive homology results from the reuse of a highly conserved set of "tool kit" genes that are repurposed in cellular interactions to produce an array of developmental outcomes. Much like the convergent wings of bats, birds, and pterosaurs that all rely on the same underlying homologous skeletal pattern, the independently evolved fly and vertebrate limbs share a very old and highly conserved suite of regulatory networks to guide their development (Carroll *et al.* 2005). The significance for anthropology is that experiments in a wide variety of model organisms are potentially relevant to understanding genetic and cellular mechanisms of development in humans and other primates. With a greater diversity of developmental data comes not only the possibility of exploring commonalities in developmental processes, but also the opportunity to identify mechanisms that produce novel phenotypes (Wagner *et al.* 2000).

Organisms are composed of functionally diverse arrays of cells, tissues, organs, and anatomical structures. Similarly, genomes and developmental processes consist of various functional constituents. DNA contains regions dedicated to coding RNA protein synthesis and gene regulation via transcription factor binding and chromatin structure. Developmental genes can be loosely classified as "selector genes" (i.e., *Hox* and *Pax6*) that activate the expression of batteries of downstream "realizator genes" that execute specific developmental operations (Pradel and White 1998). Developmental processes, such as outgrowth, typically involve the interaction of separate cell populations dedicated to mitosis (i.e., the progress zones of the limb bud and columnar zone of the growth plate) verses organization (the limb bud zone of polarizing activity (ZPA) and apical ectodermal ridge (AER) and the growth plate reserve zone and perichondrium) (Wolpert 1978; Reno *et al.* 2006). Such functional diversity is potentially linked to the complex patterns of phenotypic integration resulting in varying correlated and independent parts (Olson and Miller 1999). Morphological integration emerges, in part, from a developmental process that relies on modular, hierarchical, and nested morphogenetic fields that share patterns of gene expression and cellular interactions (Gilbert *et al.* 1996; Wagner 2007). Yet, despite this pervasive (morphological and developmental) integration, evolution has the capability to shape phenotypes with remarkable specificity, such as the dramatically elongated middle finger of the aye-aye or the reduced index finger of potto (Napier and Napier 1985). Numerous experiments demonstrate that positional information can define identity at a cell-by-cell basis through the precise action of tissue-specific genetic enhancers (DiLeone *et al.* 1998; Gompel *et al.* 2005; Guenther *et al.* 2008; Menke *et al.* 2008). Thus, the fact that homology, functional diversity, modularity, and specificity are common to phenotypes, genomes, and development is key to understanding the root of organismal complexity and the capacity for new phenotypes to evolve in all animals including humans (Wagner and Altenberg 1996).

Examples of Developmental Mechanisms Underlying Evolutionary Change

The importance of these properties for genotype-phenotype relationships can be illustrated by recent efforts using, for logistical and ethical reasons, non-primate models systems to identify the mechanistic bases for evolutionary change. Cohn and Tickle (1999) demonstrated that limblessness in snakes results from an alteration in axial *Hox* gene expression boundaries (Figure 5.1). *Hox* genes function as transcription factors that regulate downstream gene expression. They are expressed in individual overlapping domains within the paraxial and lateral plate mesoderm to assign developmental identity along the anterior/posterior vertebrate axis (Burke *et al.* 1995). There are four distinct tetrapod *Hox* clusters (named A–D), each of which can contain up to 13 paralogous genes (Duboule 2007). A key property of *Hox* genes is the spatial and temporal collinearity of their expression that results in lower-number genes typically being activated earlier and more cranially and subsequent expression of higher-number *Hox* genes caudally (Iimura *et al.* 2009). In the paraxial mesoderm, *Hox* expression has a major role in specifying vertebral identity, while in the lateral plate it specifies forelimb, flank, and hind limb position (Wellik 2009). In pythons, the paraxial domains of *Hoxc6* and *Hoxc8*, which are typically associated with thoracic identity in other tetrapods, are extended to the most anterior vertebral somites, resulting in rib-bearing vertebrae from the axis to the cloaca (Figure 5.1A) (Cohn and Tickle 1999). In limbed tetrapods, such as the chick, the lateral plate anterior expression boundaries of *Hoxb5*, *Hoxc6*, and *Hoxc8* correspond to the position of the forelimb. However, in the python these expression domains are also shifted cranially, and as Cohn and Tickle proposed, obviate the signal for the anterior limb field formation. In these ways, *Hox* genes illustrate how spatial arrangements of genetic loci, gene expression patterns, and morphogenetic fields interact to form novel coherent phenotypes with relatively simple genetic changes.

While the developmental integration of different tissues allows seemingly simple shifts in gene expression territories to result in dramatic morphological alterations, evolution also has the capacity to sculpt skeletal phenotypes with remarkable precision (Guenther *et al.* 2008). Stickleback fish have undergone repeated radiations from a common marine habitat into isolated inland freshwater lakes and streams (Bell and Foster 1994). These now isolated populations have shown a remarkable tendency for parallel phenotypic evolution in response to novel ecological conditions (predation, osmoregulation, etc.). One example is the frequent loss of the pelvic spine (Figure 5.1B). By crossing marine and freshwater populations, David Kingsley and colleagues identified a specific genetic locus near the *Pituitary homeobox 1* (*Pitx1*) gene that is highly associated with pelvic loss in these fish (Shapiro *et al.* 2004). While the coding sequence of *Pitx1* remained intact, expression was lost in the presumptive

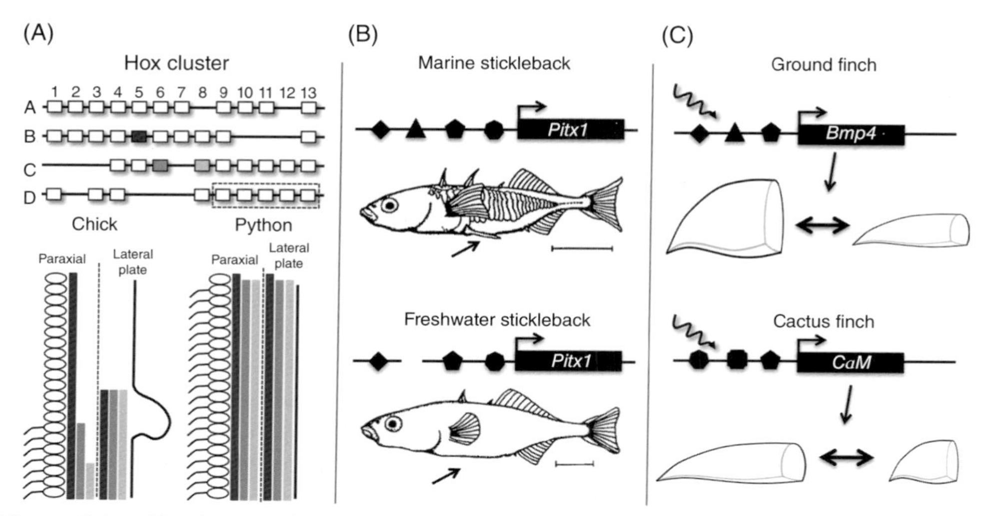

Figure 5.1. Classic examples of mechanistic change in developmental evolution. (A) Tetrapod *Hox* genes are arrayed in four separate clusters (A–D) in the genome. These genes are expressed in a pattern of temporal and spatial collinearity where lower number genes are expressed earlier and more anteriorly (cranially) along the primary axis. Note that the *Hoxd9–13* genes (dashed box) are regulated by a separate enhancer during limb development located beyond *Hoxd13* that is responsible for reversed collinearity in the tetrapod limb. In the chicken, the expression of specific *Hox* genes correspond to changes in positional identity. *Hoxc6* (medium gray) and *Hoxc8* (light gray) are expressed near the boundary between cervical and thoracic (rib-bearing) vertebrae in the paraxial mesoderm, while these genes and *Hoxb5* (dark gray) are expressed in the lateral plate near the induction of the forelimb bud. In the python, the expression boundaries of all three genes are shifted cranially to eliminate most of the cervical vertebrae and the forelimb field. (B) Marine sticklebacks have invaded multiple inland freshwater habitats resulting in parallel adaptive changes such as pelvic spine loss and lateral plate reduction. Multiple enhancers (shapes) near the *Pitx1* gene exist to activate expression in separate tissues such as the pituitary, oral cavity, and pelvic spine. In freshwater morphs, the pelvic enhancer (triangle) has been repeatedly lost in separate populations resulting in the loss of the pelvic spine. Stickleback images modified from Bell and Foster (1994). (C) Separate developmental regulatory networks specify Galapagos finch beak shape. Upstream (*trans*) factors (squiggly arrows) activate *Bmp4* expression to influence the growth of beak height and breadth. A separate network modulates beak length by regulating *CaM* expression.

pelvic region despite continued expression at other anatomical locations (i.e., the mouth and pituitary). Fine-mapping using the variation present in natural populations enabled Chan *et al.* (2010) to further isolate the locus to a 2.5-kb region approximately 30 kb away from the coding region of *Pitx1*. Interestingly, this region has undergone independent deletion in separate pelvic-reduced fish populations (Chan *et al.* 2010). To test the effect this sequence has on gene expression, the ancestral marine sequence was cloned and attached to a reporter gene. When injected into developing stickleback embryos and incorporated into the native genome, the cells that have the capacity to activate the enhancer are identified by reporter gene expression. In this case, the sequence activated specific expression in developing pelvic limb buds showing it to be a tissue specific *Pitx1* enhancer (Figure 5.1B). In addition, injection of the 2.5-kb enhancer sequence attached to the *Pitx1* coding sequence activated expression in the pelvic field and rescued spine formation in freshwater pelvic reduced fish. Many freshwater sticklebacks have undergone a targeted deletion of the pelvic enhancer while still preserving other *Pitx1 cis*-regulatory elements that drive expression in other tissues (Figure 5.1B). These experiments reveal that small regulatory changes can produce profound changes in gene expression with large phenotypic effects.

Such specificity has also been integrated into developmental networks to facilitate modular changes in morphological shape. This was illustrated using one of Darwin's classic examples of evolution, the beaks of Galapagos finches. Ground finches possess broad deep beaks well suited for crushing seeds, while cactus finches' beaks are long and pointed, useful for reaching into flowers and fruits (Figure 5.1C) (Grant 1991). Abzhanov and colleagues found that the broad deep beaks were characterized by earlier and stronger expression of *Bone morphogenetic protein 4* (*Bmp4*) in the developing dorsal beak mesenchyme (Abzhanov *et al.* 2004). Misexpression of *Bmp4* in developing chicks produced wider and deeper beaks similar to those of ground finches. In contrast, misexpression of *Noggin*, a *Bmp4* antagonist, resulted in substantial beak reduction, particularly in height and breadth. In contrast, long-beaked cactus finches displayed increased expression of *Calmodulin* (*CaM*) in the distal mesenchyme of developing beaks (Abzhanov *et al.* 2006). Similarly to *Bmp4*, misexpression of *CaM kinase kinase* (*CaMKII*, a downstream effector of *CaM*) resulted in phenotypic changes in developing chick beaks, but in this case the effects were limited to changes in length. *Bmp4* and *CaM* appear to reside in separate gene regulatory networks that target growth along distinct morphological axes of beak development (Mallarino *et al.* 2011).

These three examples from pythons, sticklebacks, and finches help demonstrate the role these highly conserved "toolkit" genes play in regulating both coordinated and specific developmental processes to construct complex phenotypes that are yet amenable to functional evolutionary change. As the sequences of these genes tend to be highly conserved, these examples also stress the importance that gene regulation often plays in phenotypic evolution (Carroll 2005). Such developmental insights are useful for addressing questions about primate developmental evolution.

Biological anthropologists have been generally receptive to the important implications of evo-devo research. Many have used observable patterns of primate phenotypic variation to explore the relationship between phenotypic and genotypic modularity (Marroig and Cheverud 2001; Hlusko 2004; Lieberman *et al.* 2004; Marroig *et al.* 2004; Young *et al.* 2010; Hlusko *et al.* 2011; Grieco *et al.* 2013). For instance, the genetic correlations shared by both mice and Old World monkeys distinguish incisor and molar size variation, and correspond to known patterns of dental homeobox gene expression patterns during development (Hlusko *et al.* 2011; Grieco *et al.* 2013). Also crucial has been the impact on functional and phylogenetic analyses through the construction of a trait classification system informed by developmental genetics and embryology to facilitate hypothesis construction regarding the developmental basis for morphological characters (Lovejoy *et al.* 1999). This methodological framework has been fundamental to many recent morphological analyses of living and fossil primates (McCollum 1999; Reno *et al.* 2000, 2008; White *et al.* 2009). However, the examples from snakes, sticklebacks, and finches, as well as others, demonstrate the power that evo-devo has to identify specific genomic and developmental mechanisms underlying morphological adaptations. Anthropology has been somewhat slow to integrate these mechanistic investigations into human and primate evolutionary variation despite frameworks defined over a decade ago (Chiu and Hamrick 2002; Carroll 2003; Lovejoy *et al.* 2003). In large part, this delay has been due to the inherent practical and ethical difficulties of conducting experimental embryological analyses on primates. Yet, the pervasive homology in genes, regulatory networks, and cellular processes that guide animal development means that analyses using more tractable laboratory models, such as the mouse, can be analogous to similar processes in primates. In addition, while there is a relative dearth of primate developmental data, we now benefit from a plethora of genomic data available for our own and other primate species (O'Bleness *et al.* 2012). As such, advances in comparative genomics have now enabled identification of human-specific genetic changes (Pollard *et al.* 2006; Prabhakar *et al.* 2006, 2008; Perry *et al.* 2007; McLean *et al.* 2011), and recent methods in gene targeting technology may make testing specific genomic variants in a transgenic mouse model within anthropological skill sets and budgets (Menke 2013).

To some conducting functional and phylogenetic analyses of primates, identifying developmental and genetic mechanisms may seem a trivial consideration. Insofar as intellectual curiosity is justification for the former, it should also be sufficient for the latter. Furthermore, development is a fundamental step within the evolutionary process. Phenotypes are not simple "readouts" of genotypic blueprints (Lovejoy *et al.* 2003). Development serves as a mediator interpreting genotypes into phenotypes, often incorporating intrinsic and extrinsic environmental inputs that are not encoded within the genotype (Hamrick 1999). For example, functionally important aspects of human knee joint shape, such as the prominent lateral lip of the patellar groove, bicondylar angle, and condyle articular surfaces, are likely to be variously influenced by differences in initial patterning and alterations in cartilage modeling in response to habitual loading behaviors (Lovejoy 2007; Tardieu 2010). Consideration of these facts can help resolve if a feature likely reflects a history of selection or indicates the performance of particular activities during life (Lovejoy *et al.* 1999). In addition, many quantifiable characters develop within the same morphogenetic fields of a more primary adaptive trait (Lovejoy *et al.* 1999) suggesting the potential for many genetic effects to be highly pleiotropic (Hlusko 2004). Patterns of morphological integration and coevolution appear to support this contention (Reno *et al.* 2008; Rolian *et al.* 2010; Young *et al.* 2010). However, as discussed above, mechanisms exist to target morphological change with remarkable specificity. Despite numerous shared genes in forelimb and hind limb development, each is specified by functionally different selector genes (*Tbx5* in the forelimb, *Tbx4* and *Pitx1* in the hind limb (Logan *et al.* 1998; Logan and Tabin 1999)), experiences key differences in gene patterning (the greater influence of *Hoxc* genes on hindlimb development (Nelson *et al.* 1996; Hostikka and Capecchi 1998; Wellik and Capecchi 2003)), and largely grows at non-homologous growth plates (proximal humerus distal radius/ulna versus distal femur and proximal tibia (Reno *et al.* 2005)). Thus, pleiotropic and co-evolutionary hypothesis should seek mechanistic validation as much as possible. Finally, the former unavailability of genetic and developmental data is not a valid argument for their lack relevance. Much in the same way that resolution of the hominoid phylogeny using genetic data has improved the precision of hypotheses concerning the nature of the human/chimpanzee last common ancestor (Richmond *et al.* 2001; Lovejoy *et al.* 2009b), identification of specific genetic and developmental mechanisms will refine the functional and phylogenetic questions asked in morphological analyses. While it is not possible to provide an encompassing review here, I describe two approaches that have helped to identify particular developmental and genetic mechanisms underlying human evolution.

Hox Genes and Coordinated Limb Evolution in Hominoids

Hominoids differ dramatically in their forelimb proportions. Highly arboreal gibbons (*Hylobates*) and orangutans (*Pongo*), and to a lesser extent chimpanzees (*Pan*), possess very long forearms and fingers, which are substantially shorter in the more terrestrial gorillas (*Gorilla*) and bipedal humans (*Homo*) (Schultz 1926). A number of functional hypotheses may be posed to explain these morphologies, yet an important primary question is what are the proximate genetic and developmental mechanisms specifying differences in skeletal proportions? Given the pervasive modular nature of both phenotypic variation and developmental processes, it is certainly possible that limb evolution is modular as well (Hallgrímsson *et al.* 2002; Rolian *et al.* 2010; Young *et al.* 2010).

As discussed above, *Hox* gene structure and function is highly conserved across animals. In particular, the basic patterns of *Hox* expression during limb development are remarkably consistent in lineages as diverse as alligators, chicks, and mice (Nelson *et al.* 1996; Reno *et al.* 2008; Vargas *et al.* 2008). Similar to axial patterning, their expression domains define regions of shared gene expression, or developmental modules, in the limb. For the *HoxD* cluster, distal limb expression is guided by a highly conserved long-range five prime (5') (nearest to *Hoxd13*) enhancer network that controls expression in the developing hands and feet (or autopods) (Gonzalez *et al.* 2007; Montavon *et al.* 2008, 2011). The impact of this separate enhancer network is that in the autopod *Hoxd* genes are expressed with reversed temporal and spatial collinearity, such that higher number genes are expressed earlier and more anteriorly (Andrey *et al.* 2013). Consequently, *Hoxd13* is the only *Hoxd* gene to be expressed in the developing thumb and big toe, whereas in the posterior (ulnar digits) *Hoxd13–Hoxd10* are expressed in overlapping domains (Figure 5.2, color plate).

Numerous experiments have shown that *Hox* genes function at later stages of development to modulate skeletal growth (Morgan and Tabin 1994; Yokouchi *et al.* 1995; Capecchi 1997; Goff and Tabin 1997; Papenbrock *et al.* 2000; Zhao and Potter 2001; Boulet and Capecchi 2002, 2004; Kmita *et al.* 2002; Wellik and Capecchi 2003). First, complete deletion of *Hoxa11* and *Hoxd11* disrupts elongation of the radius

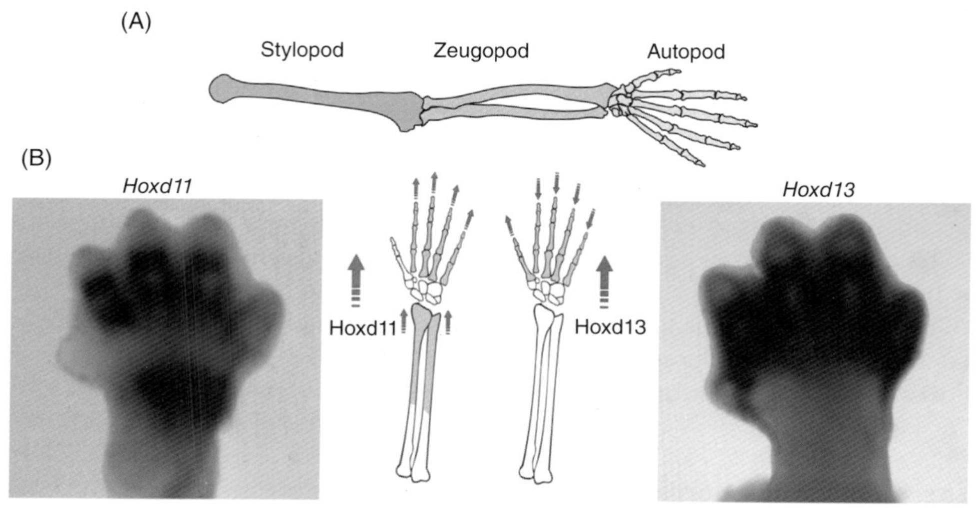

Figure 5.2. Expression of *Hoxd11* and *Hoxd13* in mammalian limb development and their proposed role in regulating hominoid forearm and digit proportions. (A) Tetrapod limbs are divided into three segments: stylopod, zeugopod, and autopod. (B) *In situ* hybridization of E13.5 mouse forelimbs indicated that *Hoxd11* is expressed in the distal zeugopod (forearm) and posterior digits, while *Hoxd13* is expressed in all five digits. Each *Hox* gene has its individual effect on skeletal growth, with *Hoxd13* promoting less growth than does *Hoxd11*. Thus, increased activity of *Hoxd13* will result in elongation in regions where it is expressed alone (thumb). However, when in competition, *Hoxd13* can act to ameliorate *Hoxd11*'s superior growth-promoting action to produce shorter fingers.

and ulna such that they resemble the short bones of the wrist and ankle (Davis *et al.* 1995), and increasing the normal dosage of *Hoxd11* expression in the distal forelimb of mice increases posterior metacarpal and phalangeal length (Boulet and Capecchi 2002). Individual *Hox* genes appear to specify different growth behaviors. For example, ectopic expression of *Hoxa13* in the mouse forearm shortens the radius and ulna. This implies that activation of the *Hoxa13* protein's targets (and subsequent downstream gene expression) yields a substantially reduced growth rate when in competition with endogenous zeugopod (forearm or leg) *Hox* genes, namely *Hoxd11* (Yokouchi *et al.* 1995; Zhao and Potter 2001). Thus it is reasonable to hypothesize that the normally overlapping *Hox* domains in the autopod serve as a mechanism underlying correlated evolution of the thumb and finger proportions during primate evolution.

Limb length is determined by the combined effect of individual growth plates that differ in growth rate, even from their counterparts residing within the same bone (Payton 1932; Bisgard and Bisgard 1935; Reno *et al.* 2000). Each growth plate has its own genetically mandated growth behavior, likely determined by residing within specific gene territories during development (i.e., *Hox*). For example, in hominoids the distal growth plates of the radius and ulna contribute twice as much growth to those bones than do the proximal growth plates (Figure 5.3A). Interestingly, the late stage expression domain of *Hoxd11* includes the distal but not the proximal zeugopod, placing the two ends within separate *Hox*-defined developmental modules.

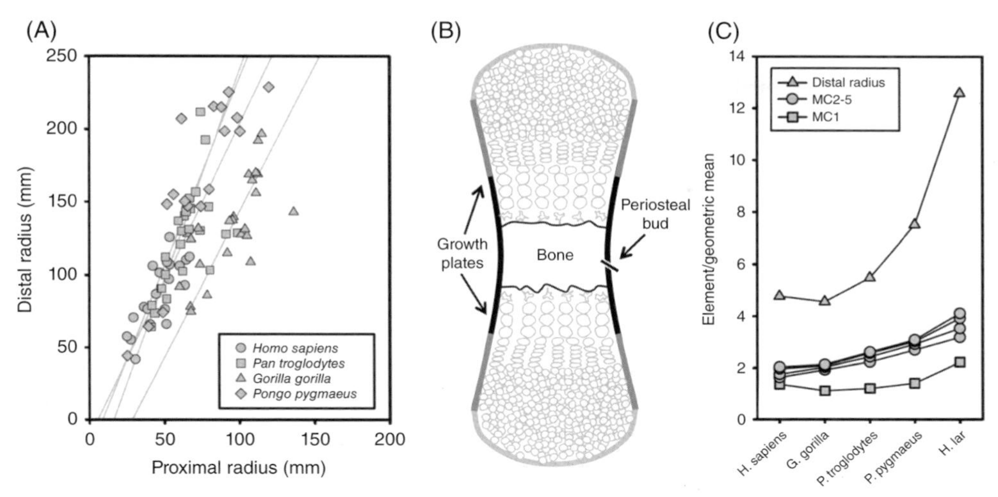

Figure 5.3. Growth plates are individually regulated to specify forearm and digit dimensions in hominoids. (A) Plot comparing proximal versus distal radial growth in juvenile great apes and humans. Proximal and distal lengths are measured to the nutrient foramen. Lines indicate the reduced major axis slopes for each species. In each case, the slopes are significantly greater than one indicating that the distal end contributes approximately twice as much to total length as proximal end. (B) During bone development, cartilaginous models are replaced by invading bone cells via the periosteal bud. The performance of each growth plate will dictate the length of the bone in each direction. The site of periosteal bud invasion can be determined by the nutrient foramen in adult bones. (C) Relative mean lengths of the distal radius and thumb and posterior (finger) metacarpals in apes. The lengths are size corrected by the geometric mean of four articular dimensions of the shoulder, elbow, and wrist joints. Note the dramatic elongation of the forearm in gibbons and orangutans. In addition, observe the longer thumbs and slight shortening of the fingers (relative to forearm length) of humans and gibbons.

To address the hypothesis that evolutionary differences in forelimb skeletal proportions correspond to *Hox* modules, we need to know the relative contribution of each growth plate to total forelimb length. Failing to assess each growth plate individually and simply relying on whole bone lengths will conflate separate growth territories and developmental modules. Fortunately, it is often possible to isolate the growth performance of individual growth plates from adult skeletons. First, the mammalian metapodials and phalanges, unlike other long bones, possess only a single growth plate such that their individual length is the result of a single growth territory (Reno *et al.* 2013a). Second, long bones form by a process of endochondral ossification in which the avascular cartilage anlagen is replaced by bone (Figure 5.3B) (Olsen *et al.* 2000). This occurs when osteogenic cells invade the cartilage model with the initial vascular supply. In adult bones, this site is marked by the nutrient foramen. Thus, in many bones (particularly the radius, ulna, and femur), the distance from the nutrient foramen to each end serves as an indicator of the relative amount of growth in each direction from the initial site of ossification (Payton 1934; Reno *et al.* 2000, 2005).

Isolating individual growth plates is quite revealing for hominoid forelimb proportions. It is now apparent that the increase in forearm length across hominoids is largely due to differences at the distal radius and ulna (Reno *et al.* 2000). In addition, the elongation of the distal forearm is highly correlated with the increased length of the fingers (Figure 5.3C). However, humans and gibbons have fingers that are somewhat shorter than would be predicted by the length of the distal forearm, and these are the two taxa with notably longer thumbs (Reno *et al.* 2008). How might these results correspond to developmental modules defined by *Hoxd* expression?

Similar to the mechanisms described above for beak evolution in Galapagos finches, we proposed that the integration of distinct developmental growth modules facilitated the evolution of hominoid forearm and digit proportions (Reno *et al.* 2008). The correlated change in distal forearm and finger length corresponds with the expression domain of *Hoxd11*, which includes the distal zeugopod and posterior digits (Figure 5.2) (Reno *et al.* 2008). As a result, it is a reasonable hypothesis that the degree of influence of *Hoxd11* (either by alteration in expression or upregulation of one or more of its target genes) results in coordinated evolution of the forearm and fingers as observed in gorillas, chimpanzees, and orangutans. The obverse effect on increased thumb length and finger reduction in humans and gibbons corresponds to the overlapping domains of *Hoxd11* and *Hoxd13*. As the most anterior digit, the thumb only lies within the *Hoxd13* territory while the more posterior digits lie within the overlapping domains of *Hoxd13* and *Hoxd11*. Experiments discussed above indicate that *Hox* genes differ with respect to their growth effects on their target tissues (Yokouchi *et al.* 1995; Goff and Tabin 1997; Zhao and Potter 2001). However, the high conservation of the DNA binding domains between different proteins suggests substantial overlap in the target sequences they recognize. Goff and Tabin have argued that the competition of the co-expressed *Hox* proteins determines the ultimate growth behavior of the bones within their domains (Goff and Tabin 1997). Their experiments suggest that while *Hoxd13* acts as a growth promoter, it promotes a lower rate of growth than does *Hoxd11*. Thus, increasing the effect of *Hoxd13* on skeletal growth in the autopod will serve to augment thumb growth, but reduce the growth of posterior digits as it is placed in competition with *Hoxd11*. Therefore, we hypothesize that humans and gibbons have increased the growth effects of *Hoxd13* to increase relative thumb/finger proportions (Reno *et al.* 2008).

A rich hominid fossil record permits the tracking of these changes in digit and forearm proportions. At 4.4 million years ago (Ma) the largely complete *Ardipithecus ramidus* skeleton provides the best indication of the ancestral hominid limb proportions with chimpanzee-like long forearm and fingers, and (despite its robusticity) a short thumb (Lovejoy *et al.* 2009a, 2009b; White *et al.* 2009). *Australopithecus afarensis* is characterized by relatively long forearms, but the evidence suggests posterior digit shortening similar to gibbons (Drapeau *et al.* 2005; Reno *et al.* 2005, 2008). In addition, the A.L. 333 composite hand skeleton indicates an increase in thumb and decrease in finger length in *Au. afarensis* to the upper range of gorilla or nearly human-like proportions (Alba *et al.* 2003; Rolian and Gordon 2013). Together, these data suggest *Au. afarensis* may have undergone an increase in *Hoxd13* influence during limb development resulting in increased thumb and intermediate finger lengths by 3.2 Ma (Reno *et al.* 2008). Relatively long forearms are maintained to 2.5 Ma as evidenced from the Bouri skeleton (Asfaw *et al.* 1999). Modern human-like forearm length, however, was first observed at 1.8 Ma in the Nariokotome *Homo erectus* skeleton (KNM WT-15000), suggesting a decreased influence of *Hoxd11* occurred within that 700,000-year span that led to shortened forearm and fingers (Walker and Leakey 1993).

The strength of such a model is that it views evolutionary diversity of primates through the lens of developmental mechanisms. It is tempting to seek adaptive explanations for each evolutionary variant (Gould and Lewontin 1979). However, it is plausible that selection in the Pliocene resulted in modification of the hand due to increased reliance on extractive foraging in terrestrial hominids, which further intensified in the Pleistocene with the systematic use of stone tools for meat acquisition (Reno *et al.* 2008). This model, however, is not intended to be all encompassing for ape and human forelimb evolution and development. The relatively short metacarpals relative to total phalangeal length observed in *Ardipithecus* (Lovejoy *et al.* 2009a) and many Miocene hominoids (Almecija *et al.* 2009) suggest parallel changes in digit proportions in African apes and humans influenced by other developmental mechanisms (Hamrick 2001; Reno 2014). In addition, while this is a plausible scenario that accords with the available morphological and developmental data, it is difficult to prove that such a mechanism underlies the evolution of hominoid limb proportions in the absence of gene expression data during ape and human limb development and/or the identification of functional genomic changes that regulate skeletal growth. The former is unlikely to become available; however, further inquiries into the interaction of the modular effects of overlapping *Hox* genes on skeletal growth in traditional models, such as the mouse, could determine downstream targets of *Hox* proteins. Thus, genomic comparisons using a new wealth of data from humans and other primates may reveal species-specific *Hoxd* regulatory variants that underlie limb evolution.

Identifying Human-Specific Genetic Changes

Since the sequencing of the full human genome, we have witnessed the completion of genomes for important model organisms such as the mouse, as well as and many of our primate relatives, including all of the great apes, gibbons, macaques, and even Neanderthals and Denisovans (Consortium 2001, 2005; Gibbs *et al.* 2007; Green *et al.* 2010; Meyer *et al.* 2012; Prado-Martinez *et al.* 2013). While it is still difficult to foresee developmental experiments involving many of these species, the availability of complete genomes makes identifying genomic differences specific to each of these species tractable. For example, Katherine Pollard and colleagues sought to determine unique regions of rapid sequence evolution in the human genome (Pollard *et al.* 2006). To find potential evolutionarily relevant loci, they used the principle that stabilizing selection conserves functional sequences in distantly related species. They then identified the subset of those regions in which humans had undergone a disproportionate rate of sequence divergence. Pollard *et al.* (2006) found that one of these sequences proved to be a non-translated RNA gene, in which the base changes resulted in altered secondary structure of the transcribed human molecule. This RNA gene is expressed in Cajal-Retzius neurons of the developing brain and the testes, two sites that have undergone significant change during human evolution.

Using a generally similar approach, James Noonan and colleagues identified a human accelerated sequence that, while failing to undergo transcription, appears to function as a gene enhancer (Prabhakar *et al.* 2006, 2008). To evaluate this possibility, they conducted a similar test as described above for the *Pitx1* regulatory site in sticklebacks. In this case, the enhancer activity of the human sequence was compared to those of the chimpanzee and rhesus macaque using mouse-primate transgenic assays. The cloned primate sequences were attached to the β-galactosidase (*lacZ*) reporter gene that, when activated, stains tissues blue. Instead of using intractable primate embryos, these three sequences were injected into three sets of mouse oocytes to observe enhancer activity during mammalian development. While the human, chimpanzee, and rhesus macaque sequences all drove gene expression at multiple anatomical sites, only the human version activated gene expression in the developing anterior portion of the distal forelimb and hind limb. So, the accelerated changes in the human sequence appear to have generated novel enhancer functions specific to developing fields of the thumb and big toe (Prabhakar *et al.* 2008). These results are intriguing given the substantial anatomical and contrasting changes in the thumb for improved manual grasping (described above) versus the big toe for pedal fulcrumation during hominid bipedalism. However, further work is necessary to determine the gene target(s) for this enhancer and its putative role in hominid limb development. Regardless, these results provide new examples of human-specific genetic changes that may impact phenotypic evolution and identify some of the mechanisms by which unique human characters may have evolved.

Through collaboration between the labs of Gill Bejerano and David Kingsley, we have sought to determine if, like in sticklebacks above (and Figure 5.1.B), genomic *deletions* played a role in human evolution (McLean *et al.* 2011). Our first task, spearheaded by Cory McLean, was to identify the set of conserved mammalian sequences that had undergone recent deletion since our divergence from the chimpanzee lineage. He first compiled a list of human-specific deletions by aligning the human, chimpanzee, and rhesus macaque reference genomes to identify the sequences that were conserved in the other two anthropoids but missing in humans. Next, he identified those chimpanzee sequences that were conserved with macaques, mice, and in some cases chickens. These two sets were subsequently merged to find human-specific conserved deletions (or hCONDELs). The intersection of the two samples constitutes cases of highly conserved chimpanzee sequences that have been lost during evolution of the human lineage.

With the locations of the individual deletions identified, the next step was to determine their functional significance for the human phenotype. One deletion resided on the X chromosome 220 kb downstream (3') from the transcriptional start site of the *Androgen Receptor* (AR) gene (Figure 5.4A, color plate) (McLean *et al.* 2011). AR is a particularly interesting candidate for human evolution because it mediates tissue specific responses to circulating androgens (testosterone and dihydrotestosterone) for the development of male specific phenotypes (George and Wilson 1994). In the human this deletion removed nearly 61 kb of sequence otherwise found in the chimpanzee. Within the chimpanzee a 4.8-kb sequence exists that maintains substantial homology among mammals including the rhesus macaque, mouse, and even the

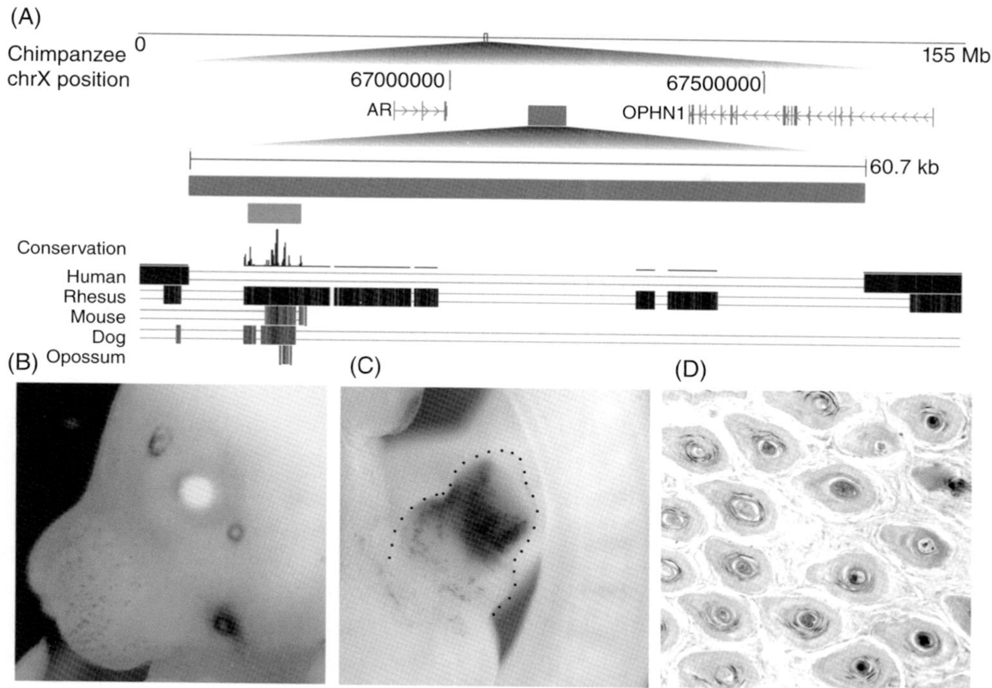

Figure 5.4. Human-specific deletions of an AR enhancer are associated with the loss of vibrissae and penile spines. (A) A 60.7-kb human-specific deletion (red bar) near AR has eliminated nearly 5 kb of highly conserved mammalian sequence. The blue bar indicates the 4.8 kb chimpanzee sequence tested for enhancer activity. The chimpanzee and homologous mouse sequences drive gene expression in embryonic (E 16.5) vibrissae (B) and genital tubercle (C). (D) The mouse enhancer is also functional in the mesodermal papillae of postnatal (day 60) penile spines.

opossum. In addition, neither the Neanderthal nor the Denisovan genomes contain sequences corresponding to the deletion, indicating its loss occurred in the hominid lineage at least 400,000–700,000 years ago (Langergraber *et al.* 2012; Reno *et al.* 2013b). Given the diversity of male secondary sexual characters in primates and their likely dependence on tissue-specific responses to circulating androgens, this genomic change could be intriguing for human phenotypic evolution (Dixson *et al.* 2005).

To determine the functional consequences of this hCONDEL near *AR*, we tested the conserved sequences for enhancer function (McLean *et al.* 2011). As the human condition represents the lack of homologous sequence, we tested the 4.8-kb chimpanzee and the homologous 7.7-kb mouse sequences for enhancer activity that was presumably lost in humans (Figure 5.4A, color plate). In the developing mouse, the chimpanzee sequence drove *lacZ* expression in the mesoderm surrounding developing facial sensory hair or vibrissae follicles, mammary glands, and distal genital tubercles at embryonic day (E) 16.5 (Figure 5.4, color plate). The mouse sequence also drove expression in these same locations, as well as in general hair follicles. To determine postnatal enhancer activity, we also established stable mouse lines expressing *lacZ* attached to the mouse enhancer. This showed that later stages of genital expression predominated in the mesodermal papillae of penile spines. Interestingly, these sites correspond to regions of known *AR* expression in the developing mammary gland and genital tubercle, and to sites where androgens are known to affect development, such as vibrissae growth and penile spine formation (Beach and Levinson 1950; Dixson 1976; Ibrahim and Wright 1983; Murakami 1987; Crocoll *et al.* 1998).

These results provide an intriguing correlation between the loss of a specific regulatory enhancer and the phenotypic loss of vibrissae and penile spines in the human lineage. All apes and every primate surveyed to date preserves the AR enhancer, and it appears to be fixed in a large sample of chimpanzees (Reno *et al.* 2013b). Vibrissae are found in nearly all mammals (humans and anteaters being the only exceptions in a large survey of mammals) (Muchlinski 2010). Even the great apes, which lack macro-vibrissae, still preserve micro-vibrissae, the shorter sensory hairs located above the lip (Muchlinski 2010). Keratinized penile spines are also found in many mammals and are widely distributed across primates, likely representing the ancestral condition for the order (Treatman-Clark 2006). Within hominoids, penile spines are observed in chimpanzees, bonobos, and gibbons (Figure 5.5) (Hill 1946; Matthews 1946; Izor *et al.* 1981). They appear to be present in juvenile gorillas but are subsequently lost by adulthood (Hill and Harrison-Matthews 1949, 1950). Orangutans do not have spines but instead have small flanges or "platelets" whose relationship to spines remains in question (Dahl 1994). Across primates, the structure of penile spines is quite diverse as they can become large and ornate in many strepsirrhines (Dixson 1987; Drea and Weil, 2008). However, these likely represent augmentations and elaborations of a more simplified ancestral spine still common in most primates, particularly in the chimpanzee (Hill 1946). The diversity in spine structure observed in mammals and primates indicates that they may have evolved to

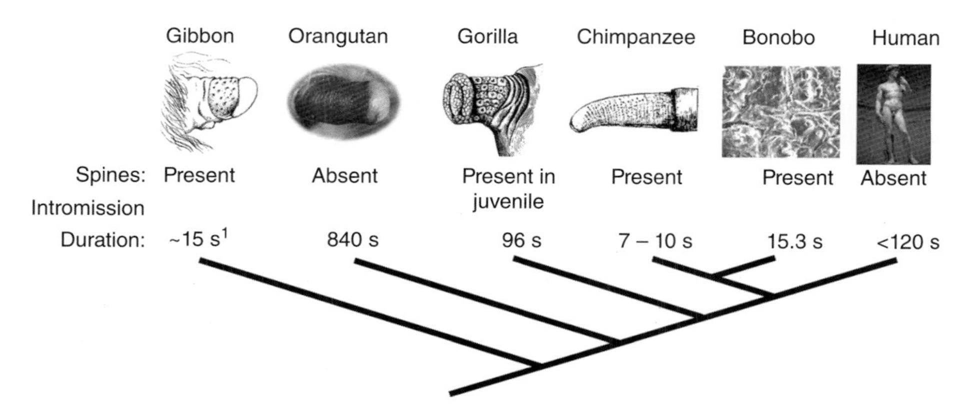

Figure 5.5. Phylogenetic distribution of penile spines and duration of intromission during copulation in hominoids. The presence of spines is likely ancestral for apes. The orangutan is the only ape with a total absence of spines, but their penises are covered by dermal platelets that may or may not be homologous to spines (Dahl 1994). Image of bonobo is a scanning electron micrograph of penile surface (Izor *et al.* 1981). Other ape images from Hill (1946); Matthews (1946); Hill and Harrison-Matthews (1950). Intromission data from Dixson (1998) except (1) from Chivers (1974); Ellefson (1974).

accommodate multiple species-specific functions (Aronson and Cooper 1967; Dixson 1998; Stockley 2002). However, the more simplified spines found in other hominoids are similar in absolute size to those of mice, rats, and some New World Monkeys (Hershkovitz 1993). Given their common association with underlying neural structures, it is reasonable to hypothesize that these spines function as male mechanosensory organs. Experimental removal of the spines in marmosets with topical depilatory treatment produces a significant increase in the duration of intromission during copulation (Dixson 1991). Copulation duration in hominoids with spines is remarkably brief, ranging from less than 10 s to 15 s in chimpanzees, bonobos, and gibbons (Figure 5.5) (Chivers 1974; Ellefson 1974; Dixson 1998). In contrast, in species lacking spines, such as orangutans, adult gorillas, and humans, copulation duration extends well over a minute (Dixson 1998; Vanden Broucke *et al.* 2007). This would suggest dramatically different reproductive behavior in ancestral hominids lacking spines compared to extant chimpanzees and bonobos. In addition, the feminization of the male canine in the Miocene and Pliocene hominids *Sahelanthropus*, *Ardipithecus*, and *Australopithecus* and low to moderate skeletal sexual dimorphism observed in the latter two suggest reproductive behaviors distinct from longer copulating, though highly dimorphic, gorillas and orangutans (Haile-Selassie 2001; Brunet *et al.* 2002; Reno *et al.* 2003, 2010, 2015; Harmon 2009; Suwa *et al.* 2009; White *et al.* 2009; but see Gordon *et al.* 2008). Several other characters in the human lineage, such as moderate sized testes with low sperm motility, permanent female mammary fat deposits, concealed ovulation from both males and females, and extended female sexual receptivity, may be explained by selection to a social structure based on pair bonding (Lovejoy 1981, 2009). Thus, the loss of penile spines adds another phenotypic change to this suite of characters that indicate an important reproductive shift during hominid evolution that may relate to increased parental care, extension of the juvenile period, cerebral expansion, and marked demographic success demonstrated by modern humans.

Conclusions

The integration of evolutionary and developmental biology provides a formal means to identify specific biological mechanisms underlying phenotypic variation and evolutionary change. Such efforts have been highly fruitful, showing that organizational and structural features such as modularity, functional diversity, specificity, and deep homology are pervasive in the genome, the developmental process, and the phenotype. This structural organization provides a mechanism to facilitate viable evolutionary change.

Initial anthropological forays into evo-devo have largely confirmed these findings. Phenotypic variation and patterns of evolutionary change in primates are highly modular (Reno *et al.* 2008; Hlusko *et al.* 2011). Identifying these modular morphological and developmental relationships in the hominoid forelimb can refine the adaptive hypotheses concerning their evolution (i.e., is the target of selection forearm or finger length?) (Reno *et al.* 2008). The extensive genomic data now available for humans and primates reveal the same basic genetic organization as in other species. Such a genome-based approach can direct attention towards previously unconsidered human-specific phenotypic changes such as penile spines (Reno *et al.* 2013b). However, in these cases more work is needed to clarify the mechanisms underlying the evolution of particular phenotypes. In the case of *Hox* genes' roles in limb proportions, specific functional regulatory changes have yet to be identified in primates. The rapidly expanding knowledge and methodologies related to the mechanisms of *Hox* regulation of target genes in the tetrapod limb, human and primate genomic variation, and computational tools available for identifying species-specific variants provide hope that such evolutionarily relevant changes will soon be identified. Further work on the AR regulatory deletion is also necessary to determine the specific functional link between the loss of this enhancer and gene expression in humans. Luckily, the depth of conservation in the genetic toolkit and regulatory networks identified in vertebrates means that relevant functional analyses can be conducted in tractable model organisms. These facts are of direct relevance to any primate functional and phylogenetic analysis. As such, evolutionary developmental anthropology is poised to make great strides towards identifying the mechanisms underlying those phenotypic changes unique to humans and our closest primate relatives.

Acknowledgements

I thank Campbell Rolian and Julia Boughner for the invitation to contribute to this volume. This work benefited from a history of collaboration with Owen Lovejoy, David Kingsley, Gill Bejerano, Melanie McCollum, Walt Horton, Terry Capellini, Marty Cohn, Maria Serrat, Burt Rosenman, Kate Guentuer, Alex Pollen, Cory McLean, and many others. I thank Kelsey Kjosness, Jasmine Hines, Campbell, and Julia for critical reading of this manuscript. The work described here was funded in part by NSF (0311768) and NIH (Ruth L. Kirschtein NRSA 1 F32 HD062137-01).

References

Abzhanov, A., Kuo, W. P., Hartmann, C. *et al.* 2006. The calmodulin pathway and evolution of elongated beak morphology in Darwin's finches. *Nature* 442:563–567.

Abzhanov, A., Protas, M., Grant, B. R. *et al.* 2004. Bmp4 and morphological variation of beaks in Darwin's finches. *Science* 305:1462–1465.

Alba, D. M., Moya-Sola, S., and Kohler, M. 2003. Morphological affinities of the *Australopithecus afarensis* hand on the basis of manual proportions and relative thumb length. *Journal of Human Evolution* 44:225–254.

Almecija, S., Alba, D. M., and Moya-Sola, S. 2009. *Pierolapithecus* and the functional morphology of Miocene ape hand phalanges: Paleobiological and evolutionary implications. *Journal of Human Evolution* 57:284–297.

Andrey, G., Montavon, T., Mascrez, B. *et al.* 2013. A switch between topological domains underlies *HoxD* genes collinearity in mouse limbs. *Science* 340:1234167.

Aronson, L. R., and Cooper, M. L. 1967. Penile spines of the domestic cat: Their endocrine-behavior relations. *Anatomical Record* 157:71–78.

Asfaw, B., White, T., Lovejoy, O. *et al.* 1999. *Australopithecus garhi*: A new species of early hominid from Ethiopia. *Science* 284:629–635.

Beach, F. A., and Levinson, G. 1950. Effects of androgen on the glans penis and mating behavior of castrated male rats. *Journal of Experimental Zoology* 114:159–171.

Bell, M. A., and Foster, S. A. 1994. Introduction to the evolutionary biology of the threespine stickleback. *In*: M. A. Bell and S. A. Foster (eds) *The evolutionary biology of the threespine stickleback*. Oxford: Oxford University Press. pp.1–27.

Bisgard, J. D., and Bisgard, M. E. 1935. Longitudinal growth of long bones. *Archives of Surgery* 31:569–587.

Boulet, A. M., and Capecchi, M. R. 2002. Duplication of the *Hoxd11* gene causes alterations in the axial and appendicular skeleton of the mouse. *Developmental Biology* 249:96–107.

Boulet, A. M., and Capecchi, M. R. 2004. Multiple roles of *Hoxa11* and *Hoxd11* in the formation of the mammalian forelimb zeugopod. *Development* 131:299–309.

Brunet, M., Guy, F., Pilbeam, D. *et al.* 2002. A new hominid from the Upper Miocene of Chad, Central Africa. *Nature* 418:145–151.

Buchanan, A. V., Sholtis, S., Richtsmeier, J., and Weiss, K. M. 2009. What are genes "for" or where are traits "from"? What is the question? *Bioessays* 31:198–208.

Burke, A. C., Nelson, C. E., Morgan, B. A., and Tabin, C. J. 1995. *Hox* gene and the evolution of vertebrate axial morphology. *Development* 121:333–346.

Capecchi, M. R. 1997. *Hox* genes and mammalian development. *Cold Spring Harbor Symposia on Quantitative Biology* 62:273–281.

Carroll, S. B. 2003. Genetics and the making of *Homo sapiens*. *Nature* 422:849–857.

Carroll, S. B. 2005. Evolution at two levels: On genes and form. *PLoS Biol* 3:e245.

Carroll, S. B. 2008. Evo-devo and an expanding evolutionary synthesis: A genetic theory of morphological evolution. *Cell* 134:25–36.

Carroll, S. B, Grenier, J. K., and Weatherbee, S. D. 2005. *From DNA to diversity*. Oxford: Blackwell.

Chan, Y. F., Marks, M. E., Jones, F. C. *et al.* 2010. Adaptive evolution of pelvic reduction in sticklebacks by recurrent deletion of a *Pitx1* enhancer. *Science* 327:302–305.

Chiu, C. H., and Hamrick, M. W. 2002. Evolution and development of the primate limb skeleton. *Evolutionary Anthropology* 11:94–107.

Chivers, D. J. 1974. *The siamang in Malaya*. Basel: S. Karger.

Cohn, M. J., and Tickle, C. 1999. Developmental basis of limblessness in snakes. *Nature* 399:474–479.

Consortium CSaA. 2005. Initial sequence of the chimpanzee genome and comparison with the human genome. *Nature* 437:69–87.

Consortium IHGS. 2001. Initial sequencing and analysis of the human genome. *Nature* 409:860–921.

Crocoll, A., Zhu, C. C., Cato, A. C., and Blum, M. 1998. Expression of androgen receptor mRNA during mouse embryogenesis. *Mechanisms of Development* 72:175–178.

Dahl, J. F. 1994. Size and form of the penis in orang-utans. *Journal of Mammalogy* 75:1–9.

Davis, A. P., Witte, D. P., Hsieh-Li, H. M. *et al.* 1995. Absence of radius and ulna in mice laking *hoxa-11* and *hoxd-11*. *Nature* 375:791–795.

DiLeone, R. J., Russell, L. B., and Kingsley, D. M. 1998. An extensive 3′ regulatory region controls expression of BMP5 in specific anatomical structures of the mouse embryo. *Genetics* 148:401–408.

Dixson, A. F. 1976. Effects of testosterone on sternal cutaneous glands and genitalia of male greater galago (*Galago crassicaudatus crassicaudatus*). *Folia Primatologica* 26:207–213.

Dixson, A. F. 1987. Observations on the evolution of genitalia and copularoty behavior in male primates. *Journal of Zoology* 213:423–443.

Dixson, A. F. 1991. Penile spines affect copulatory behaviour in a primate (*Callithrix jacchus*). *Physiology & Behavior* 49:557–562.

Dixson, A. F. 1998. *Primate sexuality: Comparative studies of the prosimians, monkeys, apes and human beings*. Oxford: Oxford University Press.

Dixson, A. F., Dixson, B., and Anderson, M. 2005. Sexual selection and the evolution of visually conspicuous sexually dimorphic traits in male monkeys, apes, and human beings. *Annual Review of Sex Research* 16:1–19.

Dobzhansky, T. 1956. What is an adaptive trait? *American Naturalist* 90:337–347.

Drapeau, M. S., Ward, C. V., Kimbel, W. H. *et al.* 2005. Associated cranial and forelimb remains attributed to *Australopithecus afarensis* from Hadar, Ethiopia. *Journal of Human Evolution* 48:593–642.

Drea, C. M., and Weil, A. 2008. External genital morphology of the ring-tailed lemur (*Lemur catta*): Females are naturally "masculinized." *Journal of Morphology* 269:451–463.

Duboule, D. 2007. The rise and fall of *Hox* gene clusters. *Development* 134:2549–2560.

Ellefson, J. O. 1974. A natural history of white-handed gibbons in the Malayan Peninsula. *In*: D. M. Rumbaugh (ed.) *Gibbon and siamang*. Basel: S. Karger. pp.1–136.

Favier, B., and Dolle, P. 1997. Developmental functions of mammalian *Hox* genes. *Molecular Human Reproduction* 3:115–131.

George, F. W., and Wilson, J. D. 1994. Sex determination and differentiation. *In*: E. Knobil and J. D. Neill (eds) *The physiology of reproduction*. New York: Raven Press.

Gibbs, R. A., Rogers, J., Katze, M. G. *et al.* 2007. Evolutionary and biomedical insights from the rhesus macaque genome. *Science* 316:222–234.

Gilbert, S. F., Opitz, J. M., and Raff, R. A. 1996. Resynthesizing evolutionary and developmental biology. *Developmental Biology* 173:357–372.

Goff, D. J., and Tabin, C. J. 1997. Analysis of *Hoxd-13* and *Hoxd-11* misexpression in chick limb bud reveals that *Hox* genes affect both bone condensation and growth. *Development* 124:627–636.

Gompel, N., Prud'homme, B., Wittkopp, P. J. *et al.* 2005. Chance caught on the wing: *Cis*-regulatory evolution and the origin of pigment patterns in *Drosophila*. *Nature* 433:481–487.

Gonzalez, F., Duboule, D., and Spitz, F. 2007. Transgenic analysis of *Hoxd* gene regulation during digit development. *Developmental Biology.* 306:847–859.

Gordon, A. D., Green, D. J., and Richmond, B. G. 2008. Strong postcranial size dimorphism in *Australopithecus afarensis*: Results from two new resampling methods for multivariate data sets with missing data. *American Journal of Physical Anthropology* 135:311–328.

Gould, S. J., and Lewontin, R. C. 1979. The spandrels of San Marco and the Panglossian paradigm: A critique of the adaptationist programme. *Proceedings of the Royal Society of London, Series B: Biological Sciences* 205:581–598.

Grant, P. R. 1991. Natural selection and Darwin's finches. *Scientific American* 265:82–87.

Green, R. E., Krause, J., Briggs, A. W. *et al.* 2010. A draft sequence of the Neandertal genome. *Science* 328:710–722.

Grieco, T. M., Rizk, O. T., and Hlusko, L. J. 2013. A modular framework characterizes micro- and macroevolution of old world monkey dentitions. *Evolution* 67:241–259.

Guenther, C., Pantalena-Filho, L., and Kingsley, D. M. 2008. Shaping skeletal growth by modular regulatory elements in the *Bmp5* gene. *PLoS Genetics* 4:e1000308.

Haile-Selassie, Y. 2001. Late Miocene hominids from the Middle Awash, Ethiopia. *Nature* 412:178–181.

Halder, G., Callaerts, P., and Gehring, W. J. 1995. Induction of ectopic eyes by targeted expression of the eyeless gene in *Drosophila*. *Science* 267:1788–1792.

Hallgrímsson, B., Willmore, K., and Hall, B. K. 2002. Canalization, developmental stability, and morphological integration in primate limbs. *American Journal of Physical Anthropology Suppl.* 35:131–158.

Hamrick, M. W. 1999. A chondral modeling theory revisited. *Journal of Theoretical Biology* 201:201–298.

Hamrick, M. W. 2001. Primate origins: Evolutionary change in digital ray patterning and segmentation. *Journal of Human Evolution* 40:339–351.

Harmon, E. 2009. Size and shape variation in the proximal femur of *Australopithecus africanus*. *Journal of Human Evolution* 56:551–559.

Hershkovitz, P. 1993. Male external genitalia of non-prehensile tailed South American monkeys. Part I. Subfamily Pitheciinae, family Cebidae. *Fieldiana: Zoology* 73:1–17.

Hill, W. C. O. 1946. Note on the male external genitalia of the chimpanzee. *Proceedings of the Zoological Society, London* 116:129–132.

Hill, W. C. O., and Harrison-Matthews, L. 1949. The male external genitalia in the gorilla, with remarks on the os penis of other Hominoidea. *Proceedings of the Zoological Society, London* 119:363–386.

Hill, W. C. O., and Harrison-Matthews, L. 1950. Supplementary note on the male external genitalia of *Gorilla*. *Proceedings of the Zoological Society, London* 120:311–315.

Hlusko, L. J. 2004. Integrating the genotype and phenotype in hominid paleontology. *Proceedings of the National Academy of Sciences, USA*. 101:2653–2657.

Hlusko, L. J., Sage, R. D., and Mahaney, M. C. 2011. Modularity in the mammalian dentition: Mice and monkeys share a common dental genetic architecture. *Journal of Experimental Zoology B (Molecular Developmental Evolution)* 316:21–49.

Hostikka, S. L., and Capecchi, M. R. 1998. The mouse *Hoxc11* gene: Genomic structure and expression pattern. *Mechanisms of Development* 70:133–145.

Ibrahim, L., and Wright, E. A. 1983. Effect of castration and testosterone propionate on mouse vibrissae. *The British Journal of Dermatology* 108:321–326.

Iimura, T., Denans, N., and Pourquie, O. 2009. Establishment of *Hox* vertebral identities in the embryonic spine precursors. *Current Topics in Developmental Biology* 88:201–234.

Izor, R. J., Walchuk, S. L., and Wilkins, L. 1981. Anatomy and systematic significance of the penis of the pygmy chimpanzee, *Pan paniscus*. *Folia Primatologica* 35:218–224.

Kmita, M., Fraudeau, N., Herault, Y., and Duboule, D. 2002. Serial deletions and duplications suggest a mechanism for the collinearity of *Hoxd* genes in limbs. *Nature* 420:145–150.

Langergraber, K. E., Prufer, K., Rowney, C. *et al.* 2012. Generation times in wild chimpanzees and gorillas suggest earlier divergence times in great ape and human evolution. *Proceedings of the National Academy of Sciences, USA* 109:15716–15721.

Lieberman, D. E., Krovitz, G. E., and McBratney-Owen, B. 2004. Testing hypotheses about tinkering in the fossil record: The case of the human skull. *Journal of Experimental Zoology B (Molecular Developmental Evolution)* 302:284–301.

Logan, M., Simon, H. G., and Tabin, C. 1998. Differential regulation of T-box and homeobox transcription factors suggests roles in controlling chick limb-type identity. *Development* 125:2825–2835.

Logan, M., and Tabin, C. J. 1999. Role of *Pitx1* upstream of *Tbx4* in specification of hindlimb identity. *Science* 283:1736–1739.

Lovejoy, C. O. 1981. The Origin of Man. *Science* 211:341–350.

Lovejoy, C. O. 2007. The natural history of human gait and posture – Part 3. The knee. *Gait and Posture* 25:325–341.

Lovejoy, C. O. 2009. Reexamining human origins in light of *Ardipithecus ramidus*. *Science* 326:74e71–78.

Lovejoy, C. O., Cohn, M. J., and White, T. D. 1999. Morphological analysis of the mammalian postcranium: A developmental perspective. *Procedings of the National Academy of Sciences, USA* 96:13247–13252.

Lovejoy, C. O., McCollum, M. A., Reno, P. L., and Rosenman, B. A. 2003. Developmental biology and human evolution. *Annual Review in Anthropology* 32:85–109.

Lovejoy, C. O., Simpson, S. W., White, T. D. *et al.* 2009a. Careful climbing in the Miocene: The forelimbs of *Ardipithecus ramidus* and humans are primitive. *Science* 326:70e71–78.

Lovejoy, C. O., Suwa, G., Simpson, S. W. *et al.* 2009b. The great divides: *Ardipithecus ramidus* reveals the postcrania of our last common ancestors with African apes. *Science* 326:100–106.

Mallarino, R., Grant, P. R., Grant, B. R. *et al.* 2011. Two developmental modules establish 3D beak-shape variation in Darwin's finches. *Proceedings of the National Academy of Sciences, USA* 108:4057–4062.

Marroig, G., and Cheverud, J. M. 2001. A comparison of phenotypic variation and covariation patterns and the role of phylogeny, ecology, and ontogeny during cranial evolution of new world monkeys. *Evolution* 55:2576–2600.

Marroig, G., De Vivo, M., and Cheverud, J. M. 2004. Cranial evolution in sakis (Pithecia, Platyrrhini). II: Evolutionary processes and morphological integration. *Journal of Evolutionary Biology* 17:144–155.

Matthews, L. H. 1946. Notes on the genital anatomy and physiology of the gibbon (*Hylobates*). *Proceedings of the Royal Society, London* 116:339–364.

Mayr, E. 1963. *Animal species and evolution*. Cambridge, MA: Harvard University Press.

McCollum, M. A. 1999. The robust australopithecine face: A morphogenetic perspective. *Science* 284:301–305.

McGinnis, W., Garber, R. L., Wirz, J. *et al.* 1984. A homologous protein-coding sequence in *Drosophila* homeotic genes and its conservation in other metazoans. *Cell* 37:403–408.

McLean, C. Y., Reno, P. L., Pollen, A. A. *et al.* 2011. Human-specific loss of regulatory DNA and the evolution of human-specific traits. *Nature* 471:216–219.

Menke, D. B. 2013. Engineering subtle targeted mutations into the mouse genome. *Genesis* 51:605–618.

Menke, D. B., Guenther, C., and Kingsley, D. M. 2008. Dual hindlimb control elements in the *Tbx4* gene and region-specific control of bone size in vertebrate limbs. *Development* 135:2543–2553.

Meyer, M., Kircher, M., Gansauge, M. T. *et al.* 2012. A high-coverage genome sequence from an archaic Denisovan individual. *Science* 338:222–226.

Montavon, T., Le Garrec, J. F., Kerszberg, M., and Duboule, D. 2008. Modeling *Hox* gene regulation in digits: Reverse collinearity and the molecular origin of thumbness. *Genes and Development* 22:346–359.

Montavon, T., Soshnikova, N., Mascrez, B. *et al.* 2011. A regulatory archipelago controls *Hox* genes transcription in digits. *Cell* 147:1132–1145.

Morgan, B. A., and Tabin, C. 1994. *Hox* genes and growth: Early and late roles in limb bud morphogenesis. *Development* Suppl. 181–186.

Muchlinski, M. N. 2010. A comparative analysis of vibrissa count and infraorbital foramen area in primates and other mammals. *Journal of Human Evolution* 58:447–473.

Murakami, R. 1987. A histological study of the development of the penis of wild-type and androgen-insensitive mice. *Journal of Anatomy* 153:223–231.

Napier, J. R., and Napier, P. H. 1985. *The natural history of primates*. Cambridge, MA: MIT Press.

Nelson, C. E., Morgan, B. A., Burke, A. C. *et al.* 1996. Analysis of *Hox* gene expression in the chick limb bud. *Development* 122:1449–1466.

O'Bleness, M., Searles, V. B., Varki, A. *et al.* 2012. Evolution of genetic and genomic features unique to the human lineage. *Nature Reviews in Genetics* 13:853–866.

Olsen, B. R., Reginato, A. M., and Wang, W. 2000. Bone development. *Annual Reviews of Cell and Developmental Biology* 16:191–220.

Olson, E. C., and Miller, R. L. 1999. *Morphological integration.* Chicago, IL: University of Chicago Press.

Papenbrock, T., Visconti, R. P., and Awgulewitsch, A. 2000. Loss of the fibula in mice overexpressing *Hoxc11*. *Mechanisms of Development* 92:113–123.

Payton, C. G. 1932. The growth in length of the long bones in the madder-fed pig. *Journal of Anatomy* 66:414–425.

Payton, C. G. 1934. The position of the nutrient foramen and direction of the nutrient canal in the long bones of the madder-fed pig. *Journal of Anatomy* 68:500–510.

Perry, G. H., Dominy, N. J., Claw, K. G. *et al.* 2007. Diet and the evolution of human amylase gene copy number variation. *Nature Genetics* 39:1256–1260.

Pollard, K. S., Salama, S. R., Lambert, N. *et al.* 2006. An RNA gene expressed during cortical development evolved rapidly in humans. *Nature* 443:167–172.

Prabhakar, S., Noonan, J. P., Pääbo, S., and Rubin, E. M. 2006. Accelerated evolution of conserved noncoding sequences in humans. *Science* 314:786.

Prabhakar, S., Visel, A., Akiyama, J. A, *et al.* 2008. Human-specific gain of function in a developmental enhancer. *Science* 321:1346–1350.

Pradel, J., and White, R. A. H. 1998. From selectors to realizors. *International Journal of Developmental Biology* 42:417–421.

Prado-Martinez, J., Sudmant, P. H., Kidd, J. M. *et al.* 2013. Great ape genetic diversity and population history. *Nature* 499:471–475.

Raff, R. A. 2000. Evo-devo: The evolution of a new discipline. *Nature Reviews in Genetics* 1:74–79.

Reno, P. L. 2014. Genetic and developmental basis for parallel evolution and its significance for hominoid evolution. *Evolutionary Anthropology* 23:188–200.

Reno, P. L., DeGusta, D., Serrat, M. A. *et al.* 2005. Plio-pleistocene hominid limb proportions – Evolutionary reversals or estimation errors? *Current Anthropology* 46:575–588.

Reno, P. L., Horton, W. E., Jr., and Lovejoy, C. O. 2013a. Metapodial or phalanx? An evolutionary and developmental perspective on the homology of the first ray's proximal segment. *Journal of Experimental Biology B (Molecular Developmental Evolution)* 320:276–285.

Reno, P. L., and Lovejoy, C. O. 2015. From Lucy to Kadanuumuu: balanced analyses of Australopithecus afarensis assemblages confirm only moderate skeletal dimorphism. *PeerJ* 3: e925 https://dx.doi.org/10.7717/peerj.925.

Reno, P. L., McBurney, D. L., Lovejoy, C. O., and Horton, W. E. 2006. Ossification of the mouse metatarsal: Differentiation and proliferation in the presence/absence of a defined growth plate. *Anatomical Record* 288A:104–118.

Reno, P. L., McCollum, M. A., Cohn, M. J. *et al.* 2008. Patterns of correlation and covariation of anthropoid distal forelimb segments correspond to *Hoxd* expression territories. *Journal of Experimental Zoology Part B (Molecular and Developmental Evolution)* 310B:240–258.

Reno, P. L., McCollum, M. A., Lovejoy, C. O., and Meindl, R. S. 2000. Adaptationism and the anthropoid postcranium: Selection does not govern the length of the radial neck. *Journal of Morphology* 246:59–67.

Reno, P. L., McCollum, M. A., Meindl, R. S., and Lovejoy, C. O. 2010. An enlarged postcranial sample confirms *Australopithecus afarensis* dimorphism was similar to modern humans.

Philosophical Transactions of the Royal Society of London Series B, Biological Sciences 365:3355–3363.

Reno, P. L., McLean, C. Y., Hines, J. E. *et al.* 2013b. A penile spine/vibrissa enhancer sequence is missing in modern and extinct humans but is retained in multiple primates with penile spines and sensory vibrissae. *PLoS One* 8:e84258.

Reno, P. L., Meindl, R. S., McCollum, M. A., and Lovejoy, C. O. 2003. Sexual dimorphism in *Australopithecus afarensis* was similar to that of modern humans. *Proceedings of the National Academy of Sciences, USA* 100:9404–9409.

Richmond, B. G., Begun, D. R., and Strait, D. S. 2001. Origin of human bipedalism: The knuckle-walking hypothesis revisited. *Yearbook of Physical Anthropology* 44:70–105.

Rolian, C., and Gordon, A. D. 2013. Reassessing manual proportions in *Australopithecus afarensis*. *American Journal of Physical Anthropology* 152:393–406.

Rolian, C., Lieberman, D. E., and Hallgrímsson, B. 2010. The coevolution of human hands and feet. *Evolution* 64:1558–1568.

Schultz, A. H. 1925. Embryological evidence of the evolution of man. *Journal of the Washington Academy of Sciences* 15:247–263.

Schultz, A. H. 1926. Fetal growth of man and other primates. *Quaterly Reviews in Biology* 1:465–521.

Shapiro, M. D., Marks, M. E., Peichel, C. L. *et al.* 2004. Genetic and developmental basis of evolutionary pelvic reduction in threespine sticklebacks. *Nature* 428:717–723.

Stockley, P. 2002. Sperm competition risk and male genital anatomy: Comparative evidence for reduced duration of female sexual receptivity in primates with penile spines. *Evolutionary Ecology* 16:123–137.

Suwa, G., Kono, R. T., Simpson, S. W. *et al.* 2009. Paleobiological implications of the *Ardipithecus ramidus* dentition. *Science* 326:94–99.

Tardieu, C. 2010. Development of the human hind limb and its importance for the evolution of bipedalism. *Evolutionary Anthropology* 19:174–186.

Treatman-Clark, K. 2006. *The evolution of the reproductive system in strepsirrhine primates*. PhD thesis. Stony Brook: Stony Brook University. p.510.

Vanden Broucke, H., Everaert, K., Peersman, W. *et al.* 2007. Ejaculation latency times and their relationship to penile sensitivity in men with normal sexual function. *Journal of Urology* 177:237–240.

Vargas, A. O., Kohlsdorf, T., Fallon, J. F. *et al.* 2008. The evolution of *HoxD-11* expression in the bird wing: Insights from *Alligator mississippiensis*. *PLoS One* 3:e3325.

Wagner, G. P. 2007. The developmental genetics of homology. *Nature Reviews in Genetics* 8:473–479.

Wagner, G. P., and Altenberg, L. 1996. Complex adaptations and the evolution of evolvability. *Evolution* 50:967–976.

Wagner, G. P., Chiu, C. H., and Laubichler, M. 2000. Developmental evolution as a mechanistic science: The inference from developmental mechanisms to evolutionary processes. *American Zoologist* 40:819–831.

Walker, A., and Leakey, R. 1993. Perspectives on the Nariokotome discovery. *In*: A. Walker and R. Leaker (eds) *The Nariokotome Homo erectus skeleton*. Cambridge, MA: Harvard University Press. p.430.

Wellik, D. M. 2009. *Hox* genes and vertebrate axial pattern. *Current Topics in Developmental Biology* 88:257–278.

Wellik, D. M., and Capecchi, M. R. 2003. *Hox10* and *Hox11* genes are required to globally pattern the mammalian skeleton. *Science* 301:363–367.

White, T. D., Asfaw, B., Beyene, Y. *et al.* 2009. *Ardipithecus ramidus* and the paleobiology of early hominids. *Science* 326:75–86.

Wolpert, L. 1978. Pattern formation in biological development. *Scientific American* 239:154–164.

Yokouchi, Y., Nakazato, S., Yamamoto, M. *et al.* 1995. Misexpression of *Hoxa-13* induces cartilage homeotic transformation and changes in cell adhesiveness in chick limb buds. *Genes and Development* 9:2509–2522.

Young, N. M., Wagner, G. P., and Hallgrímsson, B. 2010. Development and the evolvability of human limbs. *Proceedings of the National Academy of Sciences, USA* 107:3400–3405.

Zhao, Y., and Potter, S. S. 2001. Functional specificity of the *Hoxa13* homeobox. *Development* 128:3197–3207.

Chapter 6

Out on a Limb: Development and the Evolution of the Human Appendicular Skeleton

Nathan M. Young[1] and Terence D. Capellini[2]
[1] *Department of Orthopaedic Surgery, University of California, San Francisco, CA, USA*
[2] *Department of Human Evolutionary Biology, Harvard University, Cambridge, MA, USA*

Introduction

Understanding the forces that have influenced the evolution of shape, size, and proportions of the human appendages, including both limbs and girdles (scapula and pelvis), has played an important part in paleoanthropology. Primates exhibit significant variation in appendicular morphology, reflecting a diverse range of locomotor and postural adaptations (see Fleagle 2013 for a review). In the human lineage, evolutionary shifts in shape, size, and proportions are associated with some of the first and arguably most important behavioral changes in hominins, such as reduced arboreality, increased bipedal specialization, and tool use (see Aiello and Dean 2002 for a review). This variation also informs questions of phylogenetic relationships and the potential for homoplasy in apes (e.g., Larson 1998; Young 2003, 2008; Pilbeam and Young 2004) and humans (e.g., Young *et al.* 2010). In this chapter we outline a framework to address evolutionary variation in the human appendages, specifically regarding key transitions from the last common ancestor (LCA) of chimpanzees and modern humans, which combines data from extant comparative models and fossil hominins to generate testable developmental hypotheses. Because the field is so new, in many cases we hope to point out areas where more information is needed rather than providing explicit answers.

Morphospaces

For the scapula, pelvis, and then limbs, we begin by describing the relevant skeletal anatomy and development, briefly discussing the tissues, signaling centers, and genes

Developmental Approaches to Human Evolution, First Edition. Edited by
Julia C. Boughner and Campbell Rolian.

currently known to be involved in their morphogenesis. We next compare phenotypic variation in each element in a "morphospace." These statistically derived multivariate or bivariate spaces simplify variation in a biologically "agnostic" fashion to a smaller number of explanatory variables and describe the boundaries of observed variation. They can also be used to infer potential "rules" of coordinated changes (i.e., covariation), which in turn can be used to predict ancestral morphotypes and, in combination with inferred evolutionary trajectories, the potential developmental transformations involved. For this last point, we utilize the idea that evolutionary trajectories through morphospace represent developmental transformations from putative ancestral morphotypes to living species. To generate each morphospace we utilize both previously published morphometric data (scapula: Young (2004, 2006, 2008); limbs: Young *et al.* (2010)), and novel (pelvis) landmark-based analyses, and use these to test existing phylogenetic reconstructions of ancestral morphotypes.

Last Common Ancestor

While the LCA of humans and apes has been inferred at various times to be similar to any number of living or extinct primates (Wrangham and Pilbeam 2002; Pilbeam and Young 2004), genetic datasets now overwhelmingly support a tree in which chimpanzees and bonobos are our closest living relatives, with gorillas, orangutans, and gibbons successively more distant (Ruvolo 1997). The human fossil record has also greatly improved over the last two decades, particularly in regards to appendages, with numerous ancestral and collateral human species (hominins) now described into the late Miocene (Brunet *et al.* 2001; Senut *et al.* 2001). Yet despite improvements in phylogenetic resolution and the early hominin fossil record, the reconstruction of the human-chimpanzee LCA is complicated by a dearth of fossil representatives of living apes. This situation leaves open for debate whether chimpanzees are evolutionarily conservative, and thus useful proxy models of the LCA, or whether they have also experienced significant morphological and behavioral changes since the Miocene, in which case they may be too derived to be relevant. This fundamental debate has led to two alternative models that we use as the basis of our inferences of human appendicular evolution and development.

The first, which we call the "African Ape" (AA) model, posits that the LCA was essentially chimpanzee-like in morphology. The assumption of this scenario is that since divergence of African apes and humans (~7 million years ago) there has been little to no significant postcranial evolution in panins compared to hominins, or at least not sufficient that they would be identifiable as a separate genus from living chimpanzees or bonobos. A critical piece of evidence supporting this idea is that differences in the face and limbs of living taxa are largely attributable to scaling of body size (Shea 1981, 1983, 1984, 1985; Hartwig-Sherer 1993), despite the more ancient divergence of *Gorilla* from *Pan* (~10 million years ago) compared to hominins (Steiper and Young 2006). Because of these scaling relationships and the association of traits like size dimorphism with social structure and behavior, if we assume the *Pan-Homo* LCA was chimpanzee-sized, then phenotypic traits would be predicted to be similar, if not exactly identical, to those of living panins (either chimpanzees or bonobos)

(Wrangham and Pilbeam, 2001). Fossil evidence tends to support this prediction, since earlier hominins are generally more *Pan*-like while later hominins are more *Homo*-like (although see below) (White *et al.* 1994; Brunet *et al.* 2001).

The second model, which we call "Ape Convergence" (AC), predicts that the human-chimpanzee LCA is a currently unknown and more primitive Miocene ape ancestor. In this scenario African apes are too highly derived to inform potential ancestral morphotypes, and African ape similarities to each other are homoplasies. Supporting this idea, Lovejoy and colleagues (2009) have interpreted a number of postcranial features in *Ardipithecus ramidus*, an early hominin, as reflective of ancestry from an arboreal, above-branch quadruped, such as the early ape *Proconsul*. Since this scenario requires substantial homoplasy in African apes, an indirect line of evidence in support of this idea is that ape appendicular evolution appears more mosaic and homoplastic than would be inferred from extant species alone. In particular, the putative ancestor of *Pongo* known as *Sivapithecus* had a postcranium more in common with above-branch quadrupeds than with the suspensory taxa that exist today, suggesting substantial postcranial homoplasy in this lineage (Pilbeam *et al.* 1990; Pilbeam and Young 2004). We assume here that the appendicular morphology of this ancestral *ur*-Hominoid was similar to that of a large, arboreal, above-branch quadruped. For modeling purposes, we use the colobine *Nasalis* as a proxy, consistent with reconstructions of early and middle Miocene apes (e.g., *Proconsul, Nacholapithecus*) (Rose 1983, 1993; Nakatsukasa and Kunimatsu 2009).

Developmental Models

Combining our comparative analyses of living primates with hypothetical LCAs we can infer both associated hominin evolutionary trajectories through morphospace and the developmental transformations that likely drove these changes. We next return to what is known of the developmental basis of morphogenesis to assess how evolutionary variation in key hominin-specific traits might be generated. In this regard, while adults serve as the basis of our phenotypic morphospaces, they also tend to be highly integrated (i.e., we may be unable to tease out distinct developmental contributions) (e.g., Young 2004), and represent the developmental endpoint of a series of spatially and temporally overlapping and hierarchical processes (e.g., patterning, segmentation, mesenchymal condensation, endochondral bone growth, etc.) (Hallgrímsson *et al.* 2009). On the other hand, during early morphogenesis the processes that establish shape and spatial proportions are more discernible, especially for structures that contribute to the hominoid scapula (Young 2008). This suggests initial developmental processes of morphogenesis may be a substantial potential source of evolutionary variation in shape, if not size. Using this rationale, we focus on how variation in the initial processes of developmental morphogenesis may contribute to evolutionary change in the hominin lineage, what alternative trajectories predict in terms of developmental shifts in these processes, and when possible, which of these is most likely given current knowledge. We finish by suggesting ways in which the likelihood of alternative developmental transformations could be tested by future experimental and comparative approaches.

Scapula

Anatomy

The scapula is a large, flat, and triangular bone enveloped by musculature. It functions primarily to connect the muscles of the forelimb and trunk to the axial skeleton and facilitate complex movements of the limb (Figure 6.1, color plate). The dorsal surface of the blade is intersected by a spine that extends laterally to an expanded process called the acromion, which serves as another site of muscle attachment. The blade converges at its lateral edge to form the structures of the neck and then head of the scapula. The coracoid process originates cranio-laterally and serves as an anchor for shoulder muscles, while a laterally (or inferiorly) facing glenoid fossa serves as the articulation for the head of the humerus (Romer and Parsons 1986).

Development

The scapula originates from cell populations derived from at least three distinct embryonic tissues: the dermomyotome, somatopleure, and neural crest (Figure 6.2, color plate). The dermomyotome is a distinct subdivision of the somites, and cells fated to form aspects of the scapula blade arise from cervical-thoracic axial levels in the developing embryo (Huang *et al.* 2000, 2006; Valasek *et al.* 2010). The somatopleure is a medial tissue derivative of the lateral plate mesoderm (LPM), a subdomain of the mesoderm that forms early during development. Somatopleuric cells at the level of the forelimb bud contribute to the blade as well as all other scapula subdomains (Durland *et al.* 2008). Both tissues and their pre-scapula progenitor cells arise separately from one another and at distinct times, but eventually become integrated into one mesenchymal mass. This mass then undergoes coordinated chondrogenesis and osteogenesis to form distinct ossification centers in the scapula (Williams *et al.* 1995). In addition, a subset of the neural crest (post-otic), a cell population that originates from the dorsal neuroectoderm, migrates to several regions of the scapula after chondrogenesis and provides major attachment regions or anchors for the branchial muscles (e.g., the trapezius) (Matsuoka *et al.* 2005).

Molecular signals from adjacent tissues also contribute to scapular morphogenesis by inducing cell differentiation, migration, and condensation formation, and by patterning associated regions of the scapula. These tissues include the axial mesoderm, body ectoderm, dermomyotome, somatopleure, and limb bud mesenchyme (Moeller *et al.* 2003; Ehehalt *et al.* 2004; Wang *et al.* 2005). Signals from these tissues (e.g., in the form of Wingless/Integrin (WNT), Bone Morphogenetic Protein (BMP), and Fibroblast Growth Factor (FGF)) are essential regulators that stimulate the relatively naïve mesodermal mesenchyme to condense and differentiate into cartilage and bone, and may also help regulate the timing of developmental events.

Studies in mouse models have revealed several genes and hierarchical pathways with definitive roles in scapula mesenchymal condensation, cartilage differentiation, and patterning of the blade, spine/acromion, and head/neck structures. Little is known regarding the exact genes that induce the dermomytome to become scapula blade

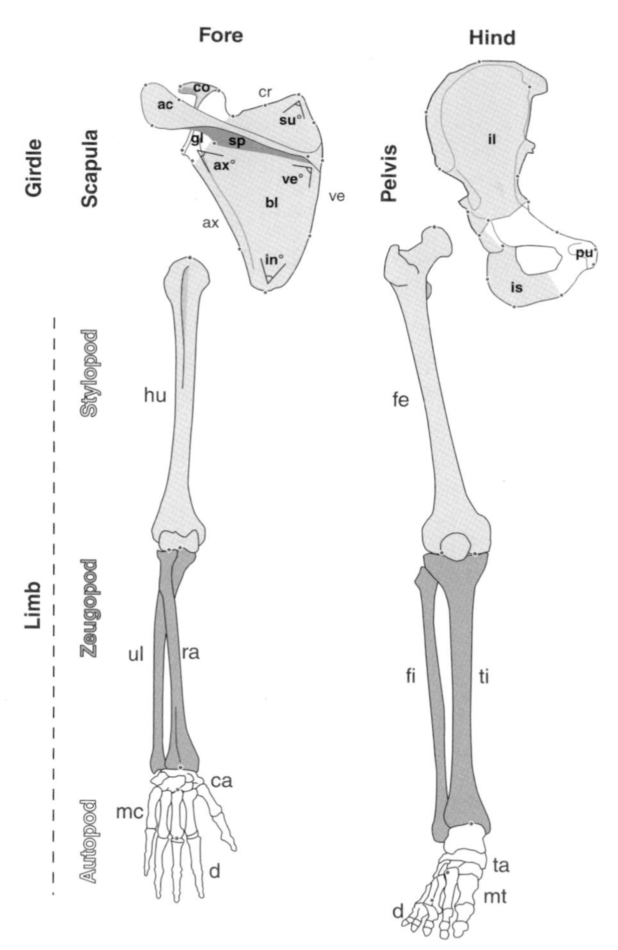

Figure 6.1. Appendicular skeletal anatomy: Primate appendages consist of a girdle attached to a limb proper. (Top-Left) The scapula (of the pectoral girdle in the forelimb) primarily consists of a roughly triangular blade (blue), traversed by a long thin spine ending in the acromion (red), a neck and head containing the glenoid (yellow), and a cranially positioned coracoid (green). The blade is bounded by three borders: (i) axillary, (ii) cranial, and (iii) vertebral. Shape can be described by several angles: inferior (in), superior (su), axillary (ax), vertebral (ve). (Top-Right) The innominate (of the pelvic girdle in the hindlimb) consists of three elements, a cranially located ilium (blue), a medial-ventral facing pubis (yellow), and a caudally facing ischium (red), all of which are fused in the middle forming circular concavity called the hip socket or acetabulum. (Bottom) Each forelimb (left) or hindlimb (right) consists of skeletal elements arranged in three "developmental" compartments. Most proximal is the stylopod (blue) containing the humerus or femur. Intermediate is the zeugopod (red) containing the radius/ulna or tibia/fibula. Most distal is the autopod (yellow) containing the carpals (ca), metacarpals (mc), and digits (d) of the hand or tarsals (ta), metatarsals (mt), and digits (d) of foot. Red circles indicate representative landmarks used to generate comparative and evolutionary morphospaces.

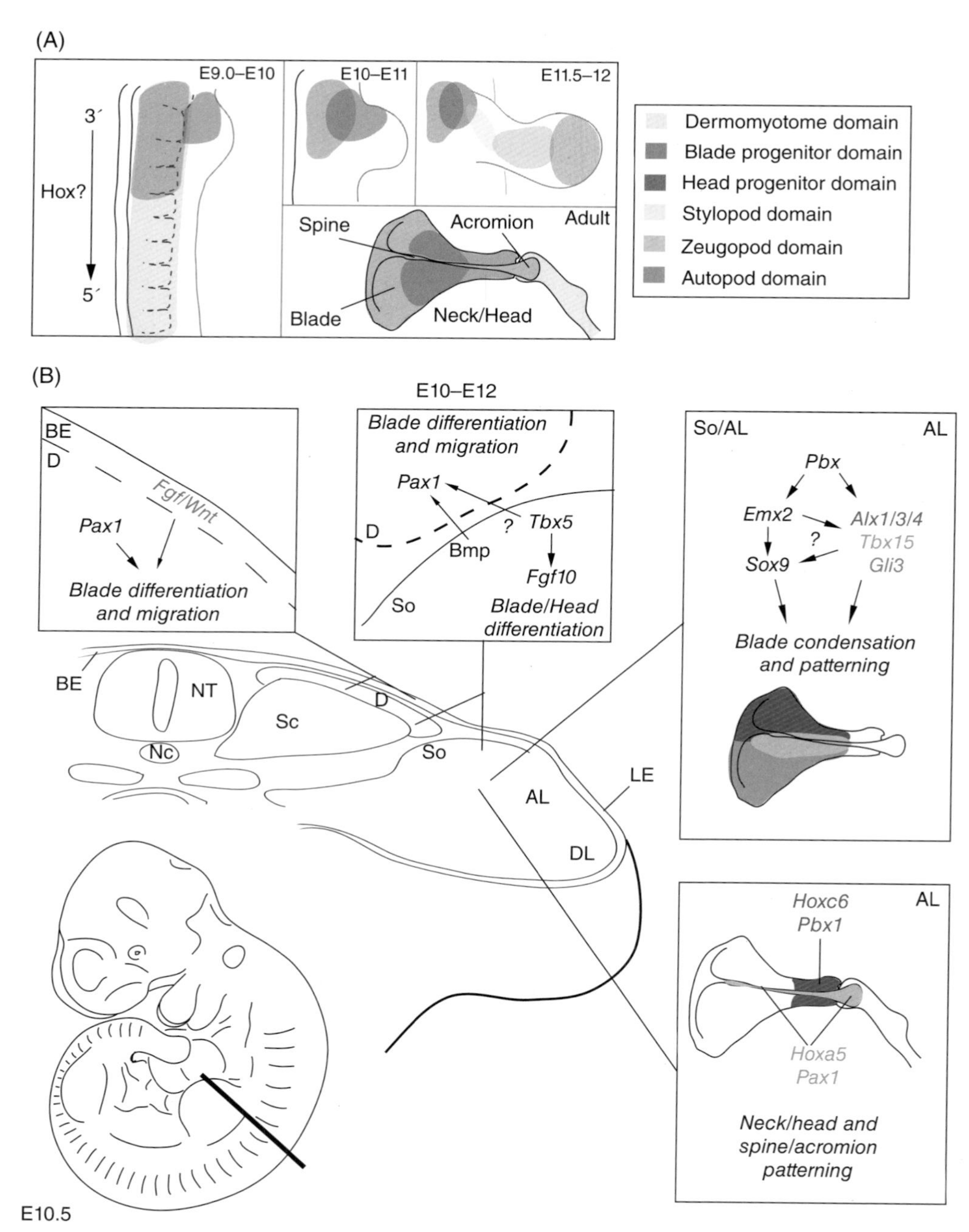

Figure 6.2. Scapular development and morphogenesis: (A) Several embryonic tissue domains give rise to the scapula during mid-gestation in the mouse. (Left) At 9–10 days gestation (E9–10) specific *Hox* gene expression in the somites (dotted features), specifically along the forelimb bud level of the dermomyotome, specify blade progenitors. Concurrently, anterior limb field tissues in the somatopleure (blue) help form head/neck progenitors as shown in the adult scapula (lower-right). Later at E10–11 and E11.5–12 tissue derivatives of the scapular blade, head, and neck form overlapping fields to give rise to the scapula proper, while

progenitors, or that cause such cells to differentiate and migrate to the somatopleuric region to form the blade, although *Hox* genes may regulate scapula position along the axis (Rancourt *et al.* 1995; Aubin *et al.* 1997; Wang *et al.* 2005). That said, once cells begin to migrate, *Sox9* and *Emx2* control formation of the blade's mesenchymal condensation (Bi *et al.* 2001; Pellegrini *et al.* 2001; Prols *et al.* 2004; Capellini *et al.* 2010). After condensation formation, regulators of endochondral ossification play key roles in scapula formation (e.g., see Long and Ornitz 2013), although the specific effectors controlling localized (i.e., scapula-specific) chondrogenesis and osteogenesis have not yet been isolated. With regard to the latter, it is likely that many genes involved in cartilage and bone cell differentiation have highly specific regulatory elements or switches that control their expression at the level of very specified anatomy (e.g., at a specific joint or long bone location). Thus, such a complex, highly modular, regulatory architecture likely governs growth control of specific elements – or even growth-plates of the scapula. For example, studies of the *cis*-regulatory architecture of *Bmp5* (Guenther *et al.* 2008) and *Gdf5* (Capellini *et al.* unpublished results) have revealed that separate enhancer elements drive gene expression and bone formation in distinct skeletal element domains including joints of the scapula.

Genetic and molecular studies using knockout mice have identified important factors patterning distinct scapula subdomains. For example, specific transcription factors pattern superior, central, and inferior blade domains and they act in the LPM, scapula mesenchyme, and/or anterior limb field to execute their effects before the onset of chondrogenesis and osteogenesis. Aristaless family genes such as *Alx1*, *Alx3*, and *Alx4* control superior blade formation (Kuijper *et al.* 2005). The T-box family member *Tbx15* controls central blade formation (Kuijper *et al.* 2005), and, along with *Gli3*, patterns the inferior portion of the blade (Johnson 1967; Hui and Joyner 1993; Kuijper *et al.* 2005), and the patterning of the vertebral border (Kuijper *et al.* 2005). Each factor's role in scapula spine development remains unclear given

Figure 6.2 (*Continued*) progenitors of the limb's three main compartments (the stylopod, zeugopod, and autopod) arise as the limb bud progresses. (B) Summary of genes and pathways involved in scapula blade, head, and neck development during mid-gestation (E10–12) depicted using a transverse section of an E10.5 mouse at the level of the anterior forelimb (AL) bud. (Top-Left box) illustrates molecular signaling from the body ectoderm (BE, purple text) via *Fgf/Wnt* to the dermomyotome (D) involving the activation of *Pax1* expression to drive early blade differentiation and migration from this tissue. (Top-Middle box) illustrates molecular interactions between the somatopleure (So) and the dermomyotome (D) to trigger blade/head differentiation in these tissues. (Top-Right box) illustrates the molecular interactions between the somatopleure (So) and anterior limb (AL); along with a depiction of those genes (in color text) that specify the unique superior (red), central (green), and inferior (blue) scapular blade domains seen below at E13.5. See text for details. (Bottom-Right box) Diagram of an E13.5 scapula demonstrating that from E9–12 specific genes function to give rise to the structures of the scapular neck and head (purple), and acromion domains (blue). In all cases, black arrows indicate activating roles, while questions marks (?) depict unclear relationships. Proximal (left); AL, anterior limb; BE, Body Ectoderm; D, Dermomyotome; DL, Distal Limb; LE, Limb Ectoderm; Nc, Notochord; NT, Neural Tube; Sc, Scleratome; So, Somatopleure.

the complex origin of the blade and spine tissues. Other genes are also involved in blade formation (*Wnt/Beta-catenin* and *Pax1*, Hill *et al.* 2006; Retinoic Acid Receptors (RARs), Lohnes *et al.* 1994; Polycomb homolog *M33*, Coré *et al.* 1997; BMP-binding protein *Cv2*, Ikeya *et al.* 2006) although their specific functions and involvement in known developmental and genetic pathways remain unclear. The *Pbx* family of transcription factors hierarchically regulates both patterning and condensation formation "pathways" (Capellini *et al.* 2010). *Pbx* genes have a wide influence on the development of all scapula components as well as the formation of the limb proper (Capellini *et al.* 2006, 2010, 2011a, 2011b). This hierarchical control is also shared with another T-box gene, *Tbx5*, whose loss in mice results in the absence of the scapula and truncations of the forelimb (Agarwal *et al.* 2003; Rallis *et al.* 2003).

Spine and acromion patterning are controlled by several factors, including *Pbx1* (Selleri *et al.* 2001), *Pax1* (Balling 1994; Timmons *et al.* 1994; Dietrich and Gruss 1995; Wilm *et al.* 1998; Adham *et al.* 2005), and *Hoxa5* (Aubin *et al.* 1997, 1998, 2002). The patterning of the scapular head and neck are controlled by several factors, including *Hoxc6* (Oliver *et al.* 1990), *Fgf10* (Min *et al.* 1998; Sekine *et al.* 1999), *Fgfr2(IIIb)* (blade and coracoid, De Moerlooze *et al.* 2000), *Pbx* (Selleri *et al.* 2001; Capellini *et al.* 2010). Importantly, upstream roles for each factor are demonstrated by the fact that overexpression or gene deletion often causes mirror image duplications or complete losses of head/neck and proximal limb structures. Interactions between these genes and those influential in limb development (e.g., *Tbx5*, *Fgf10*, among others – Min *et al.* 1998; Sekine *et al.* 1999; Ng *et al.* 2002; Agarwal *et al.* 2003) may partition the LPM into girdle and non-girdle (or limb) domains and aid in the fine-tuning of specification of these structures.

While the developmental and genetic pathways that lead to scapula development are beginning to be elucidated, specifically for aspects of the blade, spine, neck, and head, little is understood regarding how all of these structures become an integrated osseous structure during late prenatal and early postnatal life. Even less is known regarding how the specific shape of the scapula (and its subcomponents) is actualized. Nevertheless, it is during pre- and postnatal development that the morphological differences between primate scapula girdles develop, specifically, as selection has operated to produce important variation that aids in functional differences in musculo-skeletal anatomy. The next section focuses on understanding this variation in detail.

Comparative Variation

Analyzed as a comparative morphospace, most variation in anthropoid (human, ape, and monkey) scapular shape is contained in a few axes (Young 2008), and these correspond to the developmental subunits described above: (1) width and height of the scapular blade, (2) length, height, and orientation of the scapular spine, acromion, and glenoid, and (3) shape and size of the glenoid and coracoid (Figure 6.3A, color plate). For example, the first morphospace axis represents a continuum from more terrestrial cercopithecine taxa to more arboreal colobines and finally hominoids, and is largely driven by the shape and size of the scapular

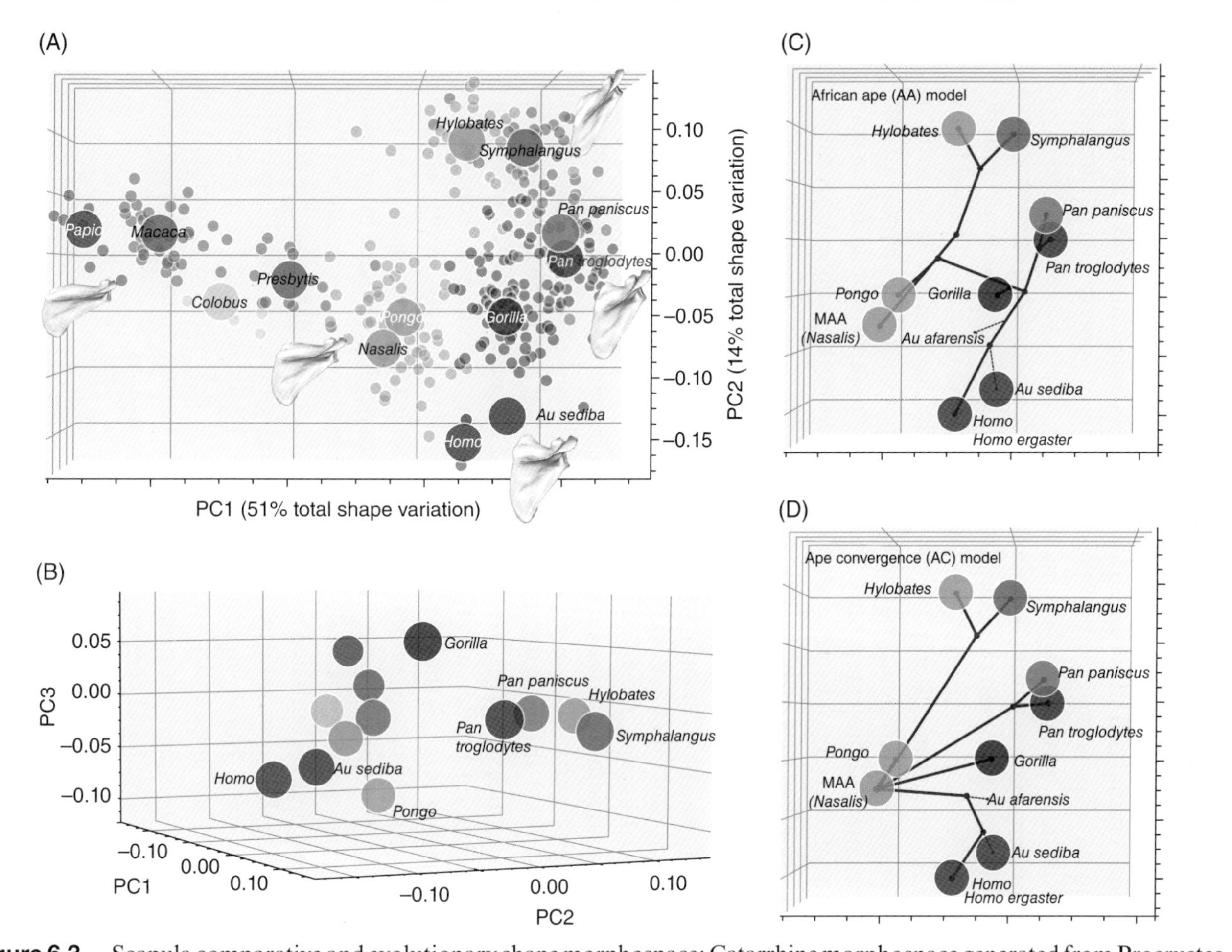

Figure 6.3. Scapula comparative and evolutionary shape morphospace: Catarrhine morphospace generated from Procrustes superimposition and principal components analysis of three-dimensional landmark data (expanded from Young (2008)). (A,B) Axis one and two primarily describe variation in blade shape and spine angle/glenoid orientation, respectively (representative scapulae illustrated), while the third axis discriminates *Gorilla* and *Pongo* based on derived characteristics of the supraspinous fossa and height of the scapular spine. (C,D) Phylogenetic reconstruction (minimum spanning tree) of the AA and AC models in the comparative morphospace including fossil hominins (dashed lines) illustrates alternative evolutionary reconstructions and trajectories in this space.

blade, relative sizes of the infraspinous and supraspinous fossae, mediolateral width of the scapular blade, cranio-caudal orientation and size of the glenoid relative to the blade, and relative size of the acromion and coracoid. The second axis describes the angle of the scapular spine and glenoid relative to the vertebral border (and also to the spinal column if one assumes a "natural" or "neutral" position of this border is parallel), from perpendicular (quadrupeds, *Pongo*, *Homo*) to obtuse and more cranial (all other apes). The third axis describes how increased supraspinous fossa volume varies within the constraints of spine angle, particularly evident in the contrast between *Gorilla* and *Pongo*.

Apes are similar to and vary from one another in a number of features relevant to our understanding of the human shoulder. In hylobatids, the fossae extend cranially or caudally a smaller distance from the scapular spine, and the angle formed by the scapular spine and the axillary border is very acute. The angle of the spine with the vertebral border is even more acute than in all other apes with the exception of *Pan*. In addition, the coracoid projects more cranially in hylobatids, similar to the coracoid projection of the glenoid and acromion. In *Gorilla* the angle between the axillary border and the spine is more obtuse and the size of the individual fossae is proportionately large and close to equal in area. The supraspinous fossa projects cranially above the level of the coracoid in African apes and humans, which distinguishes them from all other living apes. *Pan*, *Hylobates*, and *Symphalangus* differ from *Gorilla* and in general *Pongo*, in having a scapular blade that is narrower from glenoid to vertebral border. *Pongo* can be distinguished from all other apes by a relatively small supraspinous fossa (in dorsal view), a robustly developed scapular spine extending and perpendicular to the vertebral border, and a glenoid that is more cranially extended although not as pear shaped as in quadrupeds. *Pongo*'s proportionally small supraspinous fossa is mostly due to the angle of the scapular spine relative to the vertebral border when viewed in dorsal aspect. As a function of volume, this fossa is as large as that of the other apes. Interestingly, *Homo* resides at the apex of the two main axes of primate variation, reflecting a combination of typically "ape-like" and "arboreal quadruped" characteristics. Blade shape most resembles those of African apes, particularly in the well-developed superior angle of the supraspinous fossa, but is combined with a long, robust scapular spine, perpendicular to the vertebral border, as found in arboreal colobines and *Pongo*. This combination of traits in *Homo* suggests that either: (1) blade shape is a primitive retention from African apes and spine orientation is derived and convergent on quadrupeds, or (2) the spine is a primitive retention from a quadrupedal ancestor and blade shape is derived and convergent with African apes.

Evolutionary Trajectories

Phylogenetic reconstruction of the AA model evolutionary trajectory in this space suggests that an African ape–like LCA to *Homo* transition passes through a "*Gorilla*"-like morphospace along an African ape blade shape "axis." A prediction of this trajectory is that ancestral hominins would share some an African ape blade shape and spine angle features with *Gorilla*, but would lack more derived features such as a reduced height of the spine relative to a greatly expanded supraspinous fossa. The inferred AC model evolutionary trajectory is parallel to that of the terrestrial-arboreal quadrupedal axis, and would predict increasing size of the infraspinous fossa, reduced medio-lateral width, and increased glenoid and acromion size, but little to no change in spine angle and glenoid orientation relative to the vertebral border.

Until recently, the hominin fossil record could not address these predictions since only fragmentary remains of the scapula were known from *Australopithecus afarensis* (AL288-1, "Lucy") (Stern and Susman 1983) and *Australopithecus africanus* (Sts7) (Vrba 1979). However, in the last decade several well-preserved skeletons have been recovered, with the earliest known examples attributable to *Australopithecus afarensis*. An adult scapula from Woranso is described as being more similar to *Gorilla* and unlike that of *Pan* (Haile-Selassie *et al.* 2010), while the scapula of an infant (DIK-1-1) suggests similarities to *Gorilla* (Alemseged *et al.* 2006; Green and Alemseged 2012). More recently, the scapula of the later *Australopithecus sediba* (MH-2) has been described as having morphometric affinities to both *Pongo* and *Homo* (Churchill *et al.* 2013).

Phylogenetic analysis of the shoulder morphospace including *Australopithecus sediba* suggests australopithecines are intermediate with humans and African apes (Figure 6.3, color plate). To include data on other fossil hominins for which 3D shape information is not available or specimens are fragmentary, we utilized the fact that blade shape and spine angle can be accurately assessed from photographs of the dorsal aspect (e.g., Bello-Hellegouarch *et al.* 2013), and this variation is represented by the first two axes of the original 3D morphospace. We performed a reduced landmark-based shape analysis of the dorsal features of fossil hominin scapulae (results not shown), which confirms Alemseged *et al.*'s (2006) contention that the early australopithecine condition represented by DIK-1-1 is more similar to that of *Gorilla* than that of *Pan*. In contrast, the Woranso specimen is more similar to *Homo* than is either *Australopithecus sediba* or the Dikika child, suggesting the potential for significant variability within hominin taxa. This result is not surprising given relaxed functional and selective constraints in the hominoid scapula result in enhanced shape variance within even temporally bounded living species (Young 2006). As in the 3D analysis, *Australopithecus sediba* is intermediate to *Gorilla* and *Homo*, while the Nariokotome boy (*Homo ergaster*, KNM WT15000) is at the extreme of this axis, comparable to living *Homo*.

The arrangement of fossil hominin taxa in a temporal gradient parallel to the inferred evolutionary trajectory connecting extant *Pan* to *Homo* is consistent with predictions of the AA model of the LCA. It would imply that *Homo* scapular morphology evolved primarily as a change in spine orientation in concert with a more caudally oriented glenoid, while blade shape became medio-laterally wider but otherwise retained essential African ape–like qualities, such as the shape of the superior angle and cranial border. Interestingly, the trajectory within hominins is largely parallel to that between African apes and *Homo*, and as proposed in the AA model. As a consequence, for the AC model to be true, convergence on African ape blade morphology and secondary changes parallel to the AA model would have both needed to occur. In essence, the difference between models is whether the shift to the

African ape blade shape is convergently evolved in hominins and thus happened later, or is a shared primitive feature that occurred earlier in African ape evolution.

Developmental Transformations

The models discussed above make alternative predictions about the number and phylogenetic timing of phenotypic transformations, and generate distinct predictions for the developmental shifts linking ancestor and descendant taxa in hominin evolution, particularly as they correspond to regions that are known to have relatively discrete origins during initial morphogenesis. In the AA model, there are relatively fewer changes to account for since African ape blade shape characteristics are similar in the putative LCA, thus it is primarily a shift in the length/orientation of the spine and glenoid that must be explained (i.e., from PC2). When fossils are incorporated into the AC model, both shape of the blade and next spine orientation must be altered (i.e., both PC1 and PC2).

Above we described a series of genes that have localized effects on specific blade domains, while others influence the blade, and even the scapula as a whole (see Figure 6.2A, color plate). In this context, some genes influence spine presence and structure, while others may be compartmentally specific. Each gene or locus may have been the target of selection. For example, it is understood that mice harboring null mutations for *Tbx15* or *Gli3* substantially alter inferior blade dimensions but not downstream endochondral ossification programs (Kuijper *et al.* 2005). On the other hand, many members of the Aristaless gene family influence both superior blade patterning and spine morphology but have no influence on inferior blade patterning (Kuijper *et al.* 2005). At the same time, genes that shape initial condensation of the scapula (Bi *et al.* 2001; Pellegrini *et al.* 2001), and ultimately overall scapular morphology, may be targets of selection. For example, *Emx2*, along with *Pbx*, control the blade and the spine (Selleri *et al.* 2001; Capellini *et al.* 2010). Modifications – for example, in the form of cell number expansions or reductions or changes to the polarity of cell growth and adhesion – to this early developmental field may promote shifts in the blade condensation and also the position along the blade where the scapula spine originates (i.e., its root).

Given the high degree of postnatal integration between the blade and spine, and partially with the glenoid, changes in the size and/or shape of the initial anlagen likely explain interspecific differences among adults (Young 2004, 2006, 2008). Such changes may also underlie phylogenetic shifts in overall scapula morphology from those of above-branch cercopithecoids to below-branch hominoids. Shifts in axial level *Hox* expression may alter the patterning of blade progenitors that eventually migrate to the scapular blade domains. At the same time, shifts in somatopleure-dermomyotomal interaction caused by modifications to somatopleuric gene expression (e.g., *Pbx*, *Tbx*, *Hox*) may alter the number and fate of cells giving rise to the blade versus those giving rise to the head and neck. Interestingly, as the somatopleure itself is critically involved in forming neck and glenoid domains (Durland *et al.* 2008), gene expression changes in this domain may have cascading effects on the entire scapula. Changes in blade versus head/neck cell number could account for our main first axis finding; that is, that blade shape becomes narrower and structures of the neck/head (glenoid size, acromion, and possibly the coracoid) become more expansive from quadrupedal monkeys to apes. Interestingly, not only were blade shape and neck/head shape changed but so too was the orientation of the spine root in hominoids. The latter could indicate that the size of the progenitor pool has been altered, a potential consequence of *cis*-regulatory modification.

At this point only a few genes, such as *Pbx* family members and *Tbx5*, have been shown to simultaneously influence development of blade/spine and glenoid domains, specifically as assessed in global gene knockout mice (Selleri *et al.* 2001; Capellini *et al.* 2010; Agarwal *et al.* 2003; Rallis *et al.* 2003). Loss of *Pbx1* results in reductions of the superior-to-inferior blade dimensions (Selleri *et al.* 2001), while coordinated loss with other *Pbx* results in additional reductions in the proximal-to-distal extent of blade formation (Capellini *et al.* 2010). In both cases, alterations also occur in the scapula neck and head, including the glenoid. In this context, *Pbx* family members most likely function upstream of *Tbx5* expression in adjacent somatopleuric and limb domains (Capellini *et al.* unpublished data). Therefore, subtle variations in *Pbx* or *Tbx* gene expression (via *cis*- and trans-regulatory changes in each gene) may have resulted in shifts in somatopleuric patterning and dermomyotomal differentiation in coordinated ways. The differences in morphology, as seen in quadrupedal versus suspensory primates, suggests that correlated changes in multiple tissue domains have occurred during evolution, and thus, it may be more parsimonious to argue that selection has favored morphologies that result from shifts in early fields of growth (thus altering all scapular domains), rather than morphologies that result from the independent acquisition of mutations influencing each domain, and each conferring a fitness advantage. *Pbx* family members and *Tbx5* may have been the targets of selection given their composite influences on scapulogenesis.

However, it is important to point out that while initial shifts may be important for influencing multiple developmental domains, the basis for positing such shifts stems from studies on mutant mice that lack functional coding regions of genes. This means that gene loss has occurred in all domains of expression simultaneously, and little is known on how such pleiotropy influences scapula morphogenesis. Additionally, the species in question do not possess null mutations in the coding regions of these genes (unpublished results), and so like many of the most recent genomic and genetic studies suggest, they likely harbor mutations in key regulatory elements influencing morphology. There may very well be *cis*-regulatory element(s) for these early patterning genes (i.e., like *Pbx*, *Tbx*, *Hox*, *Emx2*, *Sox9*) that modulate scapula progenitor cell fate and number in domains that simultaneously influence both blade and the head/neck. This is likely considering that while different cell lineages contribute to these structures, they become integrated in the somatopleure and are thus under the influence of one potential continuous gene territory. Such a mechanism may be a more parsimonious way to evolve blade, spine, and glenoid shape along the cercopithecine to hominoid trajectory.

At the same time, we know that many developmentally important genes possess complex regulatory architectures, in which different regulatory elements (e.g., enhancers) can drive overlapping expression territories (reviewed in Davidson 2006).

In this context, multiple mutations in functionally overlapping yet distinct enhancer sequences may have been targets of selection. On the most extreme side, multiple mutations in single regulatory elements capable of driving gene expression in discrete cell populations (i.e., those giving rise to the blade versus spine versus head) may have also been targets of selection. Consistent with this line of thinking, we now know that repeated use of the underlying genetic/genomic structure can lead to the repeated evolution of similar adaptive traits (Chan *et al.* 2010; Jones *et al.* 2012). For example, around the northern hemisphere, freshwater stickleback fish, which typically lack pelvic fins, repeatedly lose function of a single *Pitx1 cis*-regulatory enhancer that in nearby marine fish functions to drive the formation of a pelvic fin (Chan *et al.* 2010). In freshwater forms, the deletion of this *Pitx1* element likely occurs via the presence of a hyper-fragile site in the genome, leading to a higher probability of loss or deletion of DNA in that region. The underlying genomic architecture at this locus thus permits such repeated change. This loss has been argued to be the result of selection acting against the presence of pelvic fins, as their presence in freshwater fish may lead to increased losses by large preying insects using them to clasp and take fish in a predatory manner. Interestingly, Jones and others (2012) reveal that this phenomenon is not limited only to the *Pitx1* locus but to many loci around the stickleback genome, indicating that in sticklebacks recurrent use of functional haplotypes may underlie repeated evolution of many biological traits. These studies force us to think differently about the likelihood of directional evolutionary change, the nature of parallelisms, and what constitutes parsimony on a real and functional level.

Our second axis (see Figure 6.3, color plate) largely described differences in the angle of the spine to that of the vertebral border, particularly as discriminating suspensory brachiating apes (e.g., hylobatids) versus African apes and finally humans. While key regulators of spine position along the scapula have yet to be found, numerous genes influence spine development, as observed in knockout mouse models. *Pbx1* (Selleri *et al.* 2001), *Pax1* (Balling 1994; Timmons *et al.* 1994; Dietrich and Gruss, 1995; Wilm *et al.* 1998; Adham *et al.* 2005), and *Hoxa5* (Aubin *et al.* 1997, 1998, 2002) all influence the extent of the spine and possibly angle, independent of roles in blade formation proper. Our third axis described specific changes between *Gorilla* and *Pongo*, particularly with respect to the volume of the supraspinous fossa and spine height. The supraspinous region is under direct control of multiple Aristaless family genes (Kuijper *et al.* 2005; Capellini *et al.* 2010), and the knockout mouse models for these genes display considerable size changes to this region independent of a major influence on central and inferior blade formation. As all of the genes above are key developmental regulators of multiple embryonic, fetal, and postnatal processes, the search for key *cis*-regulatory elements is underway (see Conclusions and Future Directions).

How we distinguish between the likelihood of the AA versus AC models using developmental data comes down to how we understand gene function, and the nature of the types of changes that are permitted by the underlying genetic and genomic architecture of the traits in question. The changes that are possible/permissible have to also be viewed in a selective and functional context. The AA model posits that spine angle shift may have been under strong selection in hominins, and as discussed above there are numerous candidate genes that may influence only spine shape/angle and not blade formation. On the other hand, the AC model posits that two seemingly separate structures must be altered as separate evolutionary events within hominins to acquire adult morphology. If the underlying genomic architecture more readily and repeatedly permits these changes, and if haplotypes harboring these regions that are prone to change are present in any frequency in an evolving population, then structural evolution may occur provided it has a fitness advantage. Thus, both models have several key features that one can look for in the genomes of extant species, to see if they display evidence of evolution. According to the AA model, we should see concerted evolutionary change in sequence (and/or key functional base-pair changes) within regulatory regions for key genes that only influence the spine, and these changes should date to particularly informative periods as supported by the fossil record. In contrast, the AC model predicts sequence evolution for regulatory regions for genes influencing both blade and spine, and such changes should be step-wise chronologically. In either case, the first step in addressing these issues is to identify the genes and their *cis*-regulatory architecture that underlies the specific traits in question (see Conclusions and Future Directions).

Pelvis

Anatomy

The pelvis is a composite of three skeletal elements: the superiorly oriented ilium, inferior-dorsally positioned ischium, and inferior-ventrally facing pubis (Figure 6.1, color plate). These elements possess numerous crests and surfaces for attachments of muscles between pelvis, lower back, and hindlimb. The two pubic elements ventrally articulate at the pubic symphysis, while each iliac blade articulates dorsally with the sacral vertebrae at the sacroiliac joints. The ilium, pubis, and ischium centrally fuse to form a lateral-facing socket called the acetabulum (hip joint), which serves as the articulation of the trunk with the hindlimb (Romer and Parsons 1986).

Development

Pelvic girdle morphogenesis is initiated approximately at hindlimb bud outgrowth, when cells of the mesodermal portion of the somatopleure become specified as pelvic progenitors (Malashichev *et al.* 2005, 2008; Durland *et al.* 2008; Pomikal and Streicher 2010) and scleratomal cells of the somites give rise to parts of the sacrum (Pomikal and Streicher 2010) (Figure 6.4, color plate). This process of specification causes somatopleuric cells to condense into a single mesenchymal mass, and likely serves to initially pattern the three main pelvic rudiments (e.g., ilium, ischium, and pubis). Eventually, this mass undergoes chondrogenesis and osteogenesis to form the pelvic bone proper (Pomikal and Streicher 2010). As with scapula morphogenesis, signaling from adjacent tissues such as the axial mesoderm, body wall ectoderm, and limb bud proper helps coordinate this process and may orchestrate the timing of formation of each individual pelvic rudiment (Malashichev *et al.* 2005, 2008).

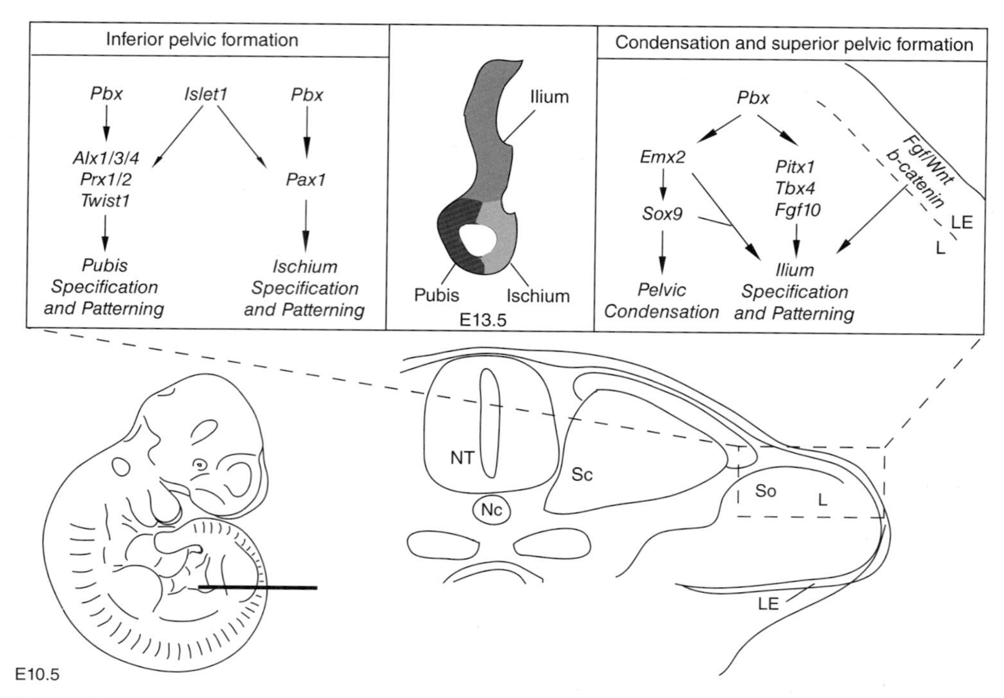

Figure 6.4. Pelvic development and morphogenesis: Diagram of a transverse section at the anterior mouse hindlimb (L) level at 10.5 days of gestation, demonstrating the major tissues, genes, and pathways known to date to be involved in pelvic formation. (Top) Major rudiments of the pelvis are shown in colors (ilium, blue; pubis, red; ischium, green) that correspond to the pathways shown in both adjacent panels. Both adjacent panels illustrate the molecular interactions between the somatopleure (So) and limb (L); along with a depiction of those genes (in color text) that specify the unique pelvic elements seen in the center diagram. (Top-Left box) illustrates the major genes and pathways, hierarchically controlled by *Pbx* and *Islet1*, which are known to date to govern inferior (ischium and pubis) formation. (Top-Right box) illustrates the major genes and molecular pathways, controlled by *Pbx*, that govern ilium formation. This includes molecular interactions likely in the form of *Fgf*/*Wnt* (purple) signaling from the limb ectoderm (LE) to the limb (L), along with those limb-determining genes (in color text) that also specify the ilium. In addition, *Pbx* via *Emx2* and their control of *Sox9* expression regulates not only ilium patterning but overall pelvic condensation and skeletal formation. See text. In all cases, black arrows indicate activating roles, while questions marks (?) depict unclear relationships. Proximal (left). L, Limb; LE, Limb Ectoderm; Nc, Notochord; NT, Neural Tube; Sc, Scleratome; So, Somatopleure.

Compared to the limb, fewer genes are currently known to influence the development of the individual pelvic rudiments. Some are associated with ilium formation ((*Emx2* (Pellegrini *et al.* 2001; Malashichev *et al.* 2005, 2008), *Fgf10* (Sekine *et al.* 1999), *Lmx1-b* (Chen *et al.* 1998), *Pitx1* (Lanctôt *et al.* 1999; Marcil *et al.* 2003), *Pbx1-3* (Selleri *et al.* 2001; Capellini *et al.* 2006, 2011a, 2011b), *Sox9* (Bi *et al.* 2001), *Tbx4* (Lanctôt *et al.* 1999), and *Tbx15* (Lausch *et al.* 2008)), while others are known to influence pubis formation (*Alx1* (Kuijper *et al.* 2005), *Alx3* (Kuijper *et al.* 2005), *Alx4* (Kuijper *et al.* 2005), *Cv2* (Ikeya *et al.* 2006), *Fgfr2-IIIb* (De Moerlooze *et al.*

2000), *Islet1* (Itou *et al.* 2012), *Msx1-2* (Lallemand *et al.* 2005), *Pbx1-3* (Capellini *et al.* 2006, 2011a, 2011b), *Prrx1* (ten Berge *et al.* 1998), *Prrx2* (ten Berge *et al.* 1998), *Twist1* (Krawchuk *et al.* 2010), *Wnt* (Lee and Behringer 2007)). To date, very few genes have been identified that are expressed and/or influence the ischium (*Islet1* (Itou *et al.* 2012), *Pax1* (Timmons *et al.* 1994; LeClair *et al.* 1999), and *Pbx* (Capellini *et al.* 2011a, 2011b). As with scapula morphogenesis, the *Pbx* family of transcription factors hierarchically regulates pelvic formation, although their roles may be to control development of only a subset of structures, such as the ilium, pubis, and acetabulum (Selleri *et al.* 2001; Capellini *et al.* 2011a, 2011b). On the other hand, inferior patterning of the pelvis (e.g., of the pubis and ischium) appears partially under the influence of *Islet1* (Itou *et al.* 2012), and in this regard two overlapping developmental zones may exist during early pelvic morphogenesis. Although *Hox* genes have been argued to be major effectors of limb development, little is known regarding their action in pelvic development. Misexpression or loss of *Hox* genes does result in mild alterations of the pelvic girdle, including defects in mice misexpressing *Hoxd12* in lateral plate derivatives (Knezevic *et al.* 1997); modest malformations of pelvic bones and sacrum in *Hoxc10* mutants (Hostikka *et al.* 2009); and lack of uterosacral ligaments in *Hoxa11* mutant mice (Connell *et al.* 2008).

Comparative Variation

Three-dimensional pelvic shape is relatively invariant among living apes (Figure 6.5, color plate). Compared to quadrupedal cercopithecoids and non-ape suspensory monkeys (here represented by the platyrrhine *Ateles*), hominoids are characterized by a modestly shorter pelvic height, and dorso-ventrally deeper pelvic outlets (PC1). At the same time, hominoids have broad and expansive ilia compared to quadrupedal monkeys and *Ateles*, with a modest and variable degree of anterior rotation at the lateral edges (PC2). A third axis largely discriminates great apes from lesser apes, and we do not consider it further here. Interestingly, differences between apes and humans are for the most part extremes of those observed between apes and quadrupedal monkeys and *Ateles*. For example, in humans the entire pelvis is further reduced in height while breadth is increased, the pelvic outlet is substantially deeper, and the large iliac blades are even more anteriorly and medially rotated into a parasagittal plane.

Evolutionary Trajectories

While other structures differ between the pelves of apes and modern humans, we focus here on how the alternative evolutionary models for the LCA affect the interpretation of the evolution of shape of the ilia (including relative size, proportions, and orientation). The AA model suggests two events, the first associated with an overall shortening of pelvic components and broad expansion of the ilia from a quadrupedal ancestor towards the living hominoid condition, and a second involving further shortening of pelvic height, including medial and anterior rotation of the ilia. In the AC model, a more direct path is predicted from an arboreal quadruped-like

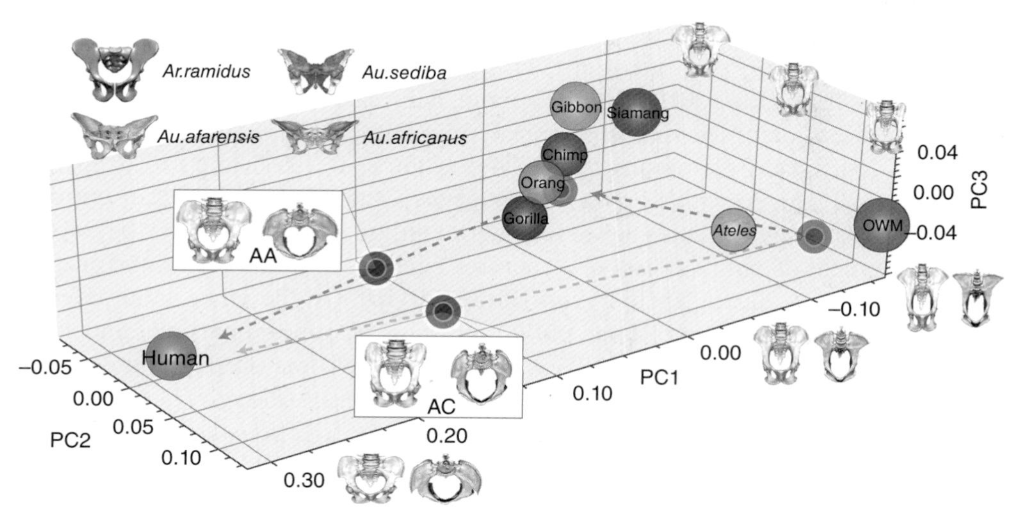

Figure 6.5. Pelvis comparative and evolutionary shape morphospace: Landmark-based three-dimensional shape analysis of the catarrhine pelvis. Axis one primarily separates humans from all other primates and describes shortening and widening of the pelvis as a whole, with anterior rotation of the ilium into a parasagittal plane. Axis two describes size of the ilium relative to the other components, and separates apes and humans from monkeys. Axis three primarily describes variation within apes. Alternative evolutionary trajectories (AA = blue, AC = green) are shown with morphological predictions calculated at the midpoint between inferred LCAs and *Homo*. These predictions can be compared to actual fossil hominin pelves shown above the reconstructed morphospace.

LCA to modern humans in which there is a coevolution of iliac rotation and expansion, all in concert with shortening, widening, and deepening of the pelvis as a whole. As a consequence, the AC model predicts intermediate ancestral morphotypes in which the ilia are smaller than that of any living hominoid. Consequently, in this model iliac expansion would be homoplastic in humans with the ape lineages. Surprisingly, *Ateles* is more similar to quadrupedal monkeys than any of the apes, suggesting a distinct suite of apomorphies (whether shared, derived or convergent) in the hominoid hindlimb.

Although fossil pelves were not included in the comparative shape analysis because of a lack of readily available 3D data, predicted intermediates of the AA and AC trajectories can be compared to reconstructions of known specimens in terms of anterior iliac blade morphology. Complete or fragmentary fossil hominin pelves are known for australopithecines such as *Australopithecus afarensis* (AL288-1) (Tague and Lovejoy 1986), *Australopithecus africanus* (Sts14), and various species of early *Homo* (e.g., KNM-WT 15000) (Ruff and Walker 1993). The earliest described hominin pelvis is from *Ardipithecus ramidus* (Haile-Selassie *et al.* 2010) while the later hominin *Australopithecus sediba* has also been recently described (Kibii *et al.* 2011) (Figure 6.5, color plate). These reconstructions tend to emphasize differences (e.g., the degree of pelvis shortening or broadening, height and curvature of the ilia, etc.) among hominins, particularly as they regard function and bipedalism. For example,

Australopithecus afarensis and *Australopithecus africanus* reconstructions suggest the ilia are "bowl-like," a characteristic not accounted for in the morphospace of living anthropoids. Yet in a broader context the ilia of fossil hominins tend to share more features than not, and in these the fossil ilia are intermediate between apes and modern humans. For example, all hominins have broad and expansive ilia on par with living apes. In particular, the highly reconstructed *Ardipthecus ramidus* pelvis is strikingly similar to the morphometric prediction of the AA trajectory, and consistent with the 2D graphical predictions of a *Pan-Homo* intermediate (Lovejoy *et al.* 1999). Given the functional relationship of iliac size to attachment areas of hip abductors/extensors (e.g., *mm. gluteus maximus*), the trajectory suggested by the AC model would suggest that either primitive forms of bipedalism relied less on these muscle groups, or that that iliac expansion occurred rapidly (in temporal terms) in hominin bipedal evolution.

Developmental Transformations

The primary axis of the comparative morphospace described key transformations between quadrupedal cercopithecines and hominoids, specifically decreases in pelvic height along with anterior rotated ilia, while the second axis described a broadening and expansion of the iliac blades in extant hominoids. Furthermore, the differences observed between apes and humans continue on this pattern and have been discussed as being at the extreme end of the observed morphological patterns. These types of changes may reflect alterations to key signaling pathways (see above) that specifically influence the formation of the ilium, although it is clear that the pubis and ischium also underwent changes. To date, two somewhat separate genetic pathways have been identified that influence subdomains in the pelvis: a pathway hierarchically regulated by *Pbx* that governs superior (ilium) formation, and another partially controlled by *Islet1* that influences inferior (pubic/ischial) development (Capellini *et al.* 2010; Itou *et al.* 2012). The extent to which these pathways are independent is still a matter of question, as the genetics of pelvic development are poorly known due to limitations in the number of studies and techniques required to investigate pelvic progenitor tissues. Most of the genes known to influence pelvis formation result in the absence of structures when the corresponding gene has been experimentally removed from mice. These data suggest that what we know of the genetics of pelvis formation likely concerns how cells are specified to become the anlagen that makes the individual rudiments. In other words, we know more about initial pelvic morphogenesis than about stages directly influencing the differential growth of each component or each of their substructures.

Regarding the types of changes we observe in the ilium, from a more quadrupedal ancestor to hominoids, shifts in gene regulation of *Pitx1*, *Pbx*, *Tbx4*, and *Fgf10* may subtly influence gene expression territories in the mesodermal anlagen that gives rise to the ilium. Experimental ablation of *Pbx* at different time points using a conditional cre-lox targeting system results in changes in the size and shape of the ilium (i.e., the earlier the conditional inactivation, the more ablation and reduction in pelvic size), suggesting that spatio-temporal shifts in gene expression territories may indeed influence cell number and thus morphology (Capellini, unpublished results). While it is likely that the shifts seen in these conditional studies are artificial, they could occur in nature if specific mutations arise in enhancers for key iliac developmental genes. In that vein, it is also possible that alterations to *cis*-regulatory elements (e.g., enhancers) controlling cartilage or bone growth of the ilium versus other pelvic parts may underlie such skeletal differences. Although for mutations to influence multiple parts of the pelvic girdle, they may need to act in the context of separate *cis*-regulatory elements or influence key base-pair motifs controlling ilial expression within a pan-regulatory element for the pelvis.

With all of the above in mind, the morphological changes observed between humans and chimps compared to the LCA point to particular places where major modifications have occurred. According to the AA model we should expect two evolutionarily distinct shifts: the shortening of pelvic components coupled with a broadening/expansion of the ilia and additional iliac height reduction with medial anterior rotation. With the AC model these changes occur in coordinated fashion with iliac rotation and expansion coupled with widening and shortening.

One scenario is that subtle shifts in the initial positional information (i.e., by changes in morphogenetic gradients vis-à-vis one mutational change) in the fields that govern pelvic morphogenesis may have occurred, which if selected for could modulate the entire shape of the pelvis and that of the proximal limb simultaneously (Lovejoy *et al.* 1999). Due to technical limitations, very few genes are known at this time to influence shape of the pelvis and limb. Altering shape is an important component of this perspective since there have been experiments in chicks that demonstrate that axial level specification of the limb and girdle can be shifted (i.e., via bead implants) but in such experiments shape remains mostly normal (Cohn *et al.* 1997). Those genes that do influence shape and normal morphogenesis are extremely critical for understanding such morphogenetic shifts and their results on morphology. An alternative is that multiple changes in gene regulation occurred either simultaneously or over time. Investigations into the genetic architecture of pelvic element shape will help reveal the nature of the mutations that have produced shape differences among primates.

From the perspective of gene function during limb and girdle development there are at least two potential mechanisms for such simultaneous shifts. One involves the alteration to some upstream transcription factor important for hindlimb development (*Pbx*, *Tbx4*, *Pitx1*, *Islet1*) that leads to simultaneous changes in expression in many target genes (Trans-effect). For example, in single *Pbx1* mutants, there is a simultaneous alteration of the proximal femur and girdle, while in compound *Pbx1;Pbx2* mutants there are more drastic effects across these and more distal limb domains (Selleri *et al.* 2001; Capellini *et al.* 2010). *Pbx* also influences axial level expression of *Hox* and limb location (Capellini *et al.* 2008). In these contexts, if *Pbx* acts (with *Hox*) as a "selector" gene that controls many downstream gene expression patterns/functions, then shifts in its expression via alterations in *cis*- and trans-regulatory networks may lead to coordinated changes in pelvic field and proximal limb specification. This is because the functional territory of the *Pbx* protein, a transcription factor, would be altered across many domains. The idea that *Pbx* does behave as a "selector" gene is evidenced not only by the near duplication of the humeral head in the forelimb but also by the transformation of branchial arch 1–derived structures into branchial arch 2–derived structures in *Pbx1* and *Pbx1;Pbx2* absence (Selleri *et al.* 2001; Capellini *et al.* 2010). Furthermore, compound mutant *Pbx* embryos (i.e., lacking functional copies of *Pbx1* and *Pbx2*) also possess severe alterations (i.e., reductions and loss of identities) across all vertebral domains, a phenotype partially related to shifts in axial expression of *Hox* and *Pax* (Capellini *et al.* 2008). In other words, even slightly modulating the expression of *Pbx* in the axis, the somatopleure, and the limb, may lead to alterations in positional information in all domains and subtle variations in morphology. While it is unlikely that *Pbx* genes are the culprits underlying the shape changes we see in hominoids, they provide an example of how shifts in transcription factor function may cause sweeping changes in morphology.

Another potential mechanism concerns some modification to the *cis*-regulatory architecture and topology of a gene important to cell number specification and/or ilium growth. For example, simultaneous pelvic changes may occur via the repositioning of individual gene promoters closer to their distal enhancers. As has been shown for *Hox*, such shifts in promoter-enhancer topology cause coordinated changes in multiple *Hox* expression domains simultaneously across the limb (Zákány *et al.* 2004; Kmita *et al.* 2005; Tarchini *et al.* 2006; Andre *et al.* 2013). In other words, a single *cis*-regulatory modification such as the one described could influence a key regulator of pelvic development in all domains. It is important to note that given the complexity and hierarchical nature of most genetic circuits, such *cis*- changes often have strong trans-effects. Overall, coordinated changes in the positional information across these domains may have led to initial patterning variations in human ancestors that became selected for and eventually led to the pelvic and limb morphologies quintessential to the modern human condition.

Finally, variation among species in these traits suggests high variability and mosaicism in the evolutionary trajectory that emphasized bipedal adaptations. From a developmental genetics perspective, this could mean that each species reflects an independent experiment with an underlying genomic/genetic architecture, one that may easily permit the reuse of select genomic regions, in similar ways to those described for the stickleback pelvic fin and a host of other traits (Chan *et al.* 2010; Jones *et al.* 2012). In this manner, mosaicism in the by-product of the underlying architecture, and simply reflects the flexibility of some genomic regions to mutate and be influenced by natural selection.

Limbs

Anatomy

Limbs consist of three developmental segments: stylopod, zeugopod, and autopod (Figure 6.1, color plate). The proximal-most segment (stylopod) of the forelimb is the humerus, while in the hindlimb it is the femur. Immediately distal to the stylopod is a pair of bones, the radius and ulna in the forelimb, and the tibia and fibula in the hindlimb. The distal segment or autopod, is composed of carpals (forelimb: wrist bones) and tarsals (hindlimb: calcaneus, talus), metacarpals and metatarsals, and phalanges (or digits) (Romer and Parsons 1986). Digits themselves may comprise a developmental and evolutionary module separate from the proximal portions of the autopod (Young 2013; Young *et al.* 2015).

Development

Limb morphogenesis is initiated when mesoderm and overlapping flank ectoderm interact to form a feedback loop that promotes localized proliferation of mesodermal cells. This proliferation leads to a distal outgrowth or "limb bud" that occurs at specific axial levels along the flank corresponding to future forelimbs and hindlimbs. As each bud grows, differential gene expression establishes three axes: (1) proximal-distal (PD), (2) anterior-posterior (AP), and dorsal-ventral (DV). These axes and their molecular effectors specify cells to become the primary compartments of the limb: stylopod, zeugopod, and autopod. Here we limit our discussion to the evolution and development of interlimb proportions, thus we focus mainly on PD patterning, segmentation, growth, and limb-specific regulation as developmental sources of evolutionary variation.

PD specification and patterning of the stylopod, zeugopod, and autopod appear similar in both limbs. Classic models of how the PD axis is patterned differ in whether the mechanism is related to timing (Tickle and Wolpert 2002) or spatial organization (Dudley *et al.* 2002; Sun *et al.* 2002), while more recent experimental evidence suggests that segments are generated via a balance between competing proximal flank signals (e.g., Retinoic Acid (RA)) and distal tip signals (e.g., members of the FGF signaling pathway) (Mercader *et al.* 2000; Mariani *et al.* 2008; Cooper *et al.* 2011; Roselló-Díez *et al.* 2011). Still others have suggested that segments form via Turing-like mechanisms, for example by activation-inhibition or reaction diffusion (Newman and Muller 2005), for which there is indirect macroevolutionary evidence (Young 2013). Regardless of the exact mechanism of axis formation, individual segments in both limbs are likely specified by sequentially expressed 5′ paralogous *HoxA* and *HoxD* family members (e.g., *Hox9-13*) (Davis *et al.* 1995; Wellik and Capecchi 2003). These systems are similar in both limbs, although there are differences in expression pattern among *Hox*, including 5′ members of *HoxC*, which appear to be expressed in hindlimb domains and may contribute to limb-specific differences, although this is unclear (Wellik and Capecchi 2003). There is evidence that *Hox* genes have many targets in the limb, and some of these target genes likely control the proliferation of chondrocytes (e.g., in forming normal growth plates) and undifferentiated mesenchymal cells (e.g., affecting condensation size) (Goff and Tabin 1997; Boulet and Capecchi 2004). Disruptions to PD patterning, either by knocking out the gene/network in mice or by physical manipulation of the limb bud itself, often leads to loss, truncation, or malformation of segments (Wolpert 2002; Zelzer and Olsen 2003). Developmental abnormalities also tend to be segmentally restricted in their presentation (e.g., hemimelias, autopodal segment duplications) (Grzeschik 2002; Wilkie *et al.* 2002). Together these data would suggest that genes associated with limb segment development have spatially restricted effects, at least after specification and/or segmentation has occurred.

After initial outgrowth stages are completed, limb buds and segments continue to elongate and morphological features such as hand/foot plates, and joints of the proximal and distal limbs (e.g., elbow, wrist, knee, and ankle) become identifiable.

During this time, cells in respective stylopod, zeugopod, and autopod compartments coalesce to form separate pre-chondrogenic mesenchymal condensations. These condensations are shaped via joint development, chondrogenesis, and osteogenesis to produce the many separate skeletal elements of the limb, although how this occurs is only poorly understood. At the same time, muscle and tendon cells begin to migrate into compartments of the limb to form its muscles. Later endochondral growth further contributes to longitudinal and proportional variation by alterations to the growth processes that generate the allometric relationships observed among primates and other mammals (Wilsman, Farnum, Green *et al.* 1996; Wilsman, Farnum, Leiferman *et al.* 1996).

Most importantly for a developmental perspective of proportions, both forelimb and hindlimb share a common origin as duplicated structures and consequently share a similar genetic and developmental system (Figure 6.6, color plate) (Ruvinsky

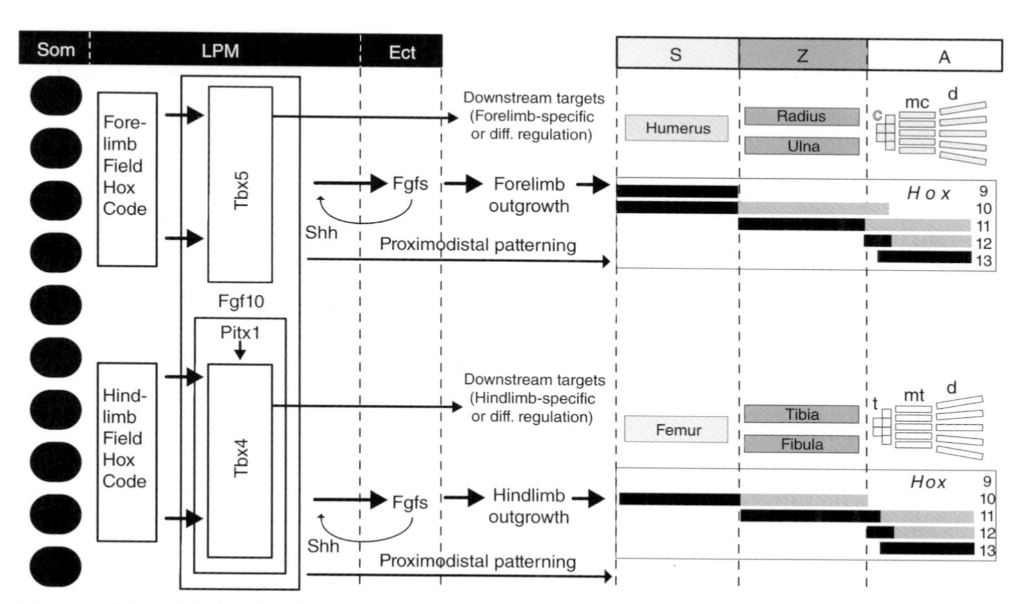

Figure 6.6. Limb development and morphogenesis: Molecular and genetic pathways involved in vertebrate appendage development. Based on expression of key genes, potentially *Hox*, among others, lateral outgrowths of tissue or buds emerge at stereotypical positions along the cranial-caudal embryonic axis. These buds experience molecular cues that trigger loss of forelimbs (*Tbx5*) or hindlimbs (*Pitx1* and *Tbx4*). Despite initial limb type divergence signals, there are shared molecular pathways that allow limbs to be patterned along the proximal-distal, anterior-posterior, and dorsal-ventral axes. For example, signaling in the posterior limb leads to *Shh* activation which not only drives anterior-posterior patterning but also drives *Fgf* signaling along the proximal-distal axis, facilitating limb bud outgrowth. The major limb compartments (stylopod, zeugopod, and autopod) form during bud initiation and outgrowth, which together with coordinated limb type specific gene expression leads to the patterning of the major skeletal elements of the limb. Likely key genes involved in this process have acquired specific *cis*-regulatory elements that mediate limb specific patterning and development. Figure adapted from Young and Hallgrímsson (2005).

and Gibson-Brown 1996; Hall 1999). This shared origin of the limbs has important consequences for their evolution (Hallgrímsson *et al.* 2002), since quantitative genetics theory predicts that genetic correlations will cause limbs to evolve in parallel, especially between homologous segments: that is, the stylopod, zeugopod, and autopod (Rolian *et al.* 2010; Young *et al.* 2010). Consequently, there must be mechanisms to differentiate limbs during their development. While the early developmental processes described are similar in both limbs, differences in how regional identity is specified presumably affect how individual elements of each compartment acquire their distinct limb-type specific features (e.g., femur versus humerus), but this process of individuation is not well understood. Several genes are known to be involved in limb specification, although it is unclear if they direct limb identity in bud tissues. For example, *Tbx5* is expressed in the region associated with the forelimb bud (Logan *et al.* 1998; Logan and Tabin 1999; Ahn *et al.* 2002; Logan 2002, 2003; Agarwal *et al.* 2003; Marcil *et al.* 2003; Rallis *et al.* 2003), while *Pitx1* and *Tbx4* are expressed in lateral plate mesoderm corresponding to the hindlimb bud, although only the former appears to be the major regulator of hindlimb identity while the latter controls outgrowth.

Comparative Variation

Primate limb proportions vary relative to body size (i.e., allometry), among segments within a limb (i.e., intramembral proportions), and between limbs (i.e., intermembral proportions). For the sake of brevity, we focus on the evolution of intermembral proportions since the human condition of relatively long legs and short arms is both distinct from all other primates and is an important behavioral correlate of bipedal adaptations in hominins.

Ape limbs are characterized by high intermembral indices, reflecting the relative elongation of the forelimb relative to the hindlimb (Figure 6.7A, color plate). The ape forelimb is also long relative to other measures of body size such as trunk length (itself reduced, for example, in the number of lumbar vertebrae and loss of the tail), while the hindlimb is relatively short. This pattern is assumed to reflect selection for suspensory behaviors associated with, among other possibilities, brachiation, vertical climbing, and/or below-branch hang-feeding (Hunt 1991; Larson 1998; Pilbeam and Young 2004). In contrast, more quadrupedal primates have similar limb lengths (i.e., intermembral proportions are closer to equivalent), and shorter limbs relative to their longer trunks. That said, there is a trend towards the ape condition related to the proportion of generalized arboreal and/or overhead suspensory behaviors utilized (e.g., in our *ur*-hominoid proxy, the large-bodied colobines *Nasalis*). Humans are decidedly the outlier among catarrhine primates, having relatively short arms and long legs relative to each other and to trunk length. This unique pattern primarily reflects selection for bipedalism on hindlimb length (Bramble and Lieberman 2004; Sockol *et al.* 2007; Pontzer *et al.* 2009), but also potentially selection for shorter forelimb elements in *Homo*, perhaps due to demands associated with increased tool use and manipulation and/or reduced reliance on arboreal behaviors.

(A)

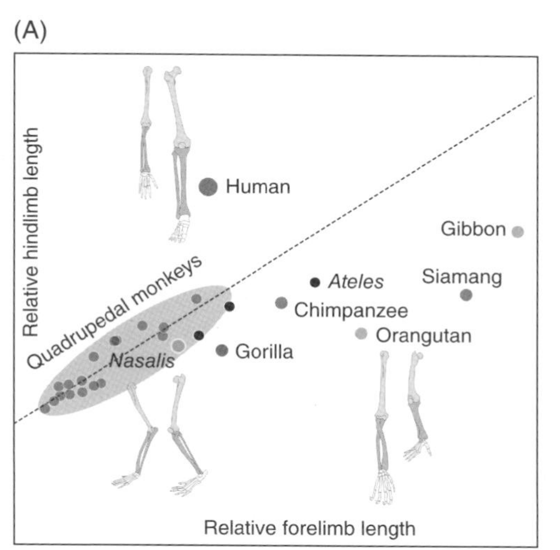

(B)

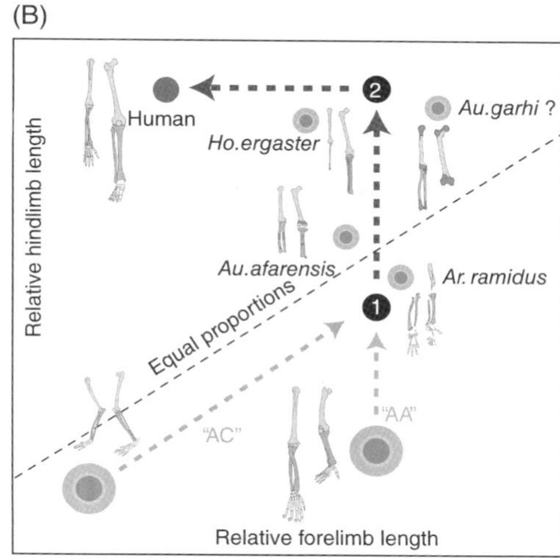

Figure 6.7. Limb comparative and evolutionary morphospace: Plots of the forelimb and hindlimb length relative to trunk length (original data from Schultz (1926)). (A) Extant primate plot shows that humans and suspensory primates differ significantly from the relationship within quadrupedal monkeys. Figures adapted from Young *et al.* (2010). Humans differ from all other primates in having relatively short forelimbs and long hindlimbs. (B) Alternative reconstructions of the LCA differ in whether this morphotype was similar to living apes ("AA"), or was more similar to a quadrupedal monkey. Regardless, fossil hominins suggest a two-step process of mosaic intermembral proportion evolution. Figure adapted from Young *et al.* (2010).

Evolutionary Trajectories

The AA model predicts an evolutionary vector orthogonal to the quadrupedal relationship in which limb lengths are similar and covary in relation to body length. The trajectory predicted by the AC model also deviates from the quadrupedal relationship but is closer to parallel, and would predict a long-backed and long-limbed ancestral morphotype. Interestingly, in both cases predicted vectors are at odds with the fossil record, suggesting a more complex series of selective episodes.

Compared to the girdles, there is substantially more information about the evolution of limb lengths and intermembral proportions (IMI), although the lack of complete skeletons generates some uncertainty in the exact estimation (Reno *et al.* 2002). Regardless, it is reasonably clear that there is a trend in early hominins such as *Australopithecus afarensis* (AL-288-1) (Jungers 1982), the partial skeleton of BOU-12/1 (potentially a specimen of *Australopithecus garhi*) (Asfaw *et al.* 1999), and *Australopithecus sediba* (MH2) (Churchill *et al.* 2013) towards increases in relative hindlimb length with smaller reductions to relative forelimb length (Figure 6.7, color plate). The estimated IMI of the earlier *Ardipithecus ramidus* suggests limbs of roughly equivalent length (Lovejoy *et al.* 2009), which, depending on how trunk length is measured, can be comparable to *Australopithecus afarensis*. Subsequently, in the *Homo* lineage, here represented by *Homo ergaster* (KNM-WT15000), relative forearm length is reduced while relative leg length increased further to modern human levels (Ruff and Walker 1993;

Reno *et al.* 2002; Richmond *et al.* 2002). These fossils suggest at least two phases of selection in hominin evolution: (i) in early hominins (e.g. *Ardipithecus*, *Australopithecus*), relative leg length increased with small to no change to primitive arm proportions, and (ii) in *Homo*, relative forearm length decreased and leg length further increased producing lower IMI (Young *et al.* 2010).

Absent fossils, neither the AA nor the AC model predicts a mosaic pattern of evolution, leaving it an open question as to which is more consistent. While fossil hominin changes are orthogonal to presumed ancestral constraints (as suggested by the AA model), this pattern is consistent with the reduced integration between limbs observed in all living apes but not found in quadrupedal anthropoids (Young *et al.* 2010). In this context, reduced integration would be "preadaptive" for selection on altered limb proportions. This trajectory also demonstrates that an equivalent intermembral proportion (i.e., where limbs are equal in length, such as in early hominins) in a fossil hominin is a predicted intermediate state of evolution from an African ape–like LCA, and thus does not have to be a primitive condition, *contra* Lovejoy *et al.* (2009). In contrast, the AC model is initially more consistent with inferred ancestral patterns of evolvability since, if one extrapolates from the quadrupedal condition, the primitive pattern of strong integration between limbs also predicts ancestral morphotypes with equal intermembral proportions. That said, as with the shoulder, the same divergent events must be explained within hominins, but this time absent the reduced integration present in an ancestral African ape similar to living African apes. Such a scenario therefore requires both an additional morphological transition and a convergent reduction in integration.

Developmental Transformations

Independent regulation of limb length that is both relative to each other and to overall body size can be viewed in part as an allometric problem regulating segment lengths to size. Studies of domesticated animals under strong artificial selection have shown that as a function of overall length of the limbs, this question may have a simple genetic basis. For example, in dogs the extreme differences in leg length between breeds appears to be regulated by variation in *Igf1* function (Sutter *et al.* 2007). But, as in humans, there is a surprisingly weak relationship between length of the limb and individual segment proportions, suggesting these are separate traits regulated by different factors (Young 2013) and the genetics of natural populations are much more complicated. It is less clear how lengths of individual limbs are regulated independently of one another, although clearly there must be a mechanism(s) given the wide diversity among primates. For example, studies from analogous serial homologues, such as insect wings, suggest that under strong directional selection it is possible to rapidly alter the intercept of the ratio between fore- and hindwing area (Frankino *et al.* 2007). Similar experimental analyses in mammalian limbs, say by performing artificial selection experiments in mouse models, would be illuminating.

As discussed previously, the genetic and developmental factors that contribute to morphological differences between limbs, in terms of shape, size, and proportions, are not as well understood, although there have been attempts to characterize overall differences in expression (Margulies *et al.* 2001; Shou *et al.* 2005). Genes associated with limb identity such as *Tbx5*, *Tbx4*, and *Pitx1* or others further upstream appear to contribute to differences by either activating downstream genes uniquely expressed in either limb or by differential regulation of genes common to the development of both limbs (Weatherbee and Carroll 1999). For example, differences in *Hox* hindlimb/forelimb proximo-distal patterning vis-à-vis the control of 5′ *HoxC* cluster in the hindlimb may be due to the different upstream transcription factors that determine fore- versus hindlimb identity and outgrowth (Margulies *et al.* 2001; Shou *et al.* 2005). Alternatively, common sets of genes expressed in both limb types may be differentially regulated in space and time; for example, by specific forelimb- versus hindlimb-specific enhancer elements. This may be the case for regulation of long bone growth plate control; such control may enable selection on differences in relative length and proportion of homologous segments as a function of postnatal (i.e., allometric) growth (see Rolian, this volume; Wilsman, Farnum, Green *et al.* 1996; Wilsman, Farnum, Leiferman *et al.* 1996).

Given the uncertainty regarding which developmental processes are involved in producing differences between limbs, the question of whether dissociated and mosaic or integrated and correlated changes are more or less likely is difficult to test without additional information. As with the scapula and pelvis, however, it is reasonable to suggest that in each model the timing of limb modifications differs, thus the identification of developmental factors and loci (i.e., functional gene sequences) that facilitate this variation may help us to understand the evolutionary timing of events and the dating of major transitions in the fossil record. In other words, are functional variants present in African apes, suggesting that they are shared and thus more ancient, or are they unique to humans, suggesting they are more recent events? Is there a genetic signature for changes in functional integration of common developmental factors (e.g., coordination of selection), and when and how do they diverge in hominoids versus humans? A comparative genomic approach that identifies both the basis of morphogenesis and the underlying genetic signature of this change in separate lineages would do much to address these alternatives.

Conclusions and Future Directions

An important goal of "evo-devo" anthropology is to help explain human evolutionary change through developmental transformations. Development is relevant to human evolution because the process by which genetic variation is "translated" into morphology is not a one-to-one correspondence. Instead, genetic variation impacts how a series of overlapping and hierarchical developmental events (e.g., patterning, segmentation, proliferation, etc.) unfold over an organism's ontogeny. On a population level, each of these events adds to, subtracts, and/or alters the magnitude and direction of phenotypic variation in "morphospace" (Hallgrímsson *et al.* 2009). The challenge of relating development to evolutionary change therefore lies in disentangling this complexity to discover both when and how during development variation is generated in complex phenotypes.

In this chapter we have attempted a systematic approach to this problem in the human girdles and limbs. We first identified key human morphological changes from alternative LCAs and derived a series of expectations regarding the evolutionary transformations of ancestral hominins. We also outlined the predicted developmental transformations associated with each model, and provided our best idea of what these imply about when and how morphogenesis changed. But for these underlying developmental transformations to have credibility and, in practical terms, be identifiable and testable in experimental settings such as the lab, clearly much work is still needed. This includes revealing the complement of genes that regulate girdle and limb morphogenesis and development, deciphering how these genes (and associated regulatory architectures) interact in networks in skeletal tissues to drive specific anatomical traits, and understanding in this context the nature in which traits such as shape, size, and proportions covary among anatomical structures and/or are independently determined. Ultimately, such an understanding allows us to put in perspective any putative functional (and heritable) mutation(s) that serves to distinguish humans from chimpanzees and other primates. Future paleontological discoveries will serve as the ultimate tests of the models used to build such expectations, and we have no doubt that, in combination with advances in technology, much of what we know now and how we test it will further evolve.

To identify the region(s) of the human or chimp genome that not only shape morphology but also have been targeted by selection in either lineage to drive the distinct morphological specializations of each species is non-trivial, involving identifying the correct mutation in the correct functional regulatory element or coding region within the genome. This is a daunting task given that current consensus derived from a number of genome sequencing projects is that upwards of 75–90% of mutations underlying phenotypic variation in humans and other vertebrates is due not to coding mutations but those influencing the regulatory portion of the genome (Jones *et al.* 2012; Grossman *et al.* 2013).

While pan-genomic projects like ENCODE (The ENCODE Project Consortium 2011; Raney *et al.* 2011) have attempted a first round of identifying function of the non-coding regions around genes, ultimately it takes thorough functional characterization using molecular, genetic, and developmental biology tools to validate an element's specific control on a nearby gene(s). Such tasks, including those performed on the genomic scale (like ENCODE), often only attempt functional connections on the level of the DNA (i.e., which pieces have functional capacity and which genes do they regulate), and most often do not focus their control on the level of anatomy or behavior. Thus, such an approach must include the assessment of a genomic region's control, and any variants within them, on a definable biological trait in order for these data to be usable to understand human evolutionary change. Assuming we understand the *cis*-regulatory architecture of a gene (i.e., all of their regulatory inputs), we can next screen each variant for interesting mutations that influence their function developmentally and thus may influence a trait's variation. A notable, landmark example of this technique is the identification of a functional *cis*-regulatory polymorphism in an enhancer for the *KitL* gene that underlies variation in (blonde) hair color in Icelandic populations (Guenther *et al.* 2014). A less concrete, but more relevant example to this chapter concerns the identification and testing of the human *HACNS1* regulatory element, which unlike the corresponding chimp and macaque element drives expression in the first digital ray and which resides near the *Gbx2* gene (Prabhakar *et al.* 2006, 2008). This type of approach, called a bottom-up approach, is useful in that it reveals meaningful biological properties regarding the regulatory topology of a gene region and how such underlies the formation of a biological trait. This is meaningful information regardless of whether we ever identify a functional variant that mediates activity. However, from an evolutionary perspective, this approach of first finding regulatory elements, then sequencing them for variants, and finally testing them, is laborious and may only reveal secondary mutations, not causal mutations that were the target of selection for the trait in question.

On the flip side, one can use pan-genomic methods (a top-down approach) to identify regions (loci) of the genome under selection (signature of selection studies), regions that associate with a trait (linkage and association studies), or regions that display mutational differences amongst humans versus other primates (comparative sequence analyses). In doing so, these methods provide very important biological data regarding the distribution of putative functional and evolutionarily relevant variation on the genomic level. There are of course many caveats to using this approach. Often these studies reflect the datasets (and populations) used in the screen and in so doing bias how we understand the biological phenomenon under study. For example, while such methods have been employed in humans or between human and chimps, they have not been used as readily on non-human primates due to technical and practical issues. This leads us to substantial biases in the way we think about our evolution. These top-down pan-genomic approaches also leave us with a list of target regions, with only the most obvious regions (i.e., those whose function is already somewhat known) often undergoing thorough functional testing; often serving as a proof of principle for the pan-genomic approach. Ultimately, most targets are never tracked down and most mutational differences between species remain untested. In addition, different genome-wide studies often disagree regarding important functional regions under selection, or associated with a trait, again often a product of technical issues.

In the best-case scenario, both approaches should highlight not only the important base pairs underlying biological traits, but their location and function within the *cis*- and trans-regulatory landscape of a gene. Such knowledge will not only benefit the human evolutionary biologist interested in identifying key traits, but a host of other researchers who can use this knowledge in other contexts, such as in medicine, basic biological science research, genomics, and so on. For these methods to be used together, multiple collaborative projects need to be formed amongst biological laboratories, each providing an experimental strength to the project. With respect to the human appendicular skeleton, the results of such collaborative work will reveal a complex genetic and genomic underpinning to limb and girdle formation, and how it evolved over the last 7 million years.

To conclude, in this chapter we have outlined a systematic approach to the "what" and "how" questions of evolutionary change. For both the limb and the girdle we have only begun to reveal the genetic and genomic architecture of their formation and we are still left with some very basic questions about how morphology is generated. For example, why does a femur look different from a humerus? What genetic factors shape morphology at the level of highly specified anatomy? How does the genetic circuitry underlying limb or girdle formation reflect the evolutionary history of their formation? These are but a few questions that we are beginning to address. Ultimately, understanding the complex genetic and genomic architecture of individual skeletal elements, of the appendages they belong to, and over evolutionary time, will inform us better on how we can model shape changes in hominoids and hominins, along with selective scenarios driving such variation.

Acknowledgements

We thank the following individuals for kindly allowing access to specimens in their care: M. Rutzmoser and J. Chupasko (Museum of Comparative Zoology), B. Randall (American Museum of Natural History), L. Gordon (National Museum of Natural History), M. Tappen (Neil Tappen Collection at the University of Minnesota), M. Schulenberg (Field Museum of Natural History), L. Jellema (Cleveland Museum of Natural History), K. Isler (University of Zürich-Irchel), O. Röhrer-Ertl (Zoologische Staatssammlung, München), W. Van Neer (Musée Royale de l'Afrique Centrale), J. Harrison (Powell-Cotton Museum), and H. Gunji (Kyoto University Primate Research Institute and the Japan Monkey Center). We thank M. D. Rose, S. Melillo, and L. Selleri for unpublished data. Additional CT data was generated from scans digitally archived online at the Kyoto University Primate Research Institute's (KUPRI) Digital Morphology Museum. Human data was generated from the Visible Human Project. The American School of Prehistoric Research, the Cora du Bois Charitable Trust, and the Alberta Ingenuity Fund (#200300516) (NMY) provided funding for this research.

References

Adham, I. M., Gille, M., Gamel, A. J. *et al.* 2005. The scoliosis (sco) mouse: A new allele of *Pax1*. *Cytogenetics Genome Research* 111(1):16–26.

Agarwal, P., Wylie, J. N., Galceran, J. *et al.* 2003. *Tbx5* is essential for forelimb bud initiation following patterning of the limb field in the mouse embryo. *Development* 130(3):623–633.

Ahn, D. G., Kourakis, M. J., Rohde, L. A. *et al.* 2002. T-box gene *tbx5* is essential for formation of the pectoral limb bud. *Nature* 417:754–758.

Aiello, L., and Dean, C. 2002. *An introduction to human evolutionary anatomy*. London: Academic Press.

Alemseged, Z., Spoor, F., Kimbel, W. H. *et al.* 2006. A juvenile early hominin skeleton from Dikika, Ethiopia. *Nature* 443:296–301.

Andrey, G., Montavon, T., Mascrez, B. *et al.* 2013. A switch between topological domains underlies *HoxD* genes collinearity in mouse limbs. *Science* 340(6137):1234167.

Asfaw, B., *et al.* 1999. *Australopithecus garhi*: A new species of early hominid from Ethiopia. *Science* 284:629–635.

Aubin, J., Lemieux, M., Moreau, J. *et al.* 2002. Cooperation of *Hoxa5* and *Pax1* genes during formation of the pectoral girdle. *Developmental Biology* 244:96–113.

Aubin, J., Lemieux, M., Tremblay, M. *et al.* 1997. Early postnatal lethality in *Hoxa-5* mutant mice is attributable to respiratory tract defects. *Developmental Biology* 192(2):432–445.

Aubin, J., Lemieux, M., Tremblay, M. *et al.* 1998. Transcriptional interferences at the *Hoxa4/Hoxa5* locus: Importance of correct *Hoxa5* expression for the proper specification of the axial skeleton. *Developmental Dynamics* 212:141–156.

Balling, R. 1994. The undulated mouse and the development of the vertebral column. Is there a human *PAX-1* homologue? *Clinical Dysmorphology* 3(3):185–191.

Bello-Hellegouarch, G., Potau, J. M., Arias-Martorel, J. *et al.* 2013. A comparison of qualitative and quantitative methodological approaches to characterizing the dorsal side of the scapula in Hominoidea and its relationship to locomotion. *International Journal of Primatology* 34:315–336.

Bi, W., Huang, W., Whitworth, D. J. *et al.* 2001. Haploinsufficiency of *Sox9* results in defective cartilage primordia and premature skeletal mineralization. *Proceedings of the National Academy of Sciences, USA* 98:6698–6703.

Boulet, A. M., and Capecchi, M. R. 2004. Multiple roles of *Hoxa11* and *Hoxd11* in the formation of the mammalian forelimb zeugopod. *Development* 131:299–309.

Bramble, D. M., and Lieberman, D. E. 2004. Endurance running and the evolution of *Homo*. *Nature* 432:345–352.

Brunet, M., Guy, F., Pilbeam, D. *et al.* 2002. A new hominid from the Upper Miocene of Chad, Central Africa. *Nature* 418:145–151.

Capellini, T. D., Di Giacomo, G., Salsi, V. *et al.* 2006. *Pbx1/Pbx2* requirement for distal limb patterning is mediated by the hierarchical control of *Hox* gene spatial distribution and *Shh* expression. *Development* 133:2263–2273.

Capellini, T. D., Handschuh, K., Quintana, L. *et al.* 2011a. Control of pelvic girdle development by genes of the *Pbx* family and *Emx2*. *Developmental Dynamics* 240(5):1173–1189.

Capellini, T. D., Vaccari, G., Ferretti, E. *et al.* 2010. Scapula development is governed by genetic interactions of *Pbx1* with its family members and with *Emx2* via their cooperative control of *Alx1*. *Development* 137:2559–2569.

Capellini, T. D., Zappavigna, V., and Selleri, L. 2011b. *Pbx* homeodomain proteins: TALEnted regulators of limb patterning and outgrowth. *Developmental Dynamics* 240(5):1063–1086.

Capellini, T. D., Zewdu, R., Di Giacomo, G. *et al.* 2008. *Pbx1/Pbx2* govern axial skeletal development by controlling *Polycomb* and *Hox* in mesoderm and *Pax1/Pax9* in sclerotome. *Developmental Biology* 321:500–514.

Chan, Y. F., Marks, M. E., Jones, F. C. *et al.* 2010. Adaptive evolution of pelvic reduction in sticklebacks by recurrent deletion of a *Pitx1* enhancer. *Science* 327(5963):302–205.

Chen, H., Lun, Y., Ovchinnikov, D. *et al.* 1998. Limb and kidney defects in *Lmx1b* mutant mice suggest an involvement of *LMX1B* in human nail patella syndrome. *Nature Genetics* 19(1):51–55.

Churchill, S. E., Holliday, T. W., Carlson, K. J. *et al.* 2013. The Upper Limb of *Australopithecus sediba*. *Science* 340(6129).

Cohn, M. J., Patel, K., Krumlauf, R. *et al.* 1997. *Hox9* genes and vertebrate limb specification. *Nature* 387(6628):97–101.

Connell, K. A., Guess, M. K., Chen, H. *et al.* 2008. *HOXA11* is critical for development and maintenance of uterosacral ligaments and deficient in pelvic prolapse. *Journal of Clinical Investigation* 118(3):1050–1055.

Cooper, K. L., Hu, J. K., ten Berge, D. *et al.* 2011. Initiation of proximal-distal patterning in the vertebrate limb by signals and growth. *Science* 332(6033):1083–1086.

Coré, N., Bel, S., Gaunt, S. J. *et al.* 1997. Altered cellular proliferation and mesoderm patterning in Polycomb-M33-deficient mice. *Development* 124(3):721–729.

Davidson, E. H. 2006. *The regulatory genome: Gene regulatory networks in development and evolution*. London: Academic Press.

Davis, A. P., Witte, D. P., Hsieh-Li, H. M. *et al.* 1995. Absence of radius and ulna in mice lacking *hoxa-11* and *hoxd-11*. *Nature* 375:791–795.

De Moerlooze, L., Spencer-Dene, B., Revest, J. M. *et al.* 2000. An important role for the *IIIb* isoform of Fibroblast Growth Factor Receptor 2 (*FGFR2*) in mesenchymal-epithelial signalling during mouse organogenesis. *Development* 127(3):483–492.

Dietrich, S., and Gruss, P. 1995. undulated phenotypes suggest a role of *Pax-1* for the development of vertebral and extravertebral structures. *Developmental Biology* 167(2):529–548.

Dudley, A. T., Ros, M. A., and Tabin, C. J. 2002. A re-examination of proximodistal patterning during vertebrate limb development. *Nature* 418:539–544.

Durland, J. L., Sferlazzo, M., Logan, M., and Burke, A. C. 2008. Visualizing the lateral somitic frontier in the *Prx1Cre* transgenic mouse. *Journal of Anatomy* 212(5):590–602.

Ehehalt, F., Wang, B., Christ, B. *et al.* 2004. Intrinsic cartilage-forming potential of dermomyotomal cells requires ectodermal signals for the development of the scapula blade. *Anatomy and Embryology.* 208(6):431–437.

The ENCODE Project Consortium. 2011. A user's guide to the encyclopedia of DNA elements (ENCODE). *PLoS Biology* 9(4).

Fleagle, J. G. 2013. *Primate evolution and adaptation*, 3rd edn. London: Academic Press.

Frankino, W. A., Stern, D. S., and Brakefield, P. M. 2007. Internal and external constraints in the evolution of a forewing-hindwing allometry. *Evolution* 61:2958–2970.

Goff, D. J., and Tabin, C. J. 1997. Analysis of *Hoxd-13* and *Hoxd-11* misexpression in chick limb buds reveals that *Hox* genes affect both bone condensation and growth. *Development* 124:627–636.

Green, D. J., and Alemseged, Z. 2012. *Australopithecus afarensis* scapular ontogeny, function, and the role of climbing in human evolution. *Science* 338(6106): 514–517.

Grossman, S. R., Andersen, K. G., Shlyakhter, I. *et al.* 2013. Identifying recent adaptations in large-scale genomic data. *Cell* 152(4):703–713.

Grzeschik, K. H. 2002. Human limb malformations; An approach to the molecular basis of development. *International Journal of Developmental Biology* 46(7):983–991.

Guenther, C., Pantalena-Filho, L., and Kingsley, D. M. 2008. Shaping skeletal growth by modular regulatory elements in the *Bmp5* gene. *PLoS Genetics* 4(12):e1000308.

Guenther, C., Tasic, B., Luo, L. *et al.* 2014. A molecular basis for classic blond hair color in Europeans. *Nature Genetics* 46:748–752.

Haile-Selassie, Y., Latimer, B. M., Alened, M. *et al.* 2010. An early *Australopithecus afarensis* postcranium from Woranso-Mille, Ethiopia. *Proceedings of the National Academy of Sciences, USA* 107(27):12121–12126.

Hall, B. K. 1999. *Evolutionary developmental biology*. London: Chapman & Hall.

Hallgrímsson, B., Jamniczky, H., Young, N. M. *et al.* 2009. Deciphering the palimpsest: Studying the relationship between morphological integration and phenotypic covariation. *Evolutionary Biology* 36:355–376.

Hallgrímsson, B., Willmore, K., and Hall, B. K. 2002. Canalization, developmental stability, and morphological integration in primate limbs. *Yearbook of Physical Anthropology* 45:131–158.

Hartwig-Scherer, S. 1993. *Allometry in hominoids: A comparative study of skeletal growth trends*. PhD dissertation, University of Zürich.

Hill, T. P., Taketo, M. M., Birchmeier, W., and Hartmann, C. 2006. Multiple roles of mesenchymal beta-catenin during murine limb patterning. *Development* 133(7):1219–1229.

Hostikka, S. L., Gong, J., and Carpenter, E. M. 2009. Axial and appendicular skeletal transformations, ligament alterations, and motor neuron loss in *Hoxc10* mutants. *International Journal of Biological Science* 5(5):397–410.

Huang, R., Christ, B., and Patel, K. 2006. Regulation of scapula development. *Anatomy and Embryology* 211(1):65–71.

Huang, R., Zhi, Q., Patel, K. *et al.* 2000. Dual origin and segmental organisation of the avian scapula. *Development* 127:3789–3794.

Hui, C. C., and Joyner, A. L. 1993. A mouse model of Greig cephalopolysyndactyly syndrome: The extra-toes[J] mutation contains an intragenic deletion of the *Gli3* gene. *Nature Genetics* 3:241–246.

Hunt, K. D. 1991. Positional behavior in the Hominoidea. *International Journal of Primatology* 12:95–118.

Ikeya, M., Kawada, M., Kiyonari, H. *et al.* 2006. Essential pro-Bmp roles of crossveinless 2 in mouse organogenesis. *Development* 133(22):4463–4473.

Itou, J., Kawakami, H., Quach, T. *et al.* 2012. *Islet1* regulates establishment of the posterior hindlimb field upstream of the *Hand2-Shh* morphoregulatory gene network in mouse embryos. *Development* 139(9):1620–1629.

Johnson, D. R. 1967. Extra-toes: A new mutant gene causing multiple abnormalities in the mouse. *Journal of Embryology and Experimental Morphology* 17:543–581.

Jones, F. C., Grabherr, M. G., Chan, Y. F. *et al.* 2012. The genomic basis of adaptive evolution in threespine sticklebacks. *Nature.* 484(7392):55–61.

Jungers, W. L. 1982. Lucy's limbs: Skeletal allometry and locomotion in *Australopithecus afarensis*. *Nature* 297:676–678.

Kibii, J. M., Churchill, S. E., Schmid, P. *et al.* 2011. A partial pelvis of *Australopithecus sediba* *Science* 333(6048):1407–1411.

Kmita, M., Tarchini, B., Zàkàny, J. *et al.* 2005. Early developmental arrest of mammalian limbs lacking *HoxA/HoxD* gene function. *Nature* 435(7045):1113–1116.

Knezevic, V., De Santo, R., Schughart, K. *et al.* 1997. *Hoxd-12* differentially affects preaxial and postaxial chondrogenic branches in the limb and regulates Sonic hedgehog in a positive feedback loop. *Development* 124(22):4523–4536.

Krawchuk, D., Weiner, S. J., Chen, Y. T. *et al.* 2010. *Twist1* activity thresholds define multiple functions in limb development. *Developmental Biology* 347(1):133–146.

Kuijper, S., Beverdam, A., Kroon, C. *et al.* 2005. Genetics of shoulder girdle formation: Roles of *Tbx15* and Aristaless-like genes. *Development* 132:1601–1610.

Lallemand, Y., Nicola, M. A., Ramos, C. *et al.* 2005. Analysis of *Msx1*; *Msx2* double mutants reveals multiple roles for Msx genes in limb development. *Development* 132(13):3003–3014.

Lanctôt, C., Moreau, A., Chamberland, M. *et al.* 1999. Hindlimb patterning and mandible development require the *Ptx1* gene. *Development* 126:1805–1810.

Larson, S. G. 1998. Parallel evolution in the hominoid trunk and forelimb. *Evolutionary Anthropology* 6:87–99.

Lausch, E., Hermanns, P., Farin, H. F. *et al.* 2008. TBX15 mutations cause craniofacial dysmorphism, hypoplasia of scapula and pelvis, and short stature in Cousin syndrome. *American Journal of Human Genetics* 83:649–655.

LeClair, E. E., Bonfiglio, L., and Tuan, R. S. 1999. Expression of the paired-box genes *Pax-1* and *Pax-9* in limb skeleton development. *Developmental Dynamics* 214:101–115.

Lee, H. H., and Behringer, R. R. 2007. Conditional expression of *Wnt4* during chondrogenesis leads to dwarfism in mice. *PLoS One* 2(5):e450.

Logan, M. 2002. SAGE profiling of the forelimb and hindlimb. *Genome Biology* 3(3): reviews1007.

Logan, M. 2003. Finger or toe: The molecular basis of limb identity. *Development* 130(26):6401–6410.

Logan, M., Simon, H. G., and Tabin, C. 1998. Differential regulation of T-box and homeobox transcription factors suggests roles in controlling chick limb-type identity. *Development* 125:2825–2835.

Logan, M, and Tabin, C. J. 1999. Role of *Pitx1* upstream of *Tbx4* in specification of hindlimb identity. *Science* 283(5408):1736–1739.

Lohnes, D., Mark, M., Mendelsohn, C. *et al.* 1994. Function of the retinoic acid receptors (RARs) during development (I). Craniofacial and skeletal abnormalities in RAR double mutants. *Development* 120(10):2723–2748.

Long, F., and Ornitz, D. M. 2013. Development of the endochondral skeleton. *Cold Spring Harbor Perspectives Biology* 5(1):a008334.

Lovejoy, C. O., Cohn, M. J., and White, T. D. 1999. Morphological analysis of the mammalian postcranium: A developmental perspective. *Proceedings Of The National Academy Of Sciences, USA*. 96:13247–13252.

Lovejoy, C. O., Suwa, G., Simpson, S. W. *et al.* 2009. The great divides: *Ardipithecus ramidus* reveals the postcrania of our last common ancestors with African apes. *Science* 326:100–106.

Malashichev, Y., Borkhvardt, V., Christ, B., and Scaal, M. 2005. Differential regulation of avian pelvic girdle development by the limb field ectoderm. *Anatomy and Embryology* 210:187–197.

Malashichev, Y., Christ, B., and Prols, F. 2008. Avian pelvis originates from lateral plate mesoderm and its development requires signals from both ectoderm and paraxial mesoderm. *Cell Tissue Research* 331:595–604.

Marcil, A., Dumontier, E., Chamberland, M. *et al.* 2003. *Pitx1* and *Pitx2* are required for development of hindlimb buds. *Development* 130:45–55.

Margulies, E. H., Kardia, S. L., and Innis, J. W. 2001. A comparative molecular analysis of developing mouse forelimbs and hindlimbs using serial analysis of gene expression (SAGE). *Genome Research* 11(10):1686–1698.

Mariani, F. V., Ahn, C. P., and Martin, G. R. 2008. Genetic evidence that FGFs have an instructive role in limb proximal-distal patterning. *Nature* 453(7193):401–405.

Matsuoka, T., Ahlberg, P. E., Kessaris, N. *et al.* 2005. Neural crest origins of the neck and shoulder. *Nature* 436:347–355.

Mercader, N., Leonardo, E., Piedra, M. E. *et al.* 2000. Opposing RA and FGF signals control proximodistal vertebrate limb development through regulation of Meis genes. *Development* 127:3961–3970.

Min, H., Danilenko, D. M., Scully, S. A. *et al.* 1998. *Fgf-10* is required for both limb and lung development and exhibits striking functional similarity to *Drosophila* branchless. *Genes and Development* 12(20):3156–3161.

Moeller, C., Swindell, E. C., Kispert, A., and Eichele, G. 2003. Carboxypeptidase Z (CPZ) modulates Wnt signaling and regulates the development of skeletal elements in the chicken. *Development* 130(21):5103–5111.

Nakatsukasa, M., and Kunimatsu, Y. 2009. *Nacholapithecus* and its importance for understanding hominoid evolution. *Evolutionary Anthropology* 18:103–119.

Newman, S. A., and Müller, G. B. 2005. Origination and innovation in the vertebrate limb skeleton: An epigenetic perspective. *Journal of Experimental Zoology B: Molecular Development and Evolution* 304(6):593–609.

Ng, J. K., Kawakami, Y., Büscher, D. *et al.* 2002. The limb identity gene *Tbx5* promotes limb initiation by interacting with *Wnt2b* and *Fgf10*. *Development* 129(22):5161–5170.

Oliver, G., De Robertis, E. M., Wolpert, L., and Tickle, C. 1990. Expression of a homeobox gene in the chick wing bud following application of retinoic acid and grafts of polarizing region tissue. *EMBO Journal* 9:3093–3099.

Pellegrini, M., Pantano, S., Fumi, M. P. *et al.* 2001. Agenesis of the scapula in *Emx2* homozygous mutants. *Developmental Biology* 232:149–156.

Pilbeam, D. R., Rose, M. D., Barry, J. C., and Shah, S. M. I. 1990. New *Sivapithecus* humeri from Pakistan and the relationship of *Sivapithecus* and *Pongo*. *Nature* 348:237–239.

Pilbeam, D. R., and Young, N. M. 2004. Hominoid evolution: synthesizing disparate data. *Comptes Rendus Palevol* 3(4):303–319.

Pomikal, C., and Streicher, J. 2010. 4D-analysis of early pelvic girdle development in the mouse (*Mus musculus*). *Journal of Morphology* 271:116–126.

Pontzer, H., Raichlen, D. A., and Sockol, M. D. 2009. The metabolic cost of walking in humans, chimpanzees, and early hominins. *Journal of Human Evolution* 56(1):43–54.

Prabhakar, S., Noonan, J. P., Pääbo, S., and Rubin, E. M. 2006. Accelerated evolution of conserved noncoding sequences in humans. *Science* 314:786.

Prabhakar, S., Visel, A., Akiyama, J. A, et al. 2008. Human-specific gain of function in a developmental enhancer. *Science* 321:1346–1350.

Prols, F., Ehehalt, F., Rodriguez-Niedenfuhr, M. *et al.* 2004. The role of *Emx2* during scapula formation. *Developmental Biology* 275:315–324.

Rallis, C., Bruneau, B. G., Del Buono, J. *et al.* 2003. *Tbx5* is required for forelimb bud formation and continued outgrowth. *Development* 130(12):2741–2751.

Rancourt, D. E., Tsuzuki, T., and Capecchi, M. R. 1995. Genetic interaction between *hoxb-5* and *hoxb-6* is revealed by nonallelic noncomplementation. *Genes and Development* 9:108–122.

Raney, B. J., Cline, M. S., Rosenbloom, K. R. *et al.* 2011. ENCODE whole-genome data in the UCSC genome browser (2011 update). *Nucleic Acids Research* 39(Database issue):D871-5.

Reno, P. L. *et al.* 2002. Plio-Pleistocene hominid limb proportions. *Current Anthropology* 46:575–588.

Richmond, B. G., Aiello, L. C., and Wood, B. A. 2002. Early hominin limb proportions. *Journal of Human Evolution* 43:529–548.

Rolian, C., Lieberman, D. and Hallgrímsson, B. 2010. The coevolution of human hands and feet. *Evolution* 64: 1558–1568.

Romer, A. S., and Parsons T. S. 1986. *The vertebrate body*. New York: Saunders College Publishing.

Rose, M. D. 1983. Miocene hominoid postcranial morphology: Monkey-like, ape like, neither, or both? *In*: R. S. Chiochon and R. S. Corruccini (eds) *New interpretations of ape and human ancestry*. New York: Plenum Press. pp.405–417.

Rose, M. D. 1993. Locomotor anatomy of Miocene hominoids. *In*: D. L. Gebo (ed.) *Postcranial adaptation in nonhuman primates*. DeKalb: Northern Illinois University Press. pp.252–272.

Roselló-Díez, A., Ros, M. A., and Torres, M. 2011. Diffusible signals, not autonomous mechanisms, determine the main proximodistal limb subdivision. *Science* 332(6033):1086–1088.

Ruff, C. B., and Walker, A. 1993. Body size and body shape. *In:* A. Walker and R. Leakey (eds) *The Nariokotome Homo erectus skeleton*. Cambridge, MA: Harvard University Press. pp 234–265.

Ruvinsky, I., and Gibson-Brown, J. J. 2000. Genetic and developmental bases of serial homology in vertebrate limb evolution. *Development* 127:5233–5244.

Ruvolo, M. 1997. Molecular phylogeny of the hominoids: Inferences from multiple independent DNA sequence data sets. *Molecular Biology and Evolution* 14:248–265.

Schultz, A. 1926. Fetal growth of man and other primates. *Quarterly Review of Biology* 1:465–521.

Sekine, K., Ohuchi, H., Fujiwara, M. *et al.* 1999. *Fgf10* is essential for limb and lung formation. *Nature Genetics* 21(1):138–141.

Selleri, L., Depew, M. J., Jacobs, Y. *et al.* 2001. Requirement for *Pbx1* in skeletal patterning and programming chondrocyte proliferation and differentiation. *Development* 128:3543–3557.

Senut, B., Pickford, M., Gommery, D. *et al.* 2001. First hominid from the Miocene (Lukeino Formation, Kenya). *Comptes Rendus de l'Académie des Sciences, Sciences de la Terre et des planètes/Earth and Planetary Sciences* 332:137–144.

Shea, B. T. 1981. Relative growth of the limbs and trunk in the African apes. *American Journal of Physical Anthropology* 56:179.

Shea, B. T. 1983. Allometry and heterochrony in the African apes. *American Journal of Physical Anthropology* 62(3):275–289.

Shea, B. T. 1984. An allometric perspective on the morphological and evolutionary relationships between pygmy (*Pan paniscus*) and common (*Pan troglodytes*) chimpanzees. *In*: R. Susman (ed.) *The pygmy chimpanzee*. New York: Plenum Press. pp.89–130.

Shea, B. T. 1985. On aspects of skull form in African apes and orangutans, with implications for hominoid evolution. *American Journal of Physical Anthropology* 68:329–342.

Shou, S., Scott, V., Reed, C. *et al.* 2005. Transcriptome analysis of the murine forelimb and hindlimb autopod. *Developmental Dynamics* 234(1):74–89.

Sockol, M. D., Raichlen, D. A., and Pontzer, H. 2007. Chimpanzee locomotor energetics and the origin of human bipedalism. *Proceedings of the National Academy of Sciences* 104(30):12265–12269.

Steiper, M. E., and Young, N. M. 2006. Primate molecular divergence dates. *Molecular Phylogenetics and Evolution* 41(2):384–394.

Stern, J. T., Jr., and Susman, R. L. 1983. The locomotor anatomy of *Australopithecus afarensis*. *American Journal of Physical Anthropology* 60:279–317.

Sun, X., Mariani, F. V., and Martin, G. R. (2002). Functions of FGF signalling from the apical ectodermal ridge in limb development. *Nature* 418:501–508.

Sutter, N. B., Bustamante, C. D., Chase, K. *et al.* 2007. A single *IGF1* allele is a major determinant of small size in dogs. *Science* 316:112–115.

Tague, R. G., and Lovejoy, C. O. 1986. The obstetric pelvis of AL 288-1 (Lucy). *Journal of Human Evolution* 15(4):237–255.

Tarchini, B., Duboule, D., and Kmita, M. 2006. Regulatory constraints in the evolution of the tetrapod limb anterior-posterior polarity. *Nature* 443(7114):985–988.

ten Berge, D., Brouwer, A., Korving, J. *et al.* 1998. *Prx1* and *Prx2* in skeletogenesis: Roles in the craniofacial region, inner ear and limbs. *Development* 125:3831–3842.

Tickle, C., and Wolpert, L. 2002. The progress zone – alive or dead? *Nature Cell Biology* 4(9):E216–217.

Timmons, P. M., Wallin, J., Rigby, P. W., and Balling, R. 1994. Expression and function of *Pax1* during development of the pectoral girdle. *Development* 120:2773–2785.

Valasek, P., Theis, S., Krejci, E. *et al.* 2010. Somitic origin of the medial border of the mammalian scapula and its homology to the avian scapula blade. *Journal of Anatomy* 216(4):482–488.

Vrba, E. 1979. A new study of the scapula of *Australopithecus africanus* from Sterkfontein. *American Journal of Physical Anthropology* 51:117–130.

Wang, B., He, L., Ehehalt, F., Geetha-Loganathan, P. *et al.* 2005. The formation of the avian scapula blade takes place in the hypaxial domain of the somites and requires somatopleure-derived BMP signals. *Developmental Biology* 287:11–18.

Weatherbee, S. D., and Carroll, S. B. 1999. Selector genes and limb identity in arthropods and vertebrates. *Cell* 97(3):283–286.

Wellik, D. M., and Capecchi, M. R. 2003. *Hox10* and *Hox11* genes are required to globally pattern the mammalian skeleton. *Science* 301:363–367.

White, T. D., Suwa, G., and Asfaw, B. 1994. *Australopithecus ramidus*, a new species of early hominid from Aramis, Ethiopia. *Nature* 371:306–312.

Wilkie, A. O., Patey, S. J., Kan, S. H. *et al.* 2002. FGFs, their receptors, and human limb malformations: Clinical and molecular correlations. *American Journal of Medical Genetics* 112(3):266–278.

Williams, P. L., Bannister, L. H., and Gray, H. 1995. *Gray's anatomy: The anatomical basis of medicine and surgery*, 38th edn. New York: Churchill Livingstone.

Wilm, B., Dahl, E., Peters, H. *et al.* 1998. Targeted disruption of *Pax1* defines its null phenotype and proves haploinsufficiency. *Development* 95(15):8692–8697.

Wilsman, N., Farnum, C. E., Green, E. M. *et al.* 1996a. Cell cycle analysis of proliferative zone chondrocytes in growth plates elongating at different rates. *Journal of Orthopedic Research* 14:562–572.

Wilsman, N., Farnum, C. E., Leiferman, E. M. *et al.* 1996b. Differential growth by growth plates as a function of multiple parameters of chondrocytic kinetics. *Journal of Orthopedic Research* 14:927–936.

Wolpert, L. 2002. The progress zone model for specifying positional information. *International Journal of Developmental Biology* 46(7):869–870.

Wrangham, R., and Pilbeam, D. 2001. African apes as time machines. *In*: B. M. F. Galdikas, N. E. Briggs, L. K. Sheeran, G. L. Shapiro, and J. Goodall (eds) *All apes great and small*, vol. 1. Springer, Berlin. pp.5–17.

Young, N. M. 2003. A reassessment of living hominoid postcranial variability: Implications for ape evolution. *Journal of Human Evolution* 45:441–464.

Young, N. M. 2004. Modularity and integration in the hominoid scapula. *Journal of Experimental Zoology, Part B: Molecular and Developmental Evolution* 302B:226–240.

Young, N. M. 2006. Function, ontogeny and canalization of shape variance in the primate scapula. *Journal of Anatomy* 209:623–636.

Young, N. M. 2008. A comparison of the ontogeny of shape variation in the anthropoid scapula: Functional and phylogenetic signal. *American Journal of Physical Anthropology* 136:247–264.

Young, N. M. 2013. Macroevolutionary diversity of amniote limb proportions predicted by developmental interactions. *Journal of Experimental Zoology, Part B: Molecular and Developmental Evolution* 32(7):420–427.

Young, N. M., and Hallgrímsson, B. 2005. Serial homology and the evolution of mammalian limb covariation structure. *Evolution* 59:2691–2704.

Young, N. M., Wagner, G. P., and Hallgrímsson, B. 2010. Development and the evolvability of human limbs. *Proceedings of The National Academy of Sciences, USA* 107:3400–3405.

Young, N. M., Winslow, B., Takkellapati, S., and Kavanagh, K. 2015. Shared rules of development predict patterns of evolution in vertebrate segmentation. *Nature Communications* 6:6690.

Zákány, J., Kmita, M., and Duboule, D. 2004. A dual role for *Hox* genes in limb anterior-posterior asymmetry. *Science* 304(5677):1669–1672.

Zelzer, E., and Olsen, B. R. 2003. The genetic basis for skeletal diseases. *Nature* 423(6937):343–348.

Chapter 7

Tinkering with Growth Plates: A Developmental Simulation of Limb Bone Evolution in Hominoids

Campbell Rolian
Faculty of Veterinary Medicine, University of Calgary, Calgary, AB, Canada

Introduction

One of the most striking interspecific differences in hominoid postcranial anatomy is the difference in their individual limb bone lengths, both in absolute terms and in relation to body mass (Schultz 1937; Aiello and Dean 1990). It has been suggested that these differences in limb bone length are functionally correlated with differences in the positional repertoire and locomotor behaviors of hominoids. For example, suspensory species such as gibbons, orangutans, and some Miocene apes (e.g., *Oreopithecus*, Jungers 1987) have forelimb bones that are relatively much longer than their hind limb bones and in relation to their body mass (Aiello and Dean 1990). From a functional perspective, longer forelimbs may have evolved in these taxa because they increase the reach between arboreal supports (Fleagle 1999). This may lower the energetic cost of brachiation by reducing the number of swings required to cover a given distance. In contrast, the hind limbs of *Homo*, which are absolutely and relatively long, are related to their committed bipedality, and may similarly have evolved by selective means as an adaptation for reducing the metabolic cost of locomotion (Sockol *et al.* 2007; Tilkens *et al.* 2007).

Adaptive evolutionary scenarios provide an explanation for *why* such distinct limb proportions were selected in the context of specific locomotor behaviors; however, they are typically less concerned with *how*, from a mechanistic perspective, this adaptive evolution is made possible (see Chapter 1 in Lynch and Walsh 1998). Evolutionary change occurs within populations, and is only possible if a trait has heritable phenotypic variation: and this applies equally to limb bone morphology. Limb bone length is a highly heritable trait (Norgard *et al.* 2008), suggesting that genetic variation is an important source of phenotypic variation in limb bone length within and among taxa, perhaps more so than environmental sources. Variation in the genetic loci that control limb developmental processes is a potentially important source of selectable phenotypic variation in limb bone size and shape. One developmental

Developmental Approaches to Human Evolution, First Edition. Edited by Julia C. Boughner and Campbell Rolian.

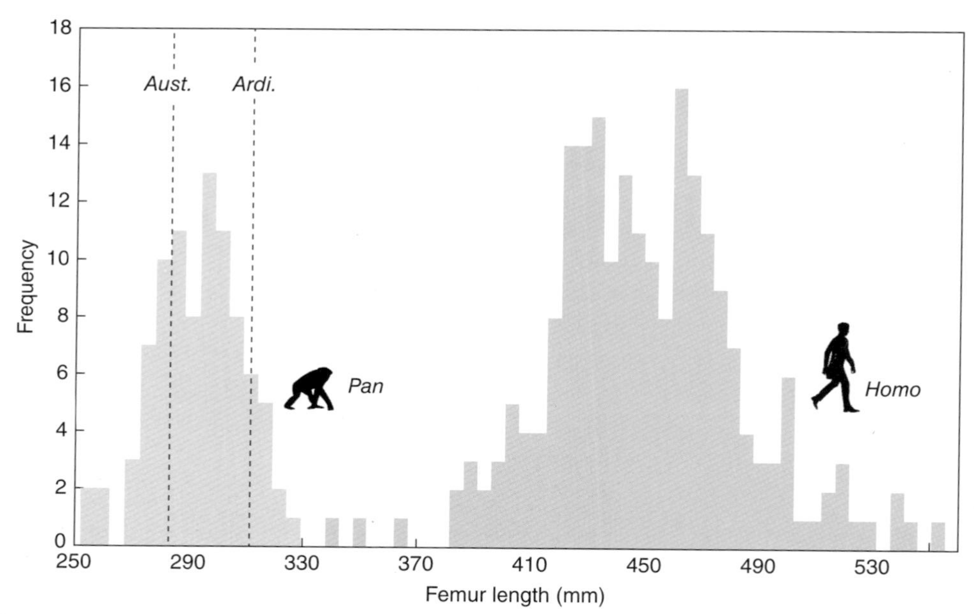

Figure 7.1. Frequency distribution of maximum absolute femur length in a sample of 89 common chimpanzees (*Pan troglodytes*), 214 modern humans (*Homo sapiens*), and two fossil hominins, the 4.4-million-year-old *Ardipithecus ramidus* (ARA-VP-6/500, see Lovejoy *et al.* 2009) and 3.2-million-year-old *Australopithecus afarensis* (A.L.288-1, "Lucy," Jungers 1982).

process of particular interest is postnatal limb growth. For example, adult femoral length is markedly different in *Pan*, *Homo*, and early hominins, being 50% longer in absolute terms in modern humans (Figure 7.1). This difference in adult length is driven by differential longitudinal postnatal growth, specifically by variation in the *rate* of longitudinal growth in the femur and other limb long bones (Figure 7.2) (Gavan 1971; Ruff 2003).

The growth pattern in Figure 7.2 suggests that genetic, molecular and cellular mechanisms that influence postnatal bone growth rates are likely effectors of evolutionary change in limb bone length among hominoids. Identifying and quantifying these mechanisms can reveal much about the evolutionary processes underlying limb diversification among primates. There are, however, both ethical and practical considerations that make it difficult to study these developmental genetic mechanisms in primates. One solution is to query common mammalian model organisms such as mice to study how differences in limb bone length originate. This is a common and fruitful approach in evolutionary developmental biology, especially when the organisms of interest are empirically intractable. It does have some limitations, however, in that the model organisms may not exhibit the same diversity of limb proportions or modes of locomotion as hominoids. For example, comparing limb proportions among mice and rats, two of the most common mammalian model organisms, may not reveal much about the developmental determinants of limb bone length beyond those associated with body size differences in these species. That said, a number of additional model rodents are becoming increasingly available, and

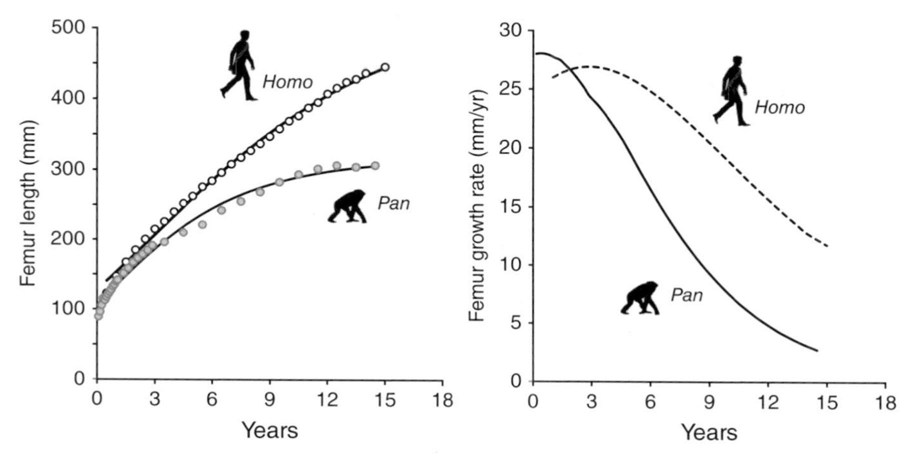

Figure 7.2. Postnatal growth (left panel) and growth rate (right panel) in the femur in chimpanzees (gray circles) and modern humans (open circles). Data for *Pan* from Gavan (1971), and for *Homo* from Ruff (2003). The growth curves were fit with a Gompertz growth curve, and the first derivatives of these were used to obtain the growth rates. The growth rates show that even though ontogeny is longer in humans, the growth rates of the femur are absolutely greater throughout much of ontogeny.

some of these have been targeted specifically because of differences in their limb proportions relative to mice and rats (Rolian 2008; Cooper *et al.* 2013).

An alternative approach that is gaining traction in evolutionary biology relies on simulations of genetic and developmental systems that evolve via natural selection (Salazar-Ciudad and Jernvall 2002; Jones *et al.* 2003, 2004, 2007; Salazar-Ciudad *et al.* 2003; Salazar-Ciudad and Marin-Riera 2013). In principle, these simulations *in silico* can be designed to model developmental systems at any level of complexity and in any anatomical context, so long as the molecular and/or cellular mechanics of development in the anatomical system in question are relatively well known. Simulation studies have the inherent advantage that they can describe developmental evolution in systems and taxa that are not amenable to experimental study, and can be useful to generate testable hypotheses regarding the types of developmental processes that underlie differences in limb proportions within and among species. Simulations of primate development thus represent a powerful tool to help characterize the developmental origins of variation in postnatal growth among hominoid limb bones.

In this study, I use an evolutionary developmental simulation of the cellular mechanisms of longitudinal growth in the limb bones, to understand how changes to these mechanisms under natural selection may have led to evolutionary divergence in limb bone length among hominoid taxa. The study seeks to answer two questions. First, what magnitude of change(s) to the cellular dynamics of limb bone longitudinal growth is necessary to cause phenotypic changes such as those observed between the femora of *Pan* and *Homo*? Second, do some of the cellular mechanisms of longitudinal bone growth change more frequently than others in response to selection acting on limb bone growth and adult length? I begin with a brief overview of the main cellular mechanisms of limb development, highlighting those mechanisms that are likely to contribute the most to variation in limb bone growth within and among taxa. These mechanisms form the basis of the subsequent developmental model and evolutionary simulation.

Limb Bone Growth

Limb long bones grow through a two-phase process, collectively known as endochondral ossification (Wagner and Karsenty 2001; Kronenberg 2003; Karsenty *et al.* 2009; Figure 7.3). The first, embryonic phase involves budding of the limbs from the body wall, and spatial patterning of the limb bones, during which the presumptive limb bones are laid down in their correct anatomical relationship (e.g., radius distal to the humerus) as condensations of mesenchymal cells. These condensations have sizes and shapes specific to each long bone. The condensation phase is followed by differentiation, in which the mesenchymal cells differentiate into two cell populations – inner and outer (Figure 7.3C) (Hall and Miyake 2000). In the inner population, mesenchymal cells become chondrocytes which proliferate and thus expand the size of the condensation. These cells eventually cease to proliferate and begin to hypertrophy, producing extra cellular matrix (ECM) that further contributes to the expansion of the condensation. The outer cell population will become the perichondrium. This sheath of connective tissue itself comprises two layers: an outer layer of fibroblasts; and an inner layer of undifferentiated mesenchymal cells that can give rise to chondrocytes and osteoblasts. In mice, condensation, differentiation, and chondrogenesis of the limb bones is complete by approximately 12 days post-conception (E12).

Osteogenesis and long bone elongation begin during the second, and much longer, phase of long bone growth. Osteogenesis begins when the hypertrophic chondrocytes in the condensation signal to the perichondrium region to trigger vascularization (Figure.7.3D). They also signal to the inner layer of cells to differentiate into osteoblasts, which begin to deposit bone near the invading blood vessels, forming a bone collar. The blood vessels carry osteoblasts into the center of the condensation, where they deposit bone on the mineralized ECM "scaffold" produced by hypertrophying chondrocytes. Waves of bone are deposited in this fashion at the metaphysis, going away from the center of the diaphysis towards the ends and periphery (cortex) of the long bone. This process establishes the primary center of ossification, which is responsible for bone elongation (Figure.7.3E). Soon after the primary center appears, additional vessels enter the condensation near the ends of the bones, creating a number of secondary centers of ossification that contribute to bone growth of the epiphyses and articular cartilage, and to the unique shape of each bone. As the bone continues to ossify, the primary and secondary ossification centers converge on one another to form the growth plates (physes) (Figure.7.3F) (Karsenty *et al.* 2009).

The cartilaginous growth plate separates the primary and secondary ossification centers. In the growth plate, chondrocytes undergo a highly orchestrated life cycle that drives longitudinal growth of the long bone. Histologically, the growth plate is divided into separate zones that reflect the life stage of the chondrocytes within them.

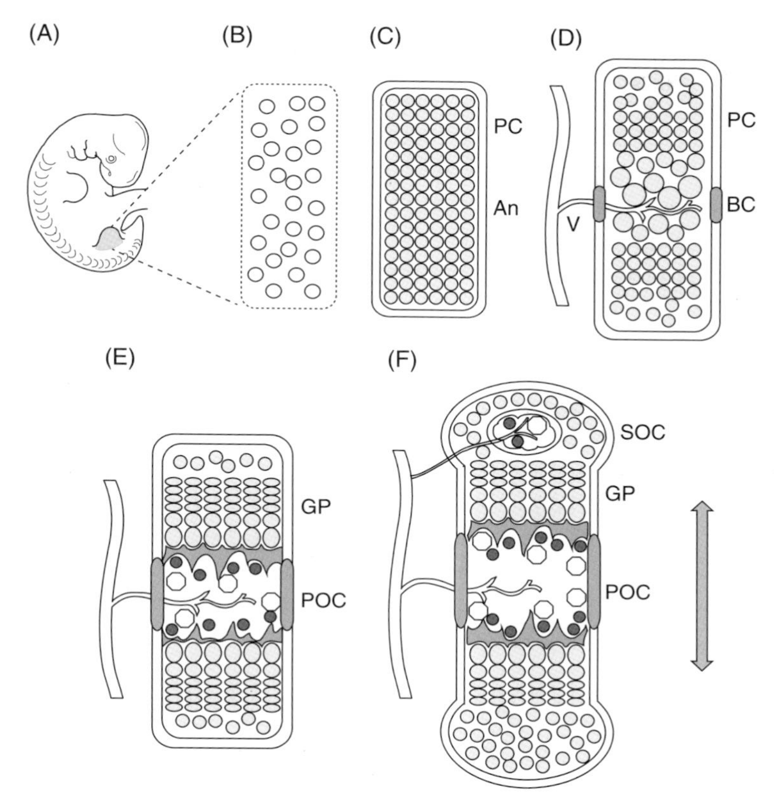

Figure 7.3. Schematic representation of limb bone formation and growth in mammals. (A) In the early embryonic stage, the limbs grow out from the body wall as undifferentiated buds of mesenchymal cells (white circles). (B) Mesenchymal cells aggregate and prospective joints begin to form in their proper anatomical positions, outlining the shape and size of the future limb bone (dashed line). (C) Mesenchymal cells condense and differentiate into chondrocytes (light gray circles), and the anlage (An) is formed, along with the perichondrium (PC). (D) Chondrocytes in the anlage hypertrophy, inducing vascular invasion (V), and differentiation of cells in the inner layer of the perichondrium (PC) into osteoblasts. These cells form a bone collar (BC) and establish the primary ossification center. At the same time, chondrocytes near the ends of the bones continue to proliferate, contributing to longitudinal growth of the bone. (E) Chondrocytes eventually become organized into growth plates (GP), with a proliferating zone (stacked flattened cells) and a hypertrophic zone closest to the bone's center. In the center, the osteoblasts (dark gray circles) lay down the primary spongiosa (dark gray) on the ECM scaffold left behind by apoptotic chondrocytes, and contribute to elongation of the bone collar. Osteoclasts (white hexagons) act in concert with osteoblasts to regulate the amount of bone deposition and resorption in the cortex and spongiosa. (F) While this process continues, the secondary centers of ossification are established in a similar manner in the periphery of the long bones. The primary and secondary ossification centers (POC and SOC) bracket the growth plates. Together, the centers contribute to growth in other directions, influencing the ultimate shape and size of the long bone.

The first zone, closest to the epiphysis, contains paired chondrocytes in a "resting" state (Abad *et al.* 2002). These cells are progenitors of cells that eventually become proliferative. Subjacent to the resting zone, the chondrocytes become actively mitotic. Histologically, the proliferative zone is recognizable by the columns of proliferating chondrocytes, which are organized parallel to the long axis of the bone. The proliferating chondrocytes within these columns undergo a number of rounds of mitosis, then mature and begin to hypertrophy.

During the maturation and hypertrophy phase, the chondrocytes enlarge partly through the absorption of water, while simultaneously producing and secreting large amounts of ECM. Histologically, the hypertrophic zone is recognized by the much larger appearance of the chondrocytes, particularly in the direction of longitudinal growth (Figure.7.3E–F). Eventually, the hypertrophic chondrocytes undergo apoptosis near the metaphysis, leaving behind a partially mineralized ECM scaffold upon which bone is deposited by osteoblasts from the primary ossification center. Through this stereotypical chondrocyte life cycle, the long bone elongates from the center towards the epiphyses. In most mammals, this process, known as endochondral ossification, continues until the growth plates senesce and the diaphysis and epiphyses fuse, marking the end of growth and skeletal maturity.

Developmental Model of Endochondral Ossification

The process of endochondral ossification highlights a number of potential cell-level sources of phenotypic variation within and among species in growth rates, and hence in adult limb bone length. For example, an increase in the rate of proliferation of the chondrocytes may help produce more hypertrophic cells per unit time, and hence increase limb bone growth over the same time period. Similarly, the rate of hypertrophy influences their final size/height, and hence how much each of those cells contributes to longitudinal growth before apoptosing. Both of these zones can also have more or fewer cells in them. Even the size of the original condensation can contribute to the adult limb bone if, for example, the condensation size influences the sizes of the original resting and proliferating zone cell populations. As long as variation in these cellular parameters has at least some genetic basis, they represent potential mechanisms underlying adaptive evolution of limb bone length within and among hominoids.

Several developmental models of endochondral ossification have been proposed to account for the regulation of longitudinal bone growth, ranging in complexity from as little as two (Stokes *et al.* 2007) or three parameters (e.g., Kember 1993; Kirkwood and Kember 1993; Rolian 2008) to as many as eight cellular parameters (Hunziker 1994; Wilsman *et al.* 1996; Farnum *et al.* 2002). Most models, however, have focused on a few cellular parameters as major regulatory determinants of long bone growth rates and ultimate bone length. These are the number and rate of mitosis of proliferating cells, and the rate of hypertrophy, or indirectly the size of the terminal hypertrophic chondrocytes. The model proposed in this study is based on these cellular processes.

The developmental model represents an idealized long bone, in which there is a single growth plate. Though many long bones have two or more growth plates, several have only one; this includes the metacarpals, metatarsals, and all phalanges, which also exhibit significant differences in relative and absolute length among hominoids. The model focuses exclusively on longitudinal, as opposed to appositional (concentric), growth. The model proposes that the growth of the long bone over an interval of time (Δt) is equal to the length contributions from two cell populations – proliferating and hypertrophic – in the growth plate over the same time interval:

$$\Delta L = \Delta L_p + \Delta L_h \tag{1}$$

Where ΔL_p and ΔL_h are the length contributions from the proliferating and hypertrophic cell populations, respectively. Each of these incremental length terms can be broken into cellular mechanisms that determine their magnitude. Let us first consider ΔL_p. The proliferative zone contains an initial population of cells, P_0. In the majority of mammals, P_0 declines over time such that by adulthood it is exhausted (Kirkwood and Kember 1993), and no more growth is possible. The number of proliferating cells present at time t, P_{pop}, was modeled as:

$$P_{pop} = P_0 \times e^{(-kt)} \tag{2}$$

where k is a decay constant whose value ensures that P_{pop} approaches 0 as growth nears completion. Assuming that all cells in this zone at time t are actively dividing (i.e., the growth fraction is = 1, Wilsman *et al.* 1996), then the number of new proliferating cells produced over Δt is:

$$P_{new} = P_{pop} \times \mu_p \times \Delta t \tag{3}$$

where μ_p is the proliferation rate (see Eq. 10, below). The increase in length produced by the creation of new proliferating cells, ΔL_p, over time interval Δt is then given by:

$$\Delta L_p = P_{new} \times P_{size} \tag{4}$$

where P_{size} is the average size of a proliferating chondrocyte, measured in the direction of longitudinal growth. ΔL_h can similarly be parsed into the cellular mechanisms that determine its magnitude. The first item to consider is how many new cells become hypertrophic over Δt, and how many of these cells apoptose near the metaphysis: in other words, how many cells transition through the hypertrophic zone over Δt. In this model, the number of *new* hypertrophic cells (H_{new}) is assumed to be equal to the number of new cells produced by mitosis in the proliferating zone (P_{new}), as well as to the number of hypertrophic cells lost to cell death over the same time (H_{lost}). Thus the zones are in a steady state with respect to each other, such that over time interval Δt:

$$P_{new} = H_{new} = H_{lost} \tag{5}$$

There is empirical evidence that cell turnover between the zones is in a steady state over at least the short term (Wilsman *et al.* 1996). In terms of growth plate kinetics, a steady state maintains a proper balance between the sizes of the proliferating and hypertrophic pools. For example, if proliferation rates were significantly greater than the turnover rates from proliferating to hypertrophic (i.e., $P_{new} >> H_{new}$), then the proliferating zone would rapidly grow to be much larger than its hypertrophic counterpart, leading to exponential growth of the limb bone. Conversely, if the rates of turnover were significantly greater ($P_{new} << H_{new}$), then this would lead to a rapid depletion of the proliferative zone chondrocytes and a premature closure of the growth plate thus limiting/stopping bone growth. Let the number of cells initially present in the hypertrophic zone be H_0. As with the proliferating zone, the size of the hypertrophic pool at time t is given by:

$$H_{pop} = H_0 \times e^{(-kt)} \quad (6)$$

where k is the same decay constant whose value ensures that the population H_{pop} approaches 0 as growth nears completion, and in parallel with the decrease in P_{pop}. The length contribution from hypertrophy, ΔL_h, is determined by the number of cells present and transitioning through the zone over Δt, multiplied by the rate at which the cells hypertrophy:

$$\Delta L_h = \left(H_{new} + H_{pop}\right) \times \mu_h \times \Delta t \quad (7)$$

where μ_h is the rate of hypertrophy, that is, the rate at which these cells grow in the direction of the longitudinal axis via physiological swelling (see Eq. 10, below). Putting it all together, and substituting P_{new} for H_{new} in Eq. 7, the length increase in a long bone with a single growth plate, over time interval Δt, is given by:

$$\Delta L = \left[P_{pop} \times \mu_p \times P_{size} + \left(P_{pop} \times \mu_p + H_{pop}\right) \times \mu_h\right] \times \Delta t \quad (8)$$

The long bone does not grow from a length of zero, but rather from the end of the condensation prior to osteogenesis. Thus, we may add a term representing the size of the original condensation, as this has been shown to influence the length of the adult bone (Hall and Miyake 2000):

$$L_{bone}\left(t + \Delta t\right) = L_{cond} + \left[P_{pop} \times \mu_p \times P_{size} + \left(P_{pop} \times \mu_p + H_{pop}\right) \times \mu_h\right] \times \Delta t \quad (9)$$

This basic model suggests that the growth rate of a long bone is primarily determined by six cellular parameters, three which are likely determined early in embryogenesis (L_{cond}, P_{pop} and H_{pop}), and two of which are rates (μ_p and μ_h, see below). To model the growth of a long bone in a comparative evolutionary context, total ontogenetic time was set to be 100% of growth, and time interval Δt was set at 1% increments. Although age at skeletal maturity and life histories differ between hominoids, Figure 7.1 suggests that it is not simply because humans have a more protracted life history that they grow longer femora than chimpanzees. The growth rates are greater in *Homo*, and indicate that this difference is rooted in the cellular parameters described above. Because growth was set to end at a value of t = 100, the decay constant k was set at 0.04, which ensures that the proliferative and hypertrophic cell population sizes decrease exponentially and approach zero as time approaches 100.

Rate Parameters

The model includes two rates, one that describes cell division, and one cell hypertrophy. Although the model accepts constant rates, this is not realistic from a biological perspective because rates are likely to change over the course of postnatal growth. For example, they may slowly increase until reaching a peak near adolescence, representing a growth spurt, and then rapidly decrease towards skeletal maturity. Linear long bone growth is often described using a sigmoid growth function, such as the Gompertz function (German *et al.* 1994; Rolian 2008). The first derivatives of theses curves describe growth rates. As can be seen in Figure 7.2, the growth rates increase to a peak during the juvenile period, whose timing may shift, and then decrease steadily towards zero. Long bone growth rates of this type can be described by underlying developmental rates of the same nature, such as proliferation or hypertrophy rates. The two rates in this model were described by the first derivative of the Richards growth function:

$$\frac{ack}{(m-1)} e^{-ct} \left(1 + ke^{-ct}\right)^{m/(1-m)} \tag{10}$$

in which the four constants, a, c, k and m, determine changes in the starting value (a), peak rate (a, c, and m), and its timing (c and k) (Figure 7.4). Equation 10 can model increases in rates over: early ontogeny; a peak (e.g., during a growth spurt); and a rapid decrease towards an effective rate of zero. Separate values of the Richards parameters a, c, k, and m were used to model the rate of proliferation μ_p and the rate of hypertrophy μ_h.

Evolutionary Simulation

The developmental model of endochondral ossification presented here relates limb bone growth primarily to the quantity, size, and behavior of the proliferative and hypertrophic cells, as in previous models (e.g., Kember 1993; Rolian 2008). Six cellular parameters are included, two of which are rates which are described by four parameters each. Thus, in total, this developmental model uses 12 variables to describe longitudinal growth of the limb bones in mammals. The objective of the evolutionary simulation is to use the model described in the previous section to test which parameters change, and by how much, in response to repeated scenarios of selection acting on an adult femoral phenotype. Previous evolutionary simulations have used two different approaches to understanding how genetic and developmental systems evolve under selection. The first focuses on evolutionary change in one or two traits, and how selection alters some of their properties (e.g., means, genetic variances, and/or covariances between traits, Burger 1993; Burger and Lande 1994; Jones *et al.* 2004). These simulations have the advantage that they model the genetic loci underlying each trait explicitly. The traits typically have arbitrary values (e.g., with a mean of zero), and hence can in principle model any type of biological system – from chemical processes to cells to tissues – if the means are adjusted by some constant that reflects a biological

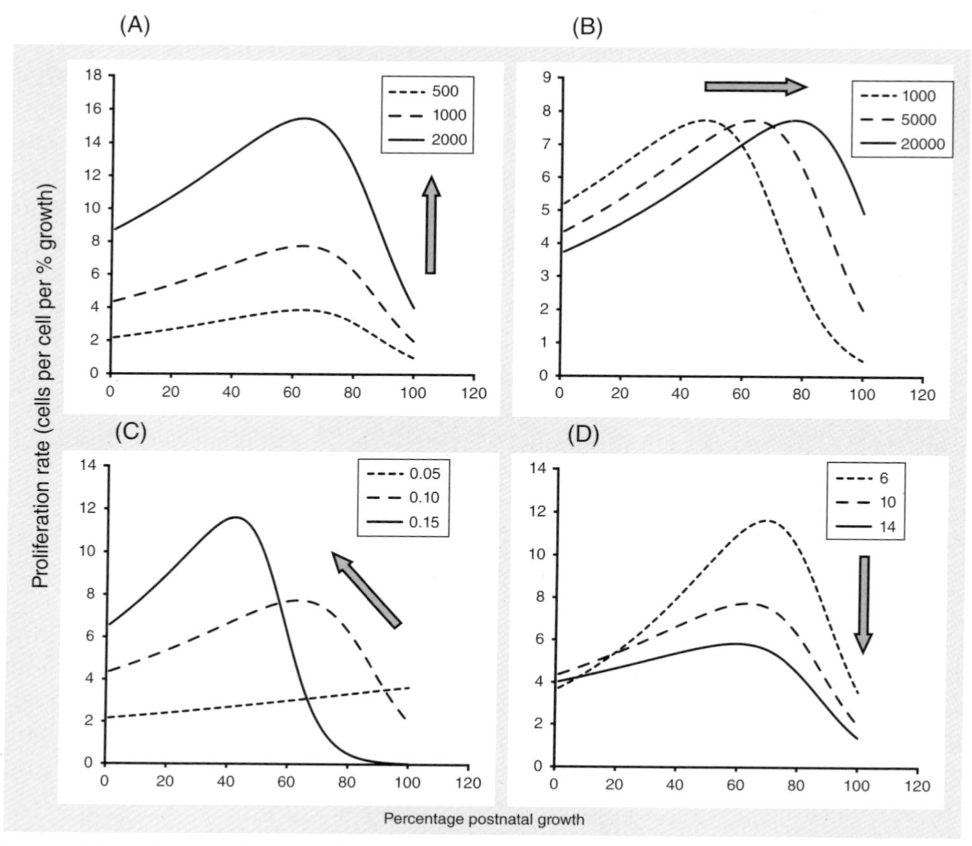

Figure 7.4. Hypothetical proliferation rate as a function of postnatal ontogeny (from 0 to 100% growth), described by the first derivative of the Richards growth function (Equation 10). The four panels illustrate the effects of different magnitudes of each of the Richards growth parameters in relation to an average value (long dashed line).

parameter (e.g., number of cells, or cell dimensions). This type of locus-based simulation has not previously been applied to the cellular mechanisms of organismal development and their inherently hierarchical nature.

In contrast, developmental evolutionary simulations, such as the one by Salazar-Ciudad and Marin-Riera (2013), model the evolution of multiple molecular and cellular developmental processes in creating a genotype-development-phenotype map for a complex morphological structure (in their case, mammal teeth). This type of model does not consider the actual allelic composition of the genotype, which is summarized only in terms of basic molecular and cellular interactions experimentally known to influence the phenotype of interest. The disadvantage of excluding genetic loci from this type of simulation is that it cannot model sexual reproduction and recombination; that is, an offspring is a genotypic and developmental clone of a single individual in the preceding generation, except for the addition of mutational effects. In such a simulation, novel genotypes (and downstream phenotypes) can only arise through mutation, as opposed to both mutation and recombination.

The evolutionary simulation presented here combines these two approaches. The 12 variables in the model of endochondral ossification are determined by a fixed number of genetic loci. For the purpose of this simulation, each trait is determined by the same number of loci, which remains invariant throughout the simulation. Furthermore, the loci are assumed to be additive, freely recombining (i.e., no linkage), non-pleiotropic, and non-epistatic. Although this may not reflect a biological reality, it is computationally simpler, largely because it circumvents the issue of determining how many loci are pleiotropic, and which shared sets of loci affect which cellular variables. The evolutionary simulation was modeled with 50 loci for each cellular variable (Jones *et al.* 2004).

Before running the simulation, a mean value for the adult phenotype (femoral length) was chosen for the starting population and as the optimum value driving the evolution of the phenotype and its developmental determinants via natural selection. Evolutionary changes were simulated as if going from a *Pan*-like postcranial anatomy in the last common ancestor (LCA) of humans and chimpanzees, towards modern *Homo* (Pilbeam 1996; Wrangham and Pilbeam 2001). In this simulation, selection was modeled for an increase in bone length, specifically, a ~50% increase in absolute femur length between a chimp-like LCA and a modern human (Figure 7.1).

The evolutionary simulation goes through the following steps each generation (Figure 7.5): (1) production of the genotypes for each individual and each of the 12 variables in the developmental model; (2) production of the adult phenotypes (i.e., bone length) using the developmental model; (3) determination of the fitness of each

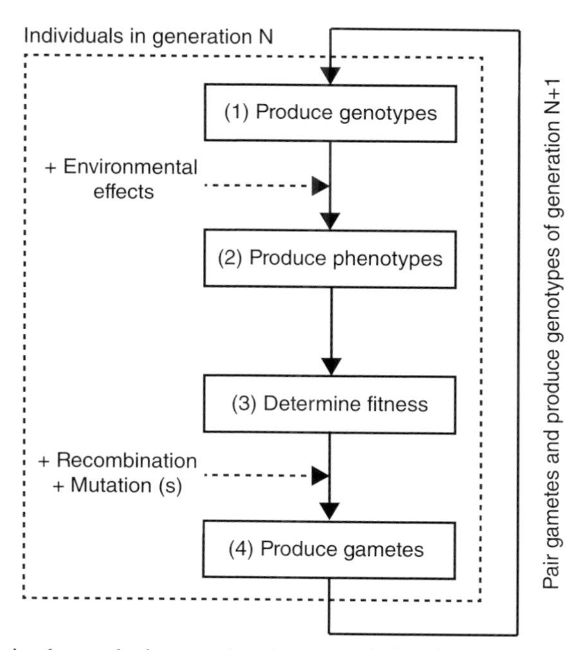

Figure 7.5. Steps in the evolutionary developmental simulation. See text for description of the steps within a generation (dashed outline). The steps are repeated for as many generations as set in the simulation.

individual relative to the adaptive optimum; (4) sexual reproduction, including random gametic phase recombination and mutation. All steps in the evolutionary simulation were computed in Matlab R2012b (Mathworks, Natick, MA). Population size was set at 500 individuals, divided equally into males and females, and did not change over the course of the evolutionary simulation.

1. *Production of the Genotypes*

   The genotypes of the first generation are produced differently from those in subsequent generations. For each individual in the first generation, a full diploid complement of 50 loci for each developmental variable was first created by drawing random numbers from a normal distribution with a mean of zero and a variance of 4×10^{-6}. This first generation is genetically heterogeneous; that is, the alleles at each locus are likely different for each individual, with as many possible alleles at a locus as there are individuals in the population. The magnitude of the variance in the first generation was chosen so that the additive effects of all the loci and developmental parameters would produce a phenotype with a mean and variance close to those of *Pan* femoral length after 1000 generations of evolution under stabilizing selection (see below).

   For each subsequent generation, the genotypes were those from the offspring of the previous generation, as determined in step (4) below. In each generation, the genotype for an individual for a given trait was obtained by summing the 50×2 allelic values for that trait. Following this, an amount of environmental variance was added to the genotypic value of each trait in each individual by drawing a random number from a normal distribution with a mean of zero and a variance equivalent to the genotypic variance. This environmental value was then multiplied by 0.5, so that the resulting narrow-sense heritability of the developmental variables in the adult phenotype would be ~0.6–0.7, in line with previously estimated heritabilities for skeletal traits (Norgard *et al.* 2008).

2. *Production of the Adult Phenotypes*

   To obtain phenotypic values for each of the developmental variables in the model, the summed genotypic and environmental value of an individual was multiplied by a constant. The magnitude of each variable's constant was determined prior to the simulation so that, when combined in the full developmental model, the resulting growth curve and adult phenotype would approximately match the growth curve and adult length for the chimpanzee femur in Figure 7.1A. The values were also selected on the basis that they yielded parameters of hypertrophy, proliferation, and chondrocyte morphology that were in agreement with published values for humans and other vertebrates (e.g., Kember and Sissons 1976; Hunziker and Schenk 1989; Hunziker 1994). The constant values did not change over the course of the evolutionary simulation, so that evolutionary change occurred only through the additive effects of the allelic values. The developmental variables for each individual were then run through the model of endochondral ossification (Eqs 1–10) to produce its adult phenotype. Each individual's growth trajectory was standardized so that it went from 0% (birth) to 100% of growth (adulthood) in increments of 1%.

3. *Determination of an Individual's Fitness and Reproductive Success*
An individual's fitness was determined by taking the absolute value of the difference between the adaptive optimum and the individual's phenotype. This value was then entered into the following fitness function:

$$\text{Fitness} = e^{\left(-be^{c}F\right)} \tag{11}$$

where F is the absolute value of the difference in femoral length between an individual's phenotype and the optimum, and b and c are positive constants that produce an exponentially decaying fitness curve (Figure 7.6). Increasing the value of c produces a more rapid drop-off in fitness (i.e., at values of F closer to the optimum), representing stronger directional or stabilizing selection. In the simulation, values of c increased as the mean phenotype moved closer to the new optimum (Figure 7.6). Lower values of c farther from the optimum (i.e., weak directional selection) ensured that early generations were not composed of only a few genotypes with extremely high fitness. In turn, higher values of c near and at the optimum (i.e., strong stabilizing selection) ensured that phenotypic variance remained similar to its value in *Homo*, and reduced the end effect of genetic drift. Note that changing the value of c in Equation 11 does not affect which developmental variables change as a result of selection, but only the rate at which the new optimum is reached.

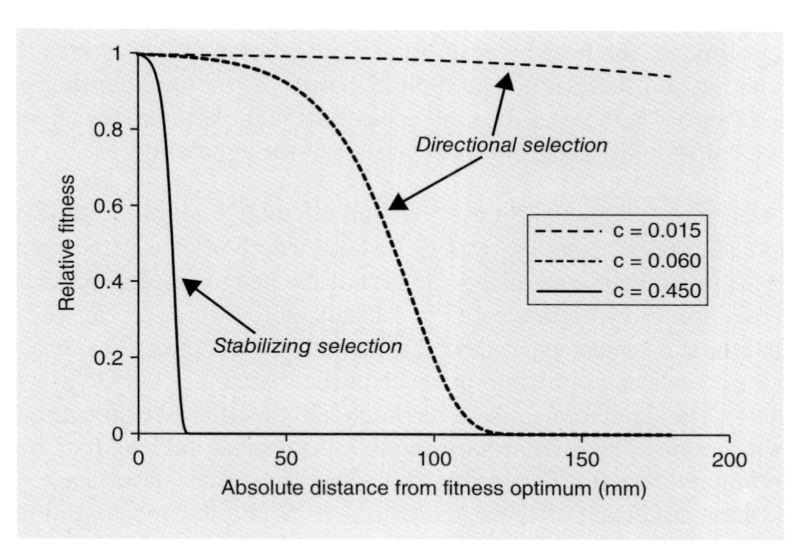

Figure 7.6. Relative fitness (between 0 and 1, where 1 is maximum fitness) as a function of the absolute distance from the new evolutionary optimum for femur length (Equation 11). At greater distances, directional selection is relatively weak, ensuring that the population is not rapidly overrepresented by a few highly fit individuals. As the population mean nears the fitness optimum, stabilizing selection takes over, and fitness drops off more rapidly as individual phenotypes are further from the optimum.

An individual's fitness value was used to determine its reproductive success following the method of Salazar-Ciudad and Marin-Riera (2013). For every individual in the following generation (i.e., the offspring), its father and mother in the current generation were chosen based on the latter's fitness value. For each offspring, two random numbers between 0 and 1 were drawn, the first for its father and the second for its mother. The father was chosen at random from the pool of potential fathers whose fitness value was greater than the first random number, and the same process was repeated for the mother. This method ensures that individuals with a higher fitness have, on average, more progeny represented in the next generation, but also allows low (non-zero) fitness individuals to leave progeny, and low-fitness individuals to mate with high-fitness individuals (Salazar-Ciudad and Marin-Riera 2013).

4. *Recombination, Reproduction, and Mutation*
In the final step of the evolutionary simulation, the haploid gametes for each parent are generated and matched to produce the diploid genotype of each offspring used in the following generation. For each developmental variable, the alleles which will be inherited from each parent are determined by constructing a gamete with one of the two allelic values at each locus, chosen at random (Jones *et al.* 2004). This process simulates recombination, but assumes that the loci are independent; that is, there is no gametic phase disequilibrium. The effects of mutations are then added to each gamete. The frequency of mutation was set at 10^{-3} per locus. If a locus was by chance mutated, a random number was drawn from a normal distribution with the same mean and variance as the original genotypic distribution and added to the genotype at that locus. Finally, the haploid genotype of the father and mother were combined to produce the diploid genotype for that offspring. The diploid genotypes of the offspring population were used in the next generation of the evolutionary simulation, repeating steps (1) to (4) for as many generations as was set in the simulation.

The evolutionary simulation was first run for 1000 generations under stabilizing selection conditions, using the average femoral length of *Pan* as the optimum (300 mm). This preliminary run allows the genetically heterogeneous population in the first generation to evolve and achieve a balance of drift, selection, and mutation, such that the subsequent experimental generations under directional selection begin with a more realistic population in which allelic frequencies vary across loci (Jones *et al.* 2004). The simulation under directional selection (the experimental simulation) was then run for 1000 generations using the average femoral length of *Homo* as the new optimum (450 mm). The preliminary simulation under stabilizing selection was run six times, and each of these was then used as the starting conditions for four iterations of the experimental simulation. The total of 24 simulation runs illustrates how starting population conditions evolve under a balance of stabilizing selection, drift, and mutation, and how each set of starting conditions influences the evolution of the developmental parameters under directional selection.

The following developmental data were used to address the two questions posed in the introduction. First, the average percentage change in the cellular and tissue level developmental variables (i.e., anlagen size, cell sizes, cell numbers, maximum and average rates of proliferation and hypertrophy) were compared against the average change in the phenotype (+50% in femur length) to determine the magnitudes of the cellular-level changes required to cause the phenotypic changes observed in the simulation. Second, the average percentage change in the developmental variables were compared amongst themselves to identify which, if any, contributes the most to differences in the rates of elongation of long bones in *Homo* and *Pan* femora. For simplicity, only changes in the actual rates of proliferation and hypertrophy are reported, rather than the changes in each of the four parameters (a, c, k, and m) that describe these rates.

Results

Preliminary Simulations

Figure 7.7 shows the average change in the developmental parameters from a genetically heterogeneous population in the first generation to a population under a balance of drift, selection, and mutation at generation 1000 in the preliminary simulation runs. The average and coefficient of variation of the developmental processes that result from the six preliminary simulations are shown in Table 7.1. After the six runs, coefficients of variation for

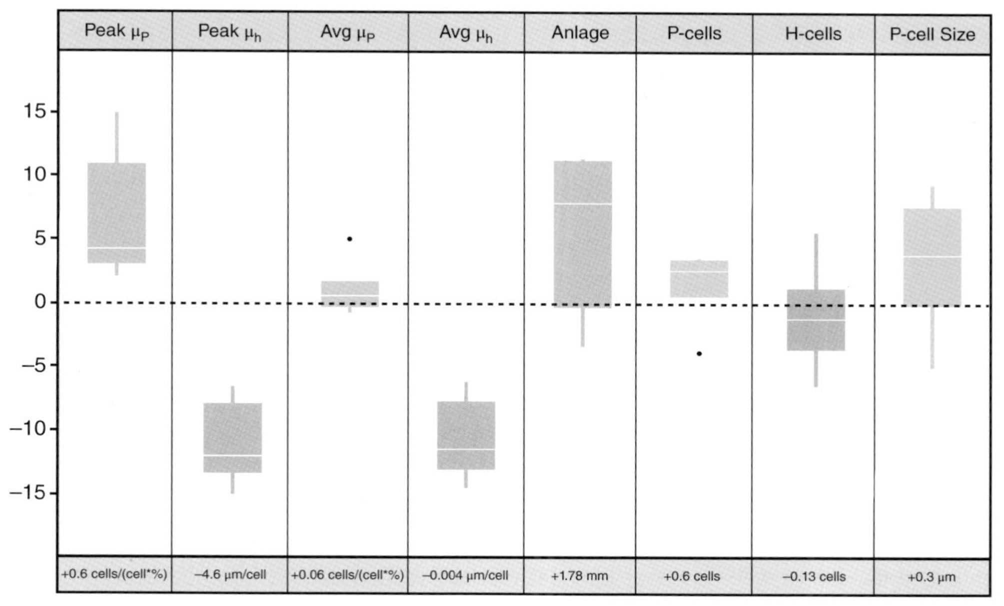

Figure 7.7. Relative changes (in percentage) in the magnitude of the starting values for the developmental parameters after 1000 generations of stabilizing selection for a chimpanzee-like optimum in femur length (Table 7.1). The lower boxes indicate what the relative changes mean in terms of the values for each parameter. Legend: μ_p, chondrocyte proliferation rate; μ_h, chondrocyte hypertrophy rate; anlage, length of the anlage; P-cells, size of the proliferating chondrocyte pool at birth; H-cells, size of the hypertrophic chondrocyte pool at birth; P-cell size, size (in μm) of the proliferating cells in the direction of longitudinal growth.

Table 7.1. Averages, coefficients of variation, and % change in the developmental parameters over the course of the preliminary simulation runs (n=6) which provide the starting conditions for the experimental runs (see also Figure 7.7).

	Parameter means (n=6) at Gen 1	Parameter means (n=6) at Gen 1000	Parameter coefficients of variation at Gen 1000	% change
Anlage size (mm)	29.99	31.77	5.74	+5.9%
Number of proliferating cells at birth	34.99	35.58	2.77	+1.7%
Number of hypertrophic cells at birth	12	11.87	3.96	−1.1%
Size of proliferating cells (μm)	10	10.34	4.82	+3.4%
Peak proliferation rate (**cells**/% **growth**)	9.29	9.88	4.63	+6.4%
Mean proliferation rate (**cells**/% **growth**)	6.54	6.60	2.04	+0.9%
Peak hypertrophy rate (μ**m**/% **growth**	41.6	36.9	3.50	−11.2%
Mean hypertrophy rate (μ**m**/% **growth**	39.7	35.4	3.42	−10.8%

each developmental variable remain small, around 5%. The peak proliferation rates are systematically higher in the end populations, and conversely the rates of hypertrophy are all slightly lower (Figure 7.7, color plate). These patterns suggest that, despite differences in genetic heterogeneity of the starting populations, the preliminary simulations converge on populations that are similar in their distribution of developmental phenotypes.

Experimental Simulations

Figure 7.8 shows the average change, after 1000 generations of selection, in the rates of proliferation and hypertrophy over ontogeny for the 24 simulations relative to the conditions at the beginning of each simulation (see also Table 7.2). Only one experimental simulation failed to reach the new phenotypic optimum, presumably for stochastic reasons related to the standing genetic variation present at the beginning of that simulation and/or to the pattern of mutation over its 1000 generations. For those simulations that did, the approximate generations at which the optimum was reached (and directional selection changed to stabilizing selection) varied from as early as 350 to as late as 920 generations. There were positive correlations between the value of the developmental parameters at generation 1 and their value after 1000 generations, especially for the rate parameters (data not shown). In other words, higher starting values for the rate parameters led to relatively higher end values. This is not unexpected, since on average, proliferation and hypertrophy rate parameters increased over the course of the simulations (Figure 7.8, color plate). However, there was no correlation between the starting values and the average change in each parameter, and several of the parameters experienced

Table 7.2. Means, coefficients of variation, and % change in the developmental parameters after the simulation runs (n=6).

	Parameter means (n=6) at Gen 1	Parameter means (n=24) at Gen 1000	Parameter coefficients of variation at Gen 1000	% change
Anlage size (mm)	31.77	33.32	5.71	+4.9%
Number of proliferating cells at birth	35.58	39.30	4.20	+10.5%
Number of hypertrophic cells at birth	11.87	12.09	7.24	+1.8%
Size of proliferating cells (μm)	10.34	10.72	8.42	+3.6%
Peak proliferation rate (cells/cell*% growth)	9.88	12.51	10.6	+27.1%
Mean proliferation rate (cells/cell*% growth)	6.60	7.53	6.6	+14.1%
Peak hypertrophy rate (μm/% growth	36.9	43.0	12.9	+16.3%
Mean hypertrophy rate (μm/% growth)	35.4	40.7	8.9	+15.0%

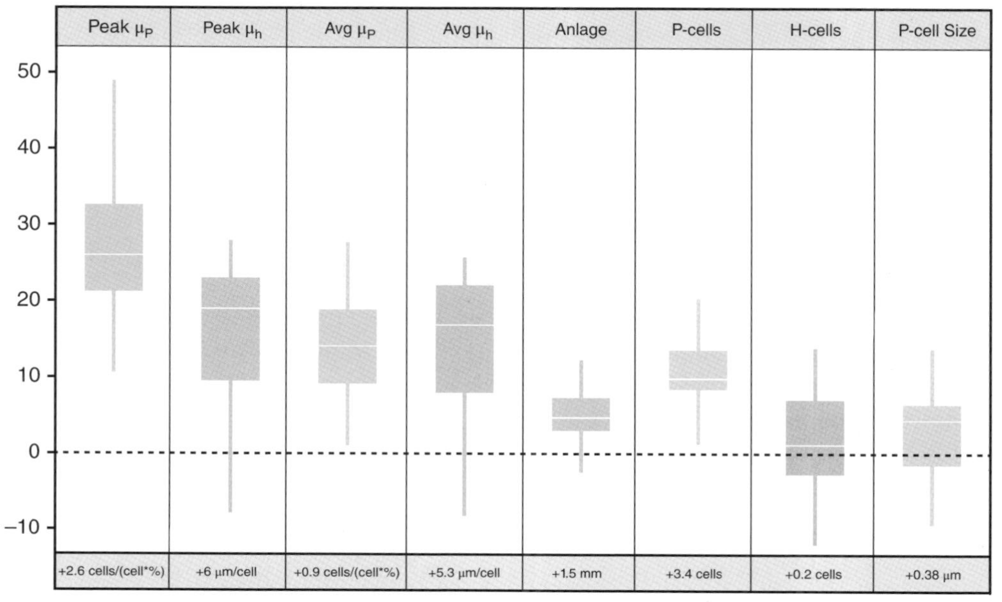

Figure 7.8. Relative changes (in percentage) in the magnitude of the developmental parameters after 1000 generations under directional selection for a new human-like optimum in femur length (Table 7.2). Legend as in Figure 7.7.

both positive and negative changes (Figure 7.8, color plate), which suggests that the starting conditions have little effect on the direction and magnitude of changes among the traits in the developmental model.

Absolute Magnitude of Changes in Cellular Parameters of Bone Growth

Regarding the magnitude of cellular changes required to produce a 50% change in absolute femur length, Figure 7.8 shows that the greatest change occurs in the chondrocyte proliferation rate, where the peak rate is on average 25% greater, while the average increase in the proliferation rate over ontogeny is ~15% (Table 7.2). Changes in the rates of hypertrophy are also notable, with both the peak and average rates increasing by 15–18%. Condensation size, the size of the original proliferative and hypertrophic chondrocyte pools, and the original size (height) of the proliferating chondrocytes do not contribute as much to increase in bone length. Thus rate appears to have the greatest influence here.

In actual numbers, differences in the cellular parameters of the growth plates before and after the simulation are much smaller than their percentage changes suggest. For example, a 10% increase in the number of cells in the proliferative zone represents an additional four proliferating chondrocytes in the human-like growth plate. Similarly, a 15% increase in the mean rate of hypertrophy represents an increase of ~5.3 microns per hypertrophic cell. Even a 25–30% increase in the peak proliferation rate does not lead to a dramatic change in the mitotic dynamics of the proliferation zone in the human-like growth plate. For example, at 10% completed growth, a human-like femoral growth plate produces ~57 more cells per percentage growth than a chimp-like growth plate. At this developmental time point, the growth plate in the human-like femur contains ~26 cells (Table 7.3). Thus, at this stage, each proliferating chondrocyte in the human-like growth plate must produce on average two additional cells per % growth. If we assume that 1% of chimpanzee or human postnatal ontogeny is approximately 55–58 days (i.e., skeletal maturity at 15–16 years old, Anderson and Green 1948; Zihlman *et al.* 2007), then each proliferating cell in a human-like femoral growth plate must produce one additional cell every 28–29 days, and its cell cycle time decreases from 9.7 to 7.7 days.

Relative Magnitude of Changes in Cellular Parameters of Bone Growth

Comparing cellular parameters to each other, the experimental simulations indicate that rates of proliferation and hypertrophy are more important and more frequently the indirect targets of selection acting on bone length in this evolutionary developmental simulation than are changes in the cell populations, condensation size, and proliferating chondrocyte height (Figure 7.8, color plate). Furthermore, the evolutionary increase in peak proliferation rates is greater than the increase in either peak or mean rates of hypertrophy, indicating that selection for increases in the peak rate of proliferation is in turn more important than selection for increases in the rate of hypertrophy. Indeed, even though both rates experience increases over the course of the simulations, increases in the rates of proliferation are inversely correlated with increases in hypertrophy (Figure 7.9). This would suggest that as chondrocytes proliferate more, they need not hypertrophy as much to achieve a similar increase in limb bone length.

Table 7.3. Differences in cellular developmental parameters between a *Pan*- and *Homo*-like femoral growth plate at 10% and 30% completed growth.

	Parameters at 10% postnatal ontogeny (mean of 24 simulation runs)		Parameters at 30% postnatal ontogeny (mean of 24 simulation runs)	
	Pan-like	*Homo*-like	*Pan*-like	*Homo*-like
Proliferating chondrocyte (P-cells) population	23.8	26.3	10.7	11.8
Proliferation rate (P-cells/cell*% growth)	5.8	7.3	7.6	10.4
New P-cells/% growth	143	200	85	124
New P-cells/day*	2.5	3.6	1.5	2.2
Cell cycle time (days*)	9.7	7.7	7.4	5.6
Hypertrophic rate (μm/% growth)	33.8	39.4	34.9	40.7
Length added (cm/% growth)	6.92	10.06	4.20	6.46
Length added (cm/day*)	0.12	0.18	0.08	0.12

*One percent postnatal ontogeny ≈56 days.

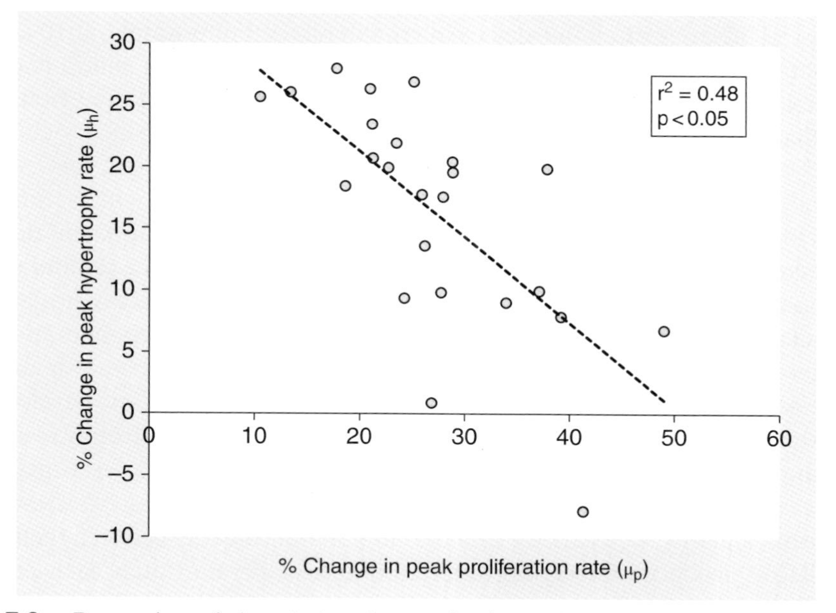

Figure 7.9. Regression of the relative changes in the peak rate of hypertrophy (μ_h) versus changes in the rate of proliferation (μ_p), illustrating the tradeoff between proliferation and hypertrophy in terms of their contributions to longitudinal bone growth.

In terms of the actual contribution of each cellular parameter to long bone elongation, even though the peak proliferation rate increases relatively more than does the rate of hypertrophy, both contribute roughly the same proportion to long bone elongation before and after the evolutionary simulation (data not shown). The fact that both proliferation and hypertrophy contribute equally to long bone elongation in the chimp- and human-like femora is a reflection that in the human-like growth plate, each additional cell that is produced in the proliferating zone still goes through the hypertrophic stage. In essence, an increase in the number of cells produced through greater proliferation always leads to a commensurate increase in the contribution to final bone length from the hypertrophic cells.

Discussion

Magnitude and Correlation of Developmental Changes

One important finding from the evolutionary simulation is how small the developmental changes within the growth plate are in relation to the magnitude of the phenotypic difference between *Pan*-like and *Homo*-like femora (Figure 7.1 and Figure 7.8, color plate). A few additional proliferating chondrocytes with a slightly faster rate of cell division, combined with a modest increase of 15–20% in the rate of hypertrophy, are enough to produce the significant differences in growth rates between sister taxa seen in Figure 7.2. The developmental model shows that any differences between the cell populations, especially in the earlier proliferation stage, are compounded over the stereotypical life cycle of chondrocytes in growth plates. For example, at 10% growth, the human-like growth plate has on average 2.5 more cells than a chimp-like growth plate (Table 7.3). Because of an increased proliferation rate at that time, however, the human-like growth plate produces 57 more cells per % growth. Since each of these cells goes through the hypertrophic stage during the same time period, this leads to an additional 3 cm (~45% more) to the femur of a human compared with a chimpanzee.

As for the direction and magnitude of change across the developmental parameters, the simulation shows that changes in anlage size, the size of the proliferating and hypertrophic chondrocyte pools, and the size of proliferative cells are not as important effectors of evolutionary change in limb bone length as are changes in the rates of proliferation and hypertrophy. The latter are broadly comparable, with an average increase of 15%. This suggests that, at least in this evolutionary developmental model, the mechanisms of chondrocyte proliferation and hypertrophy – and their underlying genetic loci – are equally likely to change under selection. Furthermore, the fact that there is a significant negative relationship between the two chondrocyte mechanisms suggests the existence of a developmental tradeoff between them – when more cells are produced through proliferation/mitosis per unit time, each cell hypertrophies less as it contributes to bone elongation. This pattern further illustrates the compounding nature of the chondrocyte life cycle in the growth plate.

Advantages and Limitations of Evolutionary Developmental Simulations

This study aimed to determine whether evolutionary developmental simulations of endochondral ossification can provide useful insights into the developmental origins of variation in limb bone length at the population level, and in models that do not lend themselves to empirical studies of a developmental genetic or cellular nature, such as primates. The simulation combined the developmental modeling of Salazar-Ciudad and Marin-Riera (2013) with the approach of Jones *et al.* (2004), in which the genetic loci for each of the developmental traits are modeled explicitly. This combined approach includes sexual reproduction and recombination, while considering the hierarchical developmental processes that link the genotype and phenotype of a complex quantitative trait such as limb bone length. Overall, the model identifies which developmental traits under genetic control change in response to natural selection acting on the phenotype, and that with large enough populations and numbers of loci, distant new fitness optima can be reached relatively rapidly, even when directional selection is weak.

As long as the developmental model is a reasonable approximation of the real cellular mechanisms of endochondral ossification, these types of simulations provide useful information on evolutionary change in genetically and developmentally intractable organisms. Of critical importance is the need for as many of the model's developmental parameters to be grounded in empirical data for the taxa of interest. In this simulation, empirical data available for humans and chimpanzees included the growth curves for their femora (Figure 7.2), as well as the approximate number of proliferating chondrocytes and size of terminal hypertrophic cells in the human distal femoral growth plate (Kember and Sissons 1976; Sissons and Kember 1977). No histomorphometric data for chimpanzee growth plates are currently available, and thus many of the starting conditions that underlie a *Pan*-like femur had to be estimated from human data.

Still, it is important to acknowledge some of the limitations of evolutionary developmental simulations. The most obvious is that it remains a theoretical enterprise. This applies not only to the developmental model, but also to evolutionary processes such as directional selection towards new optima on a fitness landscape. While the former can be validated experimentally in mammalian model organisms (see further), evolutionary change under directional selection cannot be observed directly without experimental evolution (e.g., artificial selection experiments). This limitation applies to all evo-devo studies, however, where a greater emphasis is placed on macroevolutionary changes in developmental mechanisms among distantly related organisms than on the types of microevolutionary changes illustrated in this simulation.

At the same time, this limitation is also one of the greatest strengths of evolutionary simulations. Using *in silico* approaches, one can easily change the parameters of the developmental model and/or evolutionary simulation to gauge their effects on evolutionary outcomes. For example, because the genetic loci are modeled explicitly, one can vary their number and allelic diversity to assess how much these parameters would constrain or facilitate evolutionary developmental change in morphology. Fifty loci were chosen for each developmental parameter, and the alleles for each locus were generated randomly. Despite the preliminary evolutionary simulations which likely culled some of the original allelic diversity, such a large number of loci may grossly overestimate the true number of alleles present in a real finite population (Kimura and Crow 1964), and hence the standing genetic variation and resulting rates of evolutionary change under directional selection. One could also model changes in the effect size of alleles and new mutations, to determine how often the evolution of a substantial change in morphology is associated with a small number of genes of large phenotypic effects (Chan *et al.* 2010). Finally, one could also vary the magnitude of environmental effects in each trait, reflecting varying degrees of environmental perturbations acting, for example, early *in utero* (e.g., those affecting anlage size) versus later in postnatal development (e.g., those affecting proliferation or hypertrophy). Together, these sorts of changes could offer important insights into genetic constraints on the tempo and mode of developmental and morphological evolution.

Related to this, the model presented here assumed the genetic loci for the developmental traits were not pleiotropic, and more generally that there were no correlations between the developmental traits. It is highly likely that some, if not all of the developmental traits in the model, for example, the number of cells in each population and their relationship to the original size of the anlage, are determined in part by common genetic loci. If this is the case, then these traits may coevolve. The downstream effects on rates of evolutionary change depend on the direction and magnitude of the correlation between the traits. For example, anlage size and the number of chondrocytes in the proliferation and hypertrophic zones may all be positively correlated through earlier embryonic mechanisms of mesenchymal condensation and chondrocyte differentiation (Atchley and Hall 1991; Hall and Miyake 2000). If true, then these genetic and developmental correlations could accelerate evolutionary increases in femur length through increases in the frequency of the alleles that have a positive additive effect on those traits. Simulations are thus also useful to model the impact of genetic correlations by varying the proportion of pleiotropic loci at each trait and the magnitudes of their shared effects on each developmental trait. In sum, although this evolutionary developmental simulation was relatively simple, it illustrates the utility and potential of *in silico* experiments in evo-devo (Salazar-Ciudad and Marin-Riera 2013). It also lays the foundation for more refined developmental simulations in which factors such as allelic diversity and effect size, pleiotropy, and the environment can be modeled, and their effects on the pattern and process of developmental evolution can be gauged.

Future Directions: Empirical Validations

The value of simulations for understanding the developmental basis of evolutionary differences in morphology within and between taxa can and, whenever possible, should be determined through validation against empirical data. In this case, this means validating the growth plate data against growth plate parameters in human and chimpanzee distal femora. At present these samples are not widely available for nonhuman primates, though there are suitable collections of primate embryological and fetal material in which these data could be collected, such as the Embryo Collection at the Hubrecht Laboratory in the Netherlands (Richardson and Narraway 1999).

A more important question, perhaps, is whether having these primate data would even provide an adequate measure of the value of simulations in evolutionary developmental anthropology. The evolutionary developmental simulation of limb bone growth revealed that only minimal changes to chondrocyte kinetics are required to effect significant changes in bone size within and between populations. In this model, a 10% increase in the size of the proliferating chondrocyte population translates into ~4 more cells in human-like femora (Figure 7.8, color plate). In practical terms, such a small increase probably falls within the range of error for histomorphometric measurements of primate growth plates. The same limitation applies to a 5 μm difference in terminal hypertrophic chondrocyte size. Such small differences may require impractically large sample sizes for rare primate embryological material. These limitations notwithstanding, even a few growth plates from related primates can validate whether evolutionary developmental simulations are even close when it comes to intra- and interspecific differences in chondrocyte kinetics and the organization of the growth plate. Conversely, evolutionary developmental simulations can provide information on whether it is even worth pursuing experimental studies, or whether the answers to the questions of interest will be beyond the sensitivity of current genetic and developmental assays.

A complementary approach to validating evolutionary developmental simulations is to compare them against data based on common model organisms in developmental biology. Data from rodents lend credence to the notion that significant evolutionary differences in bone length can be achieved through relatively minor tweaking of growth plate parameters. For example, gerbils have tibiae that are ~50% longer than mice at skeletal maturity, and are twice as large, but their time to skeletal maturity is roughly comparable (Rolian 2008). When one compares the developmental parameters of the distal femoral growth plate between the two species over its ontogeny, there are virtually no differences at any point in ontogeny in the numbers of cells in the proliferative zone or the size of the terminal chondrocytes (Table 7.4, Figure 7.10). In contrast, most of the length difference appears to be due to the rate of cell division in the gerbil femoral growth plate, which is up to two times as high as in the mouse (although the proliferation rate was not measured directly, see Rolian 2008 for details). In concordance with the developmental model in this study, this suggests that, rather than rates of hypertrophy or other cellular parameters, rates of proliferation are greater effectors of evolutionary change in limb bone length.

Using model organisms, one could also map observed cellular changes onto underlying genetic and molecular changes. The molecular basis of growth plate function and bone growth is increasingly well studied (reviewed in Mackie *et al.* 2008, 2011). In particular, many (though likely not all) genetic developmental pathways that control the differentiation of mesenchymal cells to chondrocytes, their proliferation, the transition from a proliferative to pre-hypertrophic state, and finally chondrocyte hypertrophy, have been documented in mice (Wagner and Karsenty 2001). In most studies, genetic engineering approaches are employed, in which genes of interest are knocked out and their effects on the organization of the growth plate and final bone morphology can be documented. High throughput genomic technology is becoming more widespread and cost-effective, and the number of fully sequenced genomes in model organisms is increasing rapidly. This means that we may soon be able to identify which of the molecular genetic pathways are differentially regulated during growth in the same tissues across related species. This would offer an additional means of empirically validating hypotheses generated from evolutionary developmental simulations.

Table 7.4. Life history, morphology, and growth plate cellular parameters in *Mus musculus* (mouse) and *Meriones unguiculatus* (Mongolian gerbil) at 7 and 28 days postnatal. See also Figure 7.10, and Rolian (2008) for details.

	Mus musculus	*Meriones unguiculatus*	*Difference (Meriones/Mus)*
Body mass at skeletal maturity (g)	23	56	2.4
Time to skeletal maturity (weeks)	12–15	12–15	–
Tibia length at skeletal maturity (mm)	11.4	16.5	1.45
P-cells at 7 days postnatal	19.5	19.5	1.00
Terminal hypertrophic cell height at 7 days (μm)	23.5	24.7	1.05
Tibial growth rate at 7 days (mm/day)	0.29	0.36	1.24
Inferred proliferation rate at 7 days postnatal (cells/day)	0.64	0.69	1.08
P-cells at 28 days postnatal	17.1	15	0.87
Terminal hypertrophic cell height at 28 days postnatal (μm)	22.7	23.3	1.03
Tibial growth rate at 7 days (mm/day)	0.13	0.23	1.77
Inferred proliferation rate at 28 days postnatal (cells/day)	0.33	0.65	1.97

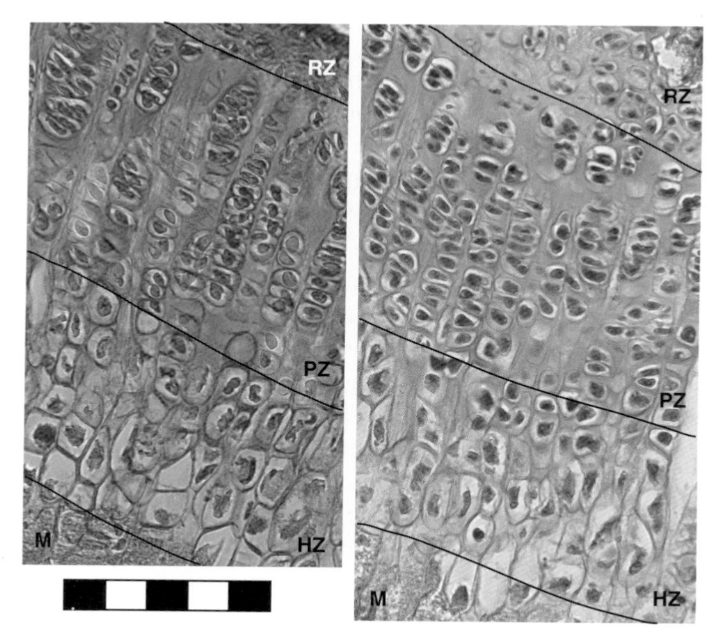

Figure 7.10. Histological sections of the distal femoral growth plates of 4-week-old gerbil (*Meriones unguiculatus*, left) and mouse (*Mus musculus*, right). The growth plates are highly similar in organization, size, and number of cells, despite the fact that the gerbil femur will grow to be 50% longer than the mouse femur over the same duration of postnatal growth (Table 7.4, see Rolian 2008). Legend: RZ, resting zone; PZ, proliferating zone; HZ, hypertrophic zone; M, metaphysis. Scale bar = 100 microns.

Conclusion

As this volume attests, there is growing interest among physical anthropologists in the genetic and developmental basis of biological differences among primates. This includes the developmental mechanisms that make humans unique among primates, not only in terms of morphology (see, e.g., Boughner, Reno, Young and Capellini, this volume) but also cognitively (see, e.g., Lalueza-Fox, Crespi and Leach, Charvet and Finlay, this volume). Yet many of the same features that make humans and non-human primates unique among vertebrates also make them empirically intractable in a developmental biology context. Fortunately, ongoing developmental research in model organisms continues to reveal the molecular and cellular mechanisms of limb bone patterning and growth with increasing detail. Here, I used this knowledge to model hominoid limb bone growth and evolution *in silico*, in order to understand how growth plate chondrocyte behavior in the long bones evolved from our last common ancestor with chimpanzees. This evolutionary developmental simulation revealed that only a small amount of "tinkering" (Jacob 1977) is necessary to produce dramatic changes in our postcranial anatomy. Although this pattern matches to some extent the available evidence from rodents, it remains to be seen how closely these simulated growth plate data will match empirical data from primates, if and when these become available. In the meantime, simulated developmental evolution provides a means of generating testable hypotheses, and narrowing down the universe of possible genetic and developmental differences that underlie skeletal diversity among primates.

Acknowledgments

Thanks to Julia Boughner and Charles Roseman for reading and providing valuable feedback on earlier versions of this manuscript. This work was supported by NSERC Discovery Grant #4181932, and the Faculty of Veterinary Medicine at the University of Calgary.

References

Abad, V., Meyers, J., Weise, M. *et al.* 2002. The role of the resting zone in growth plate chondrogenesis. *Endocrinology* 143:1851–1857.

Aiello, L., and Dean, C. 1990. *An introduction to human evolutionary anatomy*. London and San Diego: Academic Press.

Anderson, M., and Green, W. T. 1948. Lengths of the femur and the tibia – norms derived from orthoroentgenograms of children from 5 years of age until epiphysial closure. *American Journal of Diseases of Children* 75:279–290.

Atchley, W., and Hall, B. 1991. A model for development and evolution of complex morphological structures. *Biological Reviews of the Cambridge Philosophical Society* 66:101–157.

Burger, R. 1993. Predictions of the dynamics of a polygenic character under directional selection. *Journal of Theoretical Biology* 162:487–513.

Burger, R., and Lande, R. 1994. On the distribution of the mean and variance of a quantitative trait under mutation-selection-drift balance. *Genetics* 138:901–912.

Chan, Y. F., Marks, M. E., Jones, F. C. *et al.* 2010. Adaptive evolution of pelvic reduction in sticklebacks by recurrent deletion of a *Pitx1* enhancer. *Science* 327:302–305.
Cooper, K. L., Oh, S., Sung, Y. *et al.* 2013. Multiple phases of chondrocyte enlargement underlie differences in skeletal proportions. *Nature* 495:375–378.
Farnum, C. E., Lee, R., O'Hara, K., and Urban, J. P. 2002. Volume increase in growth plate chondrocytes during hypertrophy: The contribution of organic osmolytes. *Bone* 30:574–581.
Fleagle, J. G. 1999. *Primate adaptation and evolution*. San Diego: Academic Press.
Gavan, J. 1971. Longitudinal, postnatal growth in chimpanzee. *In*: G. H. Bourne (ed.) *The chimpanzee*. Basel: Karger.
German, R. Z., Hertweck, D. W., Sirianni, J. E., and Swindler, D. R. 1994. Heterochrony and sexual dimorphism in the pigtailed macaque (*Macaca nemestrina*). *American Journal of Physical Anthropology* 93:373–380.
Hall, B. K., and Miyake, T. 2000. All for one and one for all: Condensations and the initiation of skeletal development. *Bioessays* 22:138–147.
Hunziker, E. 1994. Mechanism of longitudinal bone growth and its regulation by growth plate chondrocytes. *Microscopy Research and Technique* 28:505–519.
Hunziker, E., and Schenk, R. 1989. Physiological mechanism adopted by chondrocytes in regulating longitudinal bone growth. *Journal of Physiology* 414:55–71.
Jacob, F. 1977. Evolution and tinkering. *Science* 196:1161–1166.
Jones, A. G., Arnold, S. J., and Borger, R. 2003. Stability of the G-matrix in a population experiencing pleiotropic mutation, stabilizing selection, and genetic drift. *Evolution* 57:1747–1760.
Jones, A. G., Arnold, S. J., and Burger, R. 2004. Evolution and stability of the G-matrix on a landscape with a moving optimum. *Evolution* 58:1639–1654.
Jones, A. G., Arnold, S. J., and Burger, R. 2007. The mutation matrix and the evolution of evolvability. *Evolution* 61:727–745.
Jungers, W. L. 1982. Lucy limbs – skeletal allometry and locomotion in *Australopithecus afarensis*. *Nature* 297:676–678.
Jungers, W. L. 1987. Body size and morphometric affinities of the appendicular skeleton in *Oreopithecus bambolii* (IGF 11778). *Journal of Human Evolution* 16:445–456.
Karsenty, G., Kronenberg, H. M., and Settembre, C. 2009. Genetic control of bone formation. *Annual Review of Cell and Developmental Biology* 25:629–648.
Kember, N. F. 1993. Cell kinetics and the control of bone growth. *Acta Paediatrica* 82, Suppl. 391:61–65.
Kember, N. F., and Sissons, H. A. 1976. Quantitative histology of the human growth plate. *Journal of Bone and Joint Surgery, British Volume* 58-B:426–435.
Kimura, M., and Crow, J. F. 1964. Number of alleles that can be maintained in finite population. *Genetics* 49:725–738.
Kirkwood, J. K., and Kember, N. F. 1993. Comparative quantitative histology of mammalian growth plates. *Journal of Zoology* 231:543–562.
Kronenberg, H. 2003. Developmental regulation of the growth plate. *Nature* 423:332–336.
Lovejoy, C. O., Suwa, G., Spurlock, L. *et al.* 2009. The pelvis and femur of *Ardipithecus ramidus*: The emergence of upright walking. *Science* 326(5949):71, 71e1–71e6.
Lynch, M., and Walsh, B. 1998. *Genetics and analysis of quantitative traits*. Sunderland, MA: Sinauer.
Mackie, E. J., Ahmed, Y. A., Tatarczuch, L. *et al.* 2008. Endochondral ossification: How cartilage is converted into bone in the developing skeleton. *International Journal of Biochemistry and Cell Biology* 40:46–62.

Mackie, E. J., Tatarczuch, L., and Mirams, M. 2011. The skeleton: A multi-functional complex organ. *The growth plate chondrocyte and endochondral ossification. Journal of Endocrinology* 211:109–121.

Norgard, E. A., Roseman, C. C., Fawcett, G. L. *et al.* 2008. Identification of quantitative trait loci affecting murine long bone length in a two-generation intercross of LG/J and SM/J mice. *Journal of Bone and Mineral Research* 23:887–895.

Pilbeam, D. 1996. Genetic and morphological records of the Hominoidea and hominid origins: A synthesis. *Molecular Phylogenetics and Evolution* 5:155–168.

Richardson, M. K., and Narraway, J. 1999. A treasure house of comparative embryology. *International Journal of Developmental Biology* 43:591–602.

Rolian, C. 2008. Developmental basis of limb length in rodents: Evidence for multiple divisions of labor in mechanisms of endochondral bone growth. *Evolution & Development* 10:15–28.

Ruff, C. 2003. Ontogenetic adaptation to bipedalism: Age changes in femoral to humeral length and strength proportions in humans, with a comparison to baboons. *Journal of Human Evolution* 45:317–349.

Salazar-Ciudad, I., and Jernvall, J. 2002. A gene network model accounting for development and evolution of mammalian teeth. *Proceedings of the National Academy of Sciences, USA* 99:8116–8120.

Salazar-Ciudad, I., Jernvall, J., and Newman, S. A. 2003. Mechanisms of pattern formation in development and evolution. *Development* 130:2027–2037.

Salazar-Ciudad, I., and Marin-Riera, M. 2013. Adaptive dynamics under development-based genotype-phenotype maps. *Nature* 497:361–364.

Schultz, A. 1937. Proportions, variability and asymmetries of the long bones of the limbs and the clavicle in man and apes. *Human Biology* 9:281–328.

Sissons, H. A., and Kember, N. F. 1977. Longitudinal bone-growth of human femur. *Postgraduate Medical Journal* 53:433–437.

Sockol, M. D., Raichlen, D. A., and Pontzer, H. 2007. Chimpanzee locomotor energetics and the origin of human bipedalism. *Proceedings of the National Academy of Sciences, USA* 104:12265–12269.

Stokes, I. A. F., Clark, K. C., Farnum, C. E., and Aronsson, D. D. 2007. Alterations in the growth plate associated with growth modulation by sustained compression or distraction. *Bone* 41:197–205.

Tilkens, M. J., Wall, C., Weaver, T. D., and Steudel-Numbers, K. 2007. The effects of body proportions on thermoregulation: An experimental approach. *Journal of Human Evolution* 53:286–291.

Wagner, E., and Karsenty, G. 2001. Genetic control of skeletal development. *Current Opinions in Genetics and Development* 11:527–532.

Wilsman, N., Farnum, C., Leiferman, E. *et al.* 1996. Differential growth by growth plates as a function of multiple parameters of chondrocytic kinetics. *Journal of Orthopaedic Research* 14:927–936.

Wrangham, R., and Pilbeam, D. 2001. African apes as time machines. *In*: B. Galdikas, N. Briggs, L. Sheeran, and G. Shapiro (eds) *All apes great and small*. New York: Kluwer Academic.

Zihlman, A. L., Bolter, D. R., and Boesch, C. 2007. Skeletal and dental growth and development in chimpanzees of the Taï National Park, Cote d'Ivoire. *Journal of Zoology* 273:63–73.

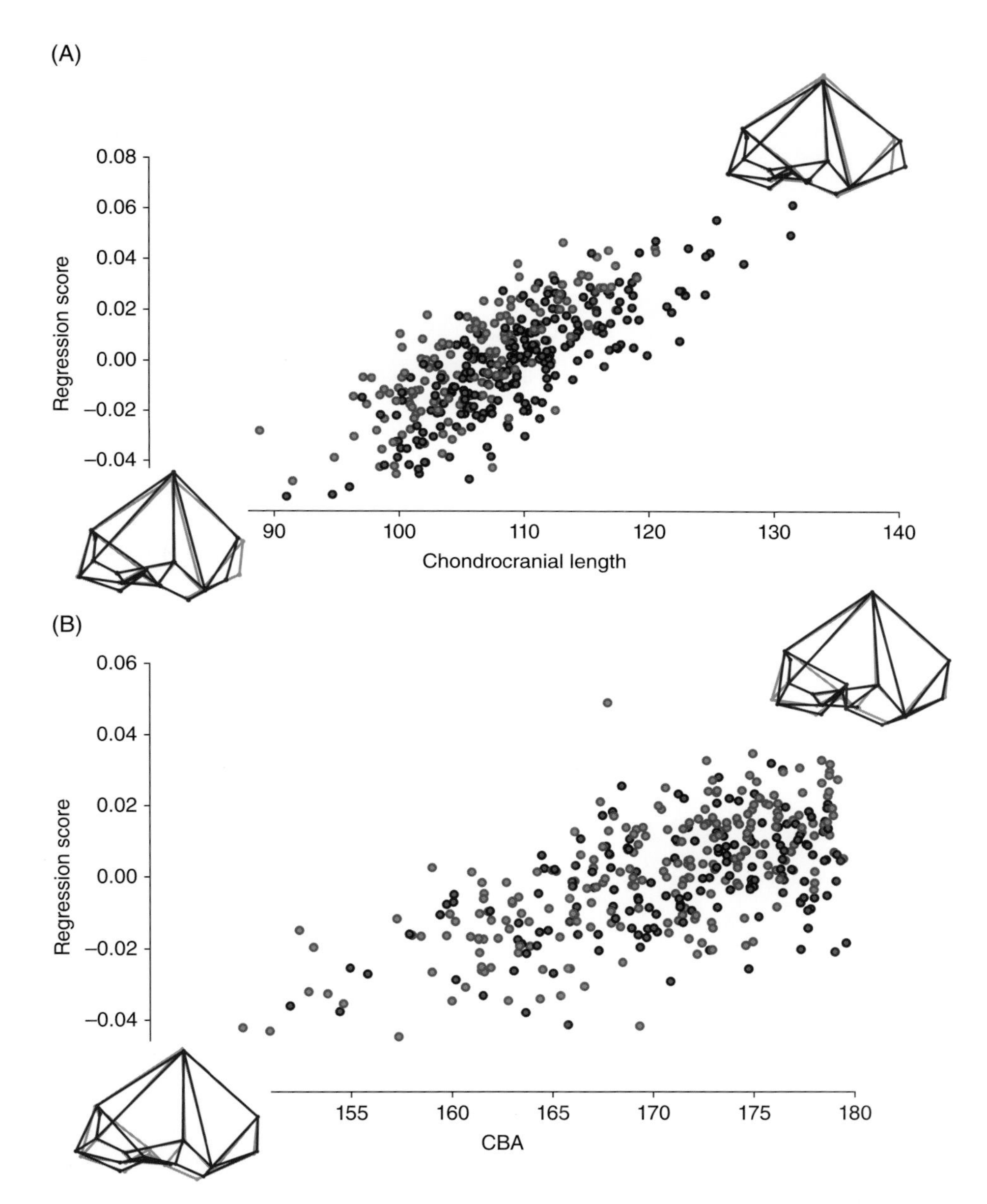

Figure 2.2. Multivariate regression of chondrocranial length (A) and cranial base angle (CBA) (B) on cranial shape in a modern human population. Regressions are pooled within sexes (orange females, blue males).

Developmental Approaches to Human Evolution, First Edition. Edited by
Julia C. Boughner and Campbell Rolian.

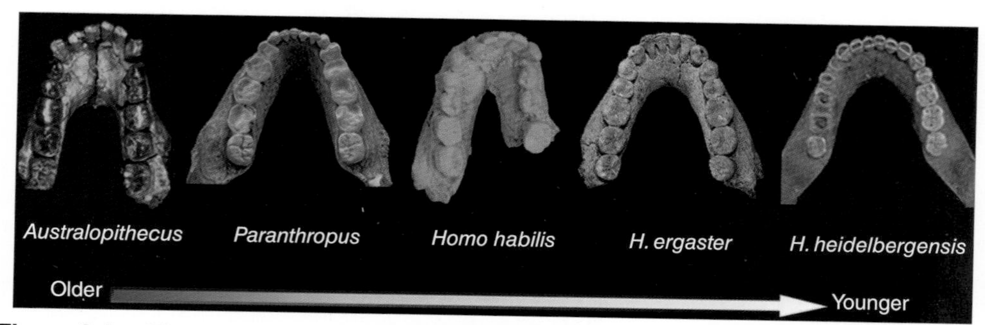

Figure 3.1. The trend in hominins towards more gracile jaws and smaller post-canine teeth, particularly from *Homo ergaster* (and *H. erectus*, not shown) onwards. Many factors, including ecological, physiological, and socio-cultural, sculpted human craniodental morphology. Explaining at the genetic level how change in tooth size and jaw robusticity occur may help to structure plausible explanations of how or why this trend happened. Images used with permission from the Smithsonian Institute and *Science* (295:15 Feb) 2002.

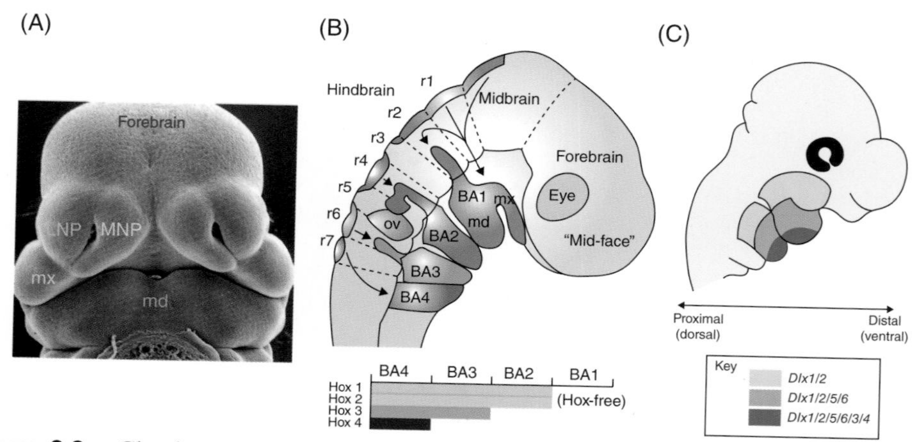

Figure 3.3. Classic stages of jaw morphogenesis in a mouse embryo model (embryonic day 10–11). The maxillary and mandibular prominences form (A) via cranial neural crest cell migration and proliferation (B). The identity of each prominence is patterned by genes (e.g., nested expression of *Dlx1-6*) (C) to eventually form distinct upper and lower jaw skeletons. Adapted with permission from (A) Lippincott Williams & Wilkins (2007); (B) *Nature Reviews Neuroscience* 8 (2007); (C) Minoux and Rijli, *Development* (2010). Legend: BA, brachial arch; LNP, lateral nasal prominence; md, mandibular prominence; MNP, medial nasal prominence; mx, maxillary prominence; r, rhombomere.

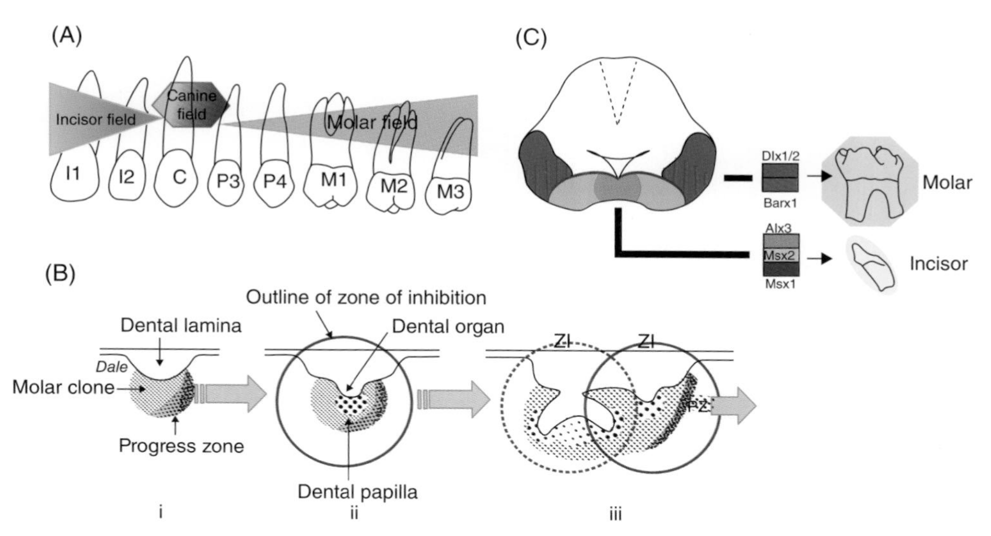

Figure 3.4. Progression of the models crafted to explain how tooth identity as well as initiation time are determined, first via the Field Theory (A), the Clone Model (B), and more recently the Molecular Field Model (C). Based on Ten Cate (1994) and Cobourne and Sharpe (2003), published in Boughner and Hallgrímsson 2008.

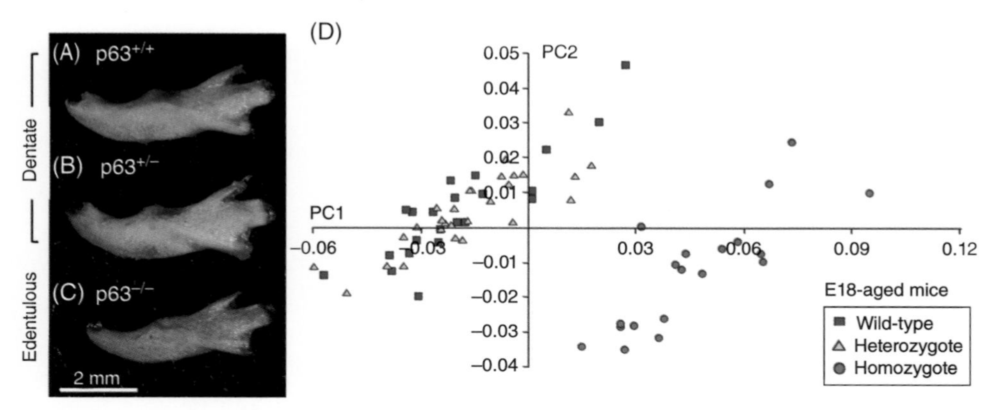

Figure 3.5. Compared to dentate (A [$p63^{+/+}$], B [$p63^{+/-}$]) mice, mandible (i.e. dentary bone) morphogenesis occurs virtually unperturbed in toothless (C [$p63^{-/-}$]) mice by embryonic day (E) 18 (birth occurs around E21). The toothless mandibles lack alveolar bone, which either never forms or is resorbed soon after forming. After a principal component analysis (PCA), the mandible shapes of the toothless $p63^{-/-}$ mice appear to match the mandible shapes of dentate mice that are 1–2 days advanced in their development (PC1): thus an E18-aged $p63^{-/-}$ mandible looks like a normal E16–17-aged mandible. Otherwise shape variance is comparable (PC2) to that of toothed mice littermates. Adapted from Paradis *et al.* (2013).

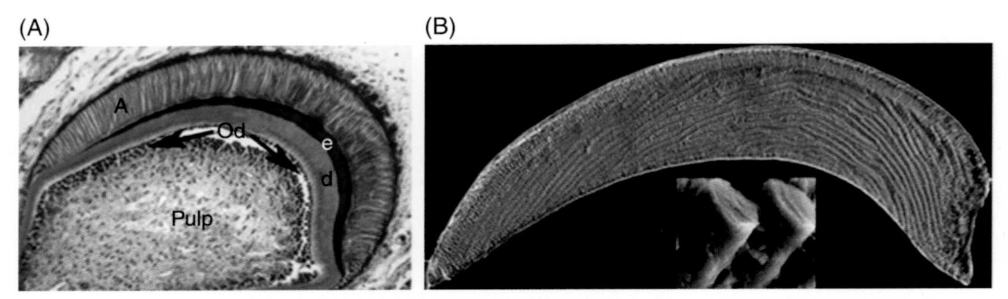

Figure 4.1. (A) Coronal section of a mouse incisor at postnatal day 2 showing the secretory ameloblast layer (A), enamel (e), odontoblast layer (Od), dentine (d), and pulp chamber. (B) Scanning electron micrograph of a coronal section of a mouse incisor at approximately 3 weeks of age showing the architecture of the mineralized enamel. Inset in B shows detail of enamel prisms (or rods) which contain the enamel crystallites.

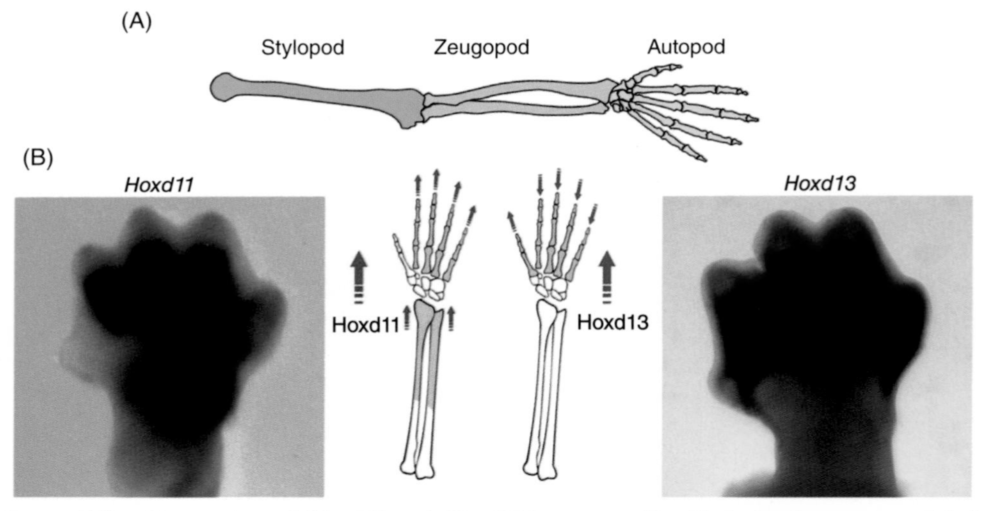

Figure 5.2. Expression of *Hoxd11* and *Hoxd13* in mammalian limb development and their proposed role in regulating hominoid forearm and digit proportions. (A) Tetrapod limbs are divided into three segments: stylopod, zeugopod, and autopod. (B) *In situ* hybridization of E13.5 mouse forelimbs indicated that *Hoxd11* is expressed in the distal zeugopod (forearm) and posterior digits, while *Hoxd13* is expressed in all five digits. Each *Hox* gene has its individual effect on skeletal growth, with *Hoxd13* promoting less growth than does *Hoxd11*. Thus, increased activity of *Hoxd13* will result in elongation in regions where it is expressed alone (thumb). However, when in competition, *Hoxd11* can act to ameliorate *Hoxd11*'s superior growth-promoting action to produce shorter fingers.

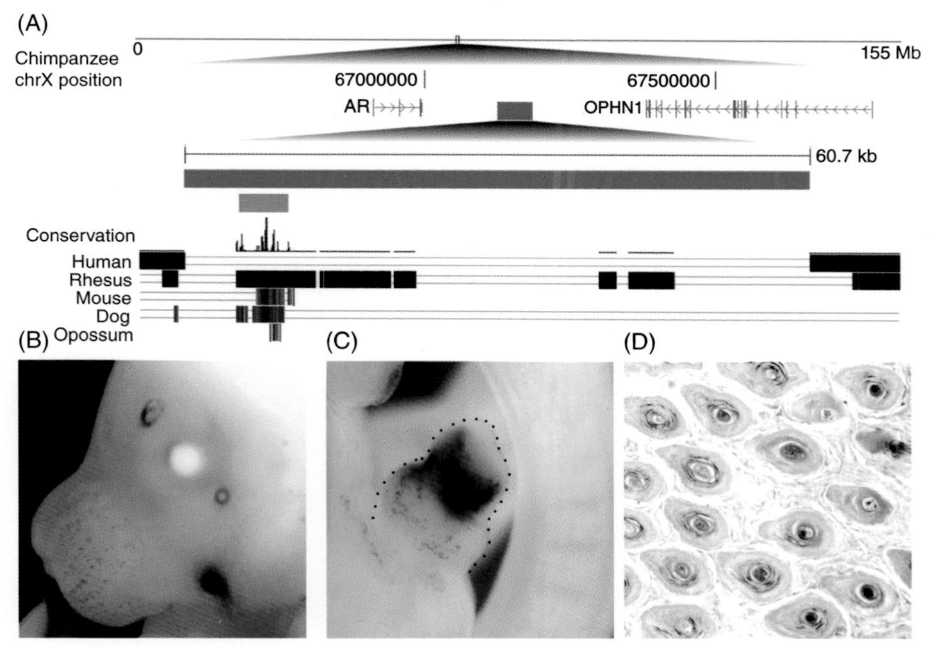

Figure 5.4. Human-specific deletions of an AR enhancer are associated with the loss of vibrissae and penile spines. (A) A 60.7-kb human-specific deletion (red bar) near AR has eliminated nearly 5 kb of highly conserved mammalian sequence. The blue bar indicates the 4.8 kb chimpanzee sequence tested for enhancer activity. The chimpanzee and homologous mouse sequences drive gene expression in embryonic (E 16.5) vibrissae (B) and genital tubercle (C). (D) The mouse enhancer is also functional in the mesodermal papillae of postnatal (day 60) penile spines.

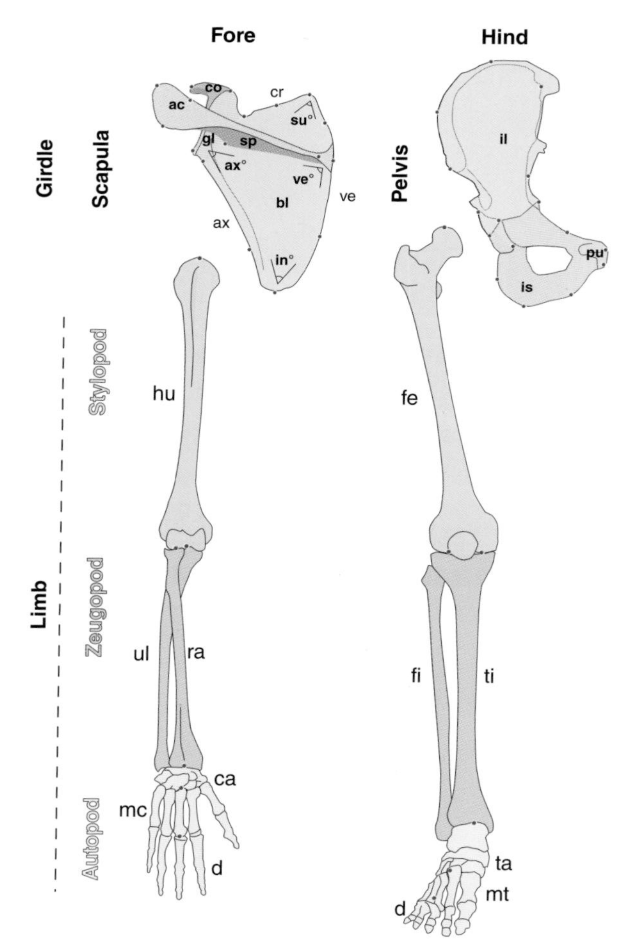

Figure 6.1. Appendicular skeletal anatomy: Primate appendages consist of a girdle attached to a limb proper. (Top-Left) The scapula (of the pectoral girdle in the forelimb) primarily consists of a roughly triangular blade (blue), traversed by a long thin spine ending in the acromion (red), a neck and head containing the glenoid (yellow), and a cranially positioned coracoid (green). The blade is bounded by three borders: (i) axillary, (ii) cranial, and (iii) vertebral. Shape can be described by several angles: inferior (in), superior (su), axillary (ax), vertebral (ve). (Top-Right) The innominate (of the pelvic girdle in the hindlimb) consists of three elements, a cranially located ilium (blue), a medial-ventral facing pubis (yellow), and a caudally facing ischium (red), all of which are fused in the middle forming circular concavity called the hip socket or acetabulum. (Bottom) Each forelimb (left) or hindlimb (right) consists of skeletal elements arranged in three "developmental" compartments. Most proximal is the stylopod (blue) containing the humerus or femur. Intermediate is the zeugopod (red) containing the radius/ulna or tibia/fibula. Most distal is the autopod (yellow) containing the carpals (ca), metacarpals (mc), and digits (d) of the hand or tarsals (ta), metatarsals (mt), and digits (d) of foot. Red circles indicate representative landmarks used to generate comparative and evolutionary morphospaces.

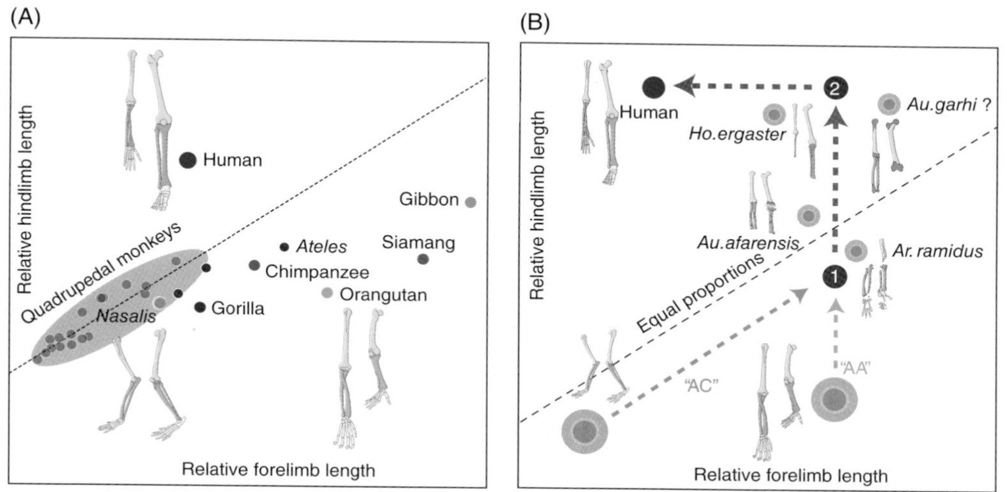

Figure 6.7. Limb comparative and evolutionary morphospace: Plots of the forelimb and hindlimb length relative to trunk length (original data from Schultz (1926)). (A) Extant primate plot shows that humans and suspensory primates differ significantly from the relationship within quadrupedal monkeys. Figures adapted from Young *et al.* (2010). Humans differ from all other primates in having relatively short forelimbs and long hindlimbs. (B) Alternative reconstructions of the LCA differ in whether this morphotype was similar to living apes ("AA"), or was more similar to a quadrupedal monkey. Regardless, fossil hominins suggest a two-step process of mosaic intermembral proportion evolution. Figure adapted from Young *et al.* (2010).

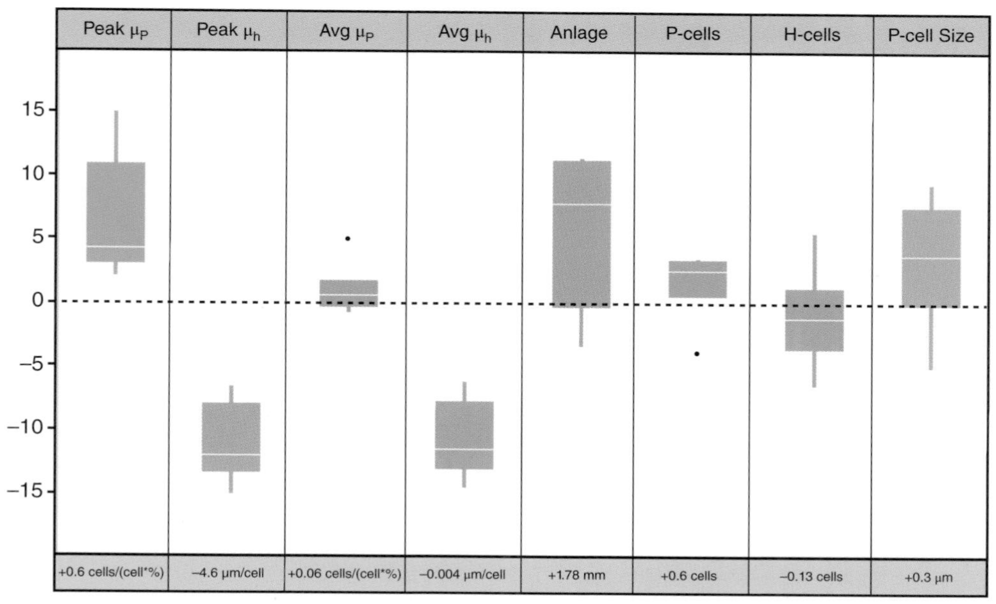

Figure 7.7. Relative changes (in percentage) in the magnitude of the starting values for the developmental parameters after 1000 generations of stabilizing selection for a chimpanzee-like optimum in femur length (Table 7.1). The lower boxes indicate what the relative changes mean in terms of the values for each parameter. Legend: μ_p, chondrocyte proliferation rate; μ_h, chondrocyte hypertrophy rate; anlage, length of the anlage; P-cells, size of the proliferating chondrocyte pool at birth; H-cells, size of the hypertrophic chondrocyte pool at birth; P-cell size, size (in μm) of the proliferating cells in the direction of longitudinal growth.

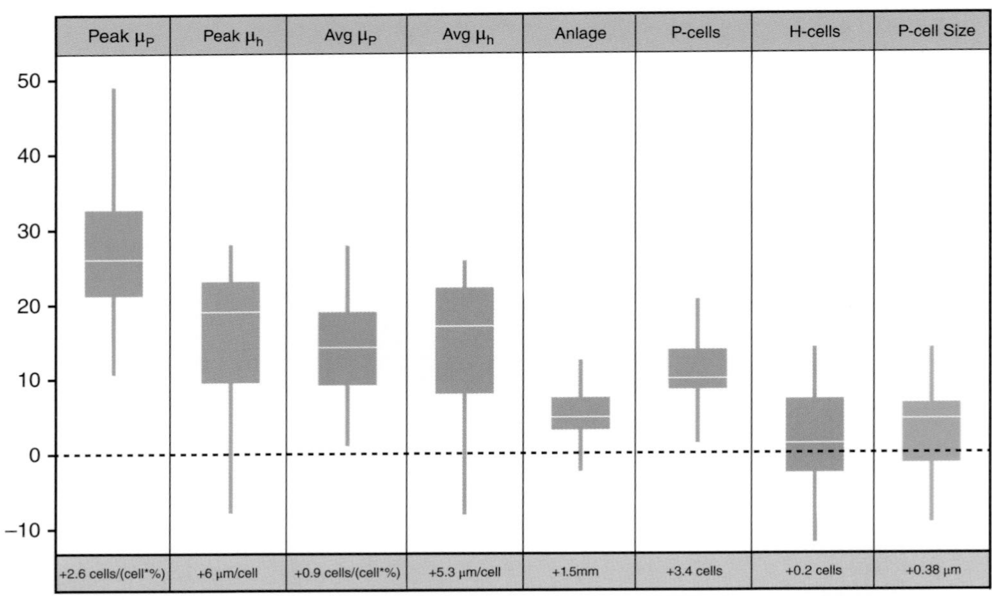

Figure 7.8. Relative changes (in percentage) in the magnitude of the developmental parameters after 1000 generations under directional selection for a new human-like optimum in femur length (Table 7.2). Legend as in Figure 7.7.

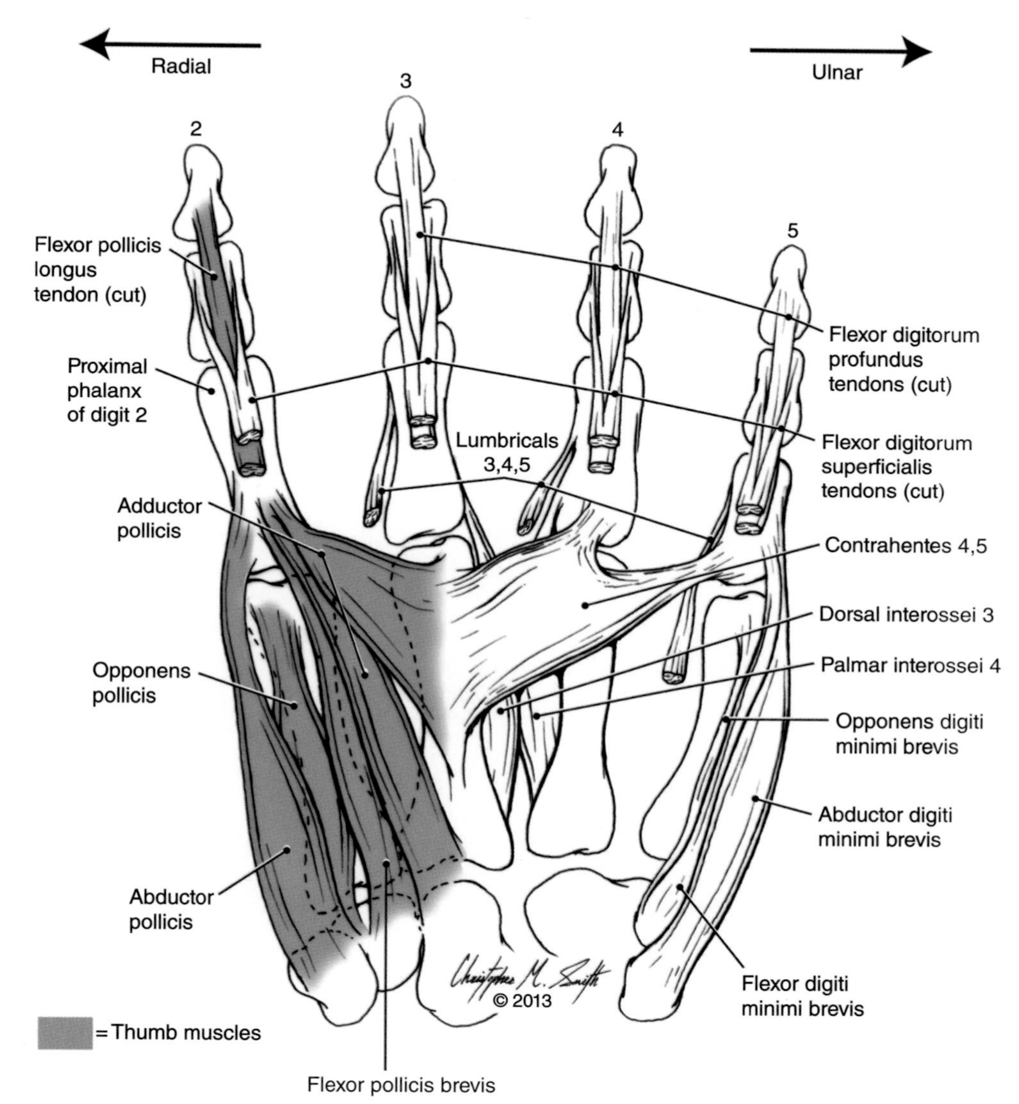

Figure 8.4. Four-digit hand of a trisomy 18 modern human newborn, showing that the muscles that usually attach to the thumb in karyotypically normal modern humans now attach to the digit that has a homeotic identity of digit 2 and develops from the anlage of digit 2, but that is now the most radial digit due to the loss of the thumb. Illustration by Christopher Smith (Johns Hopkins University, University School of Medicine Department of Art as Applied to Medicine).

(A)

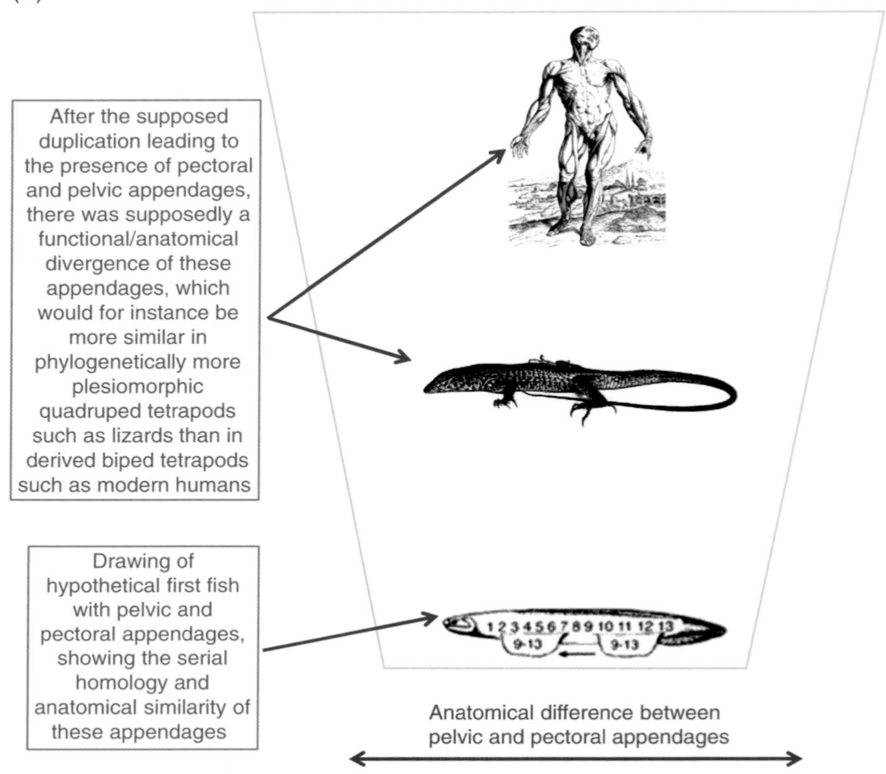

Figure 8.5. (A) Simplified scheme illustrating the "serial homology followed by functional/anatomical divergence" hypothesis often shown in textbooks and followed in more technical papers, particularly within the fields of developmental biology and evo-devo. The picture of the hypothetical fish is modified from Shubin *et al.*'s (1997) scheme showing the origin and evolution of paired appendages. According to that scheme, establishment of serially homologous appendages was proposed to result from gene cooption during the evolution of Paleozoic vertebrates. That is, *Hox* genes were initially involved in specifying regional identities along the primary body axis, particularly in caudal segments, and then during the origin of jawed fish there was a cooption of similar nested patterns of expression of *HoxD* genes in the development of both sets of paired appendages (numbers shown within the fish body). According to this scheme the cooption may have happened in both appendages simultaneously, or *Hox* expression could have been initially present in a pelvic appendage and been coopted in the development of an existing pectoral outgrowth (arrow below fish body; the pictures of the other taxa are modified from Diogo and Abdala 2010 and references therein). (B) The evolutionary history of the pelvic and pectoral appendages was more complex than the "serial homology followed by functional/anatomical divergence" hypothesis suggests. This is because it was more likely the result of a complex interplay between ontogenetic, functional, topological, and phylogenetic constraints leading to cases of anatomical divergence followed by cases of anatomical convergence ("similarity bottlenecks"). This is exemplified in this simplified scheme of the evolutionary muscle transitions leading to modern humans. Figure reused with permission from Diogo and Ziermann (2015).

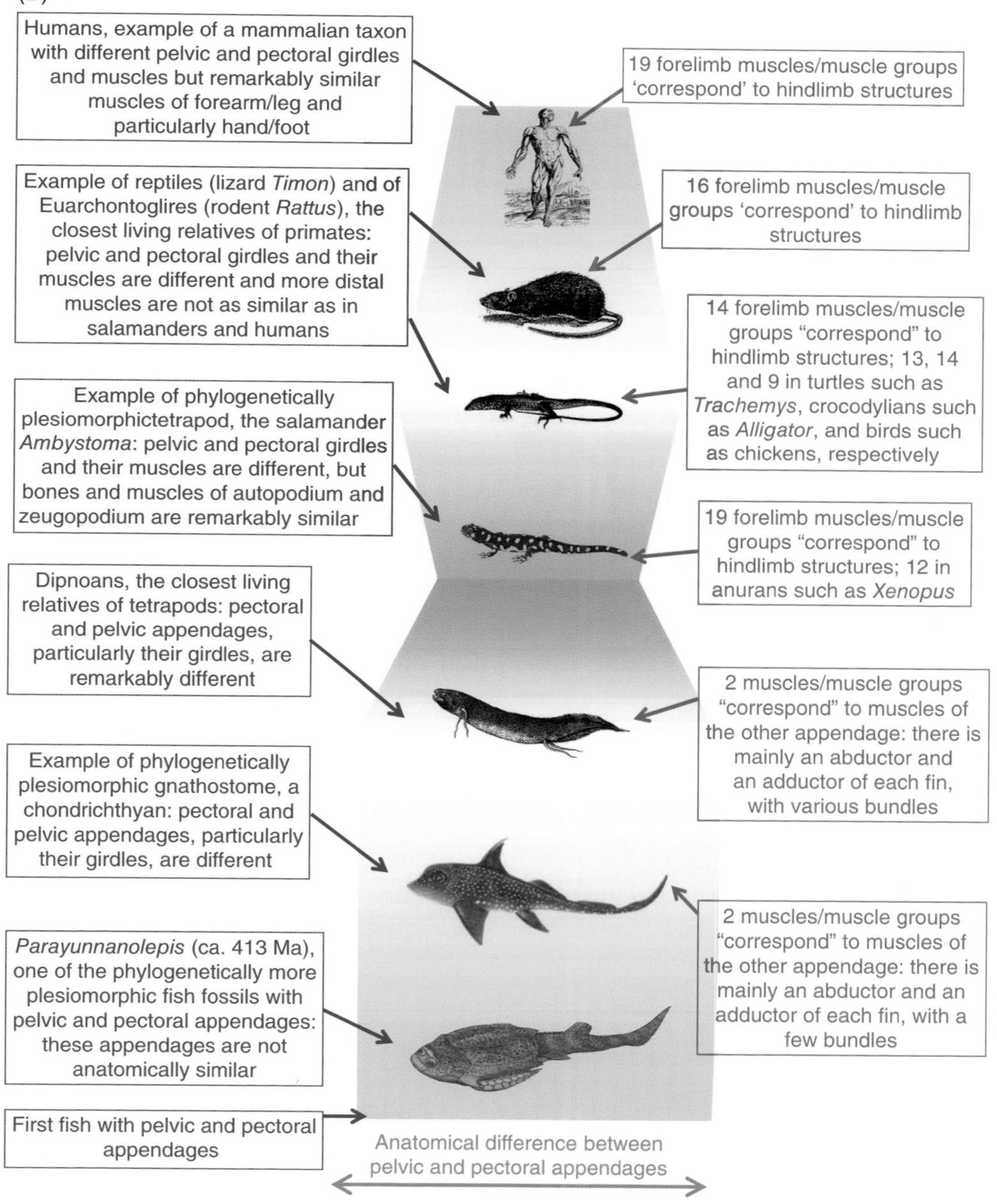

Figure 8.5. (*Continued*).

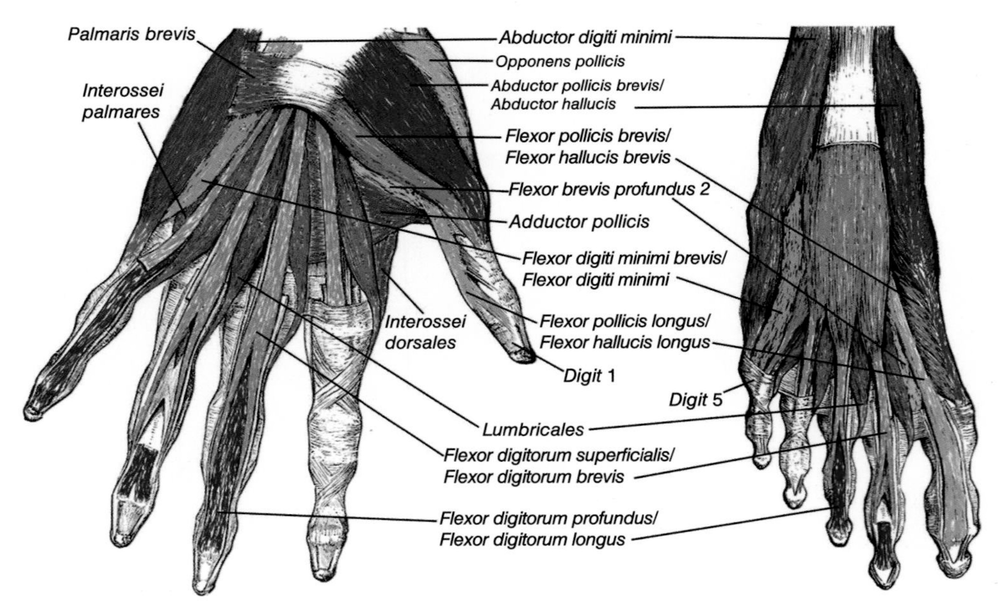

Figure 8.6. Superficial musculature of the modern human hand seen in palmar view (on the left) and of the modern human foot seen in plantar view (on the right). There are striking similarities between the muscles of the autopodium (hand/foot) of the forelimb and hindlimb of modern humans. Modified from Diogo *et al.* 2013c.

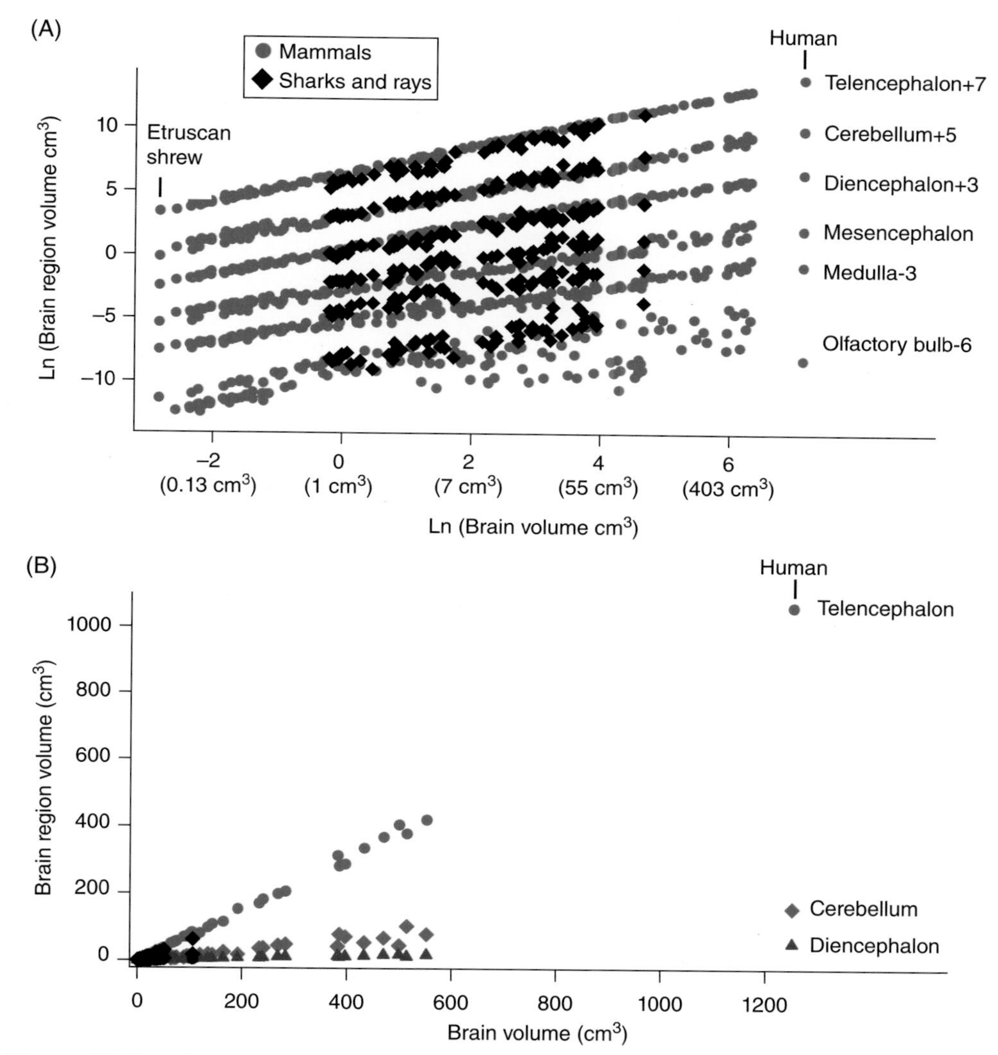

Figure 10.2. (A) The natural-logged values of brain region volumes are regressed against the natural-logged values of overall brain size in mammals, as well as in sharks and rays. Major brain regions strongly covary with overall brain size. One important exception is the olfactory bulb, which varies relatively independently of other brain regions. Brain regions also expand allometrically with respect to the other brain regions. In particular, the telencephalon and cerebellum become disproportionately enlarged relative to other brain regions, such as the medulla, as brains expand. For example, the Etruscan shrew has a brain volume around 50 mm^3. Its telencephalon and its medulla occupy approximately 51% and 20% of its brain, respectively. The human has a brain volume of approximately 1250 cm^3. Its telencephalon occupies approximately 85% of its brain but its medulla occupies less than 1% of its brain. (B) To appreciate the allometry of brain region volumes, the telencephalon, cerebellum, and diencephalon volumes are plotted against overall brain size in mammals and in sharks. The telencephalon becomes disproportionately larger than the cerebellum, which in turn becomes disproportionately larger than the diencephalon. A constant was added to the brain region volume in (A) to distinguish brain region volumes on the same graph. These constants are listed adjacent to each named structure. Data are from Stephan *et al.* 1981; Reep *et al.* 2007; and Yopak *et al.* 2009.

Figure 10.3. Neuron numbers per mm^2 of cortical surface area in a galago (*Otelemur crassicaudatus*) show that neuron numbers under a unit of cortical surface area increase across its rostro-caudal axis. Data are from Collins *et al.* 2010.

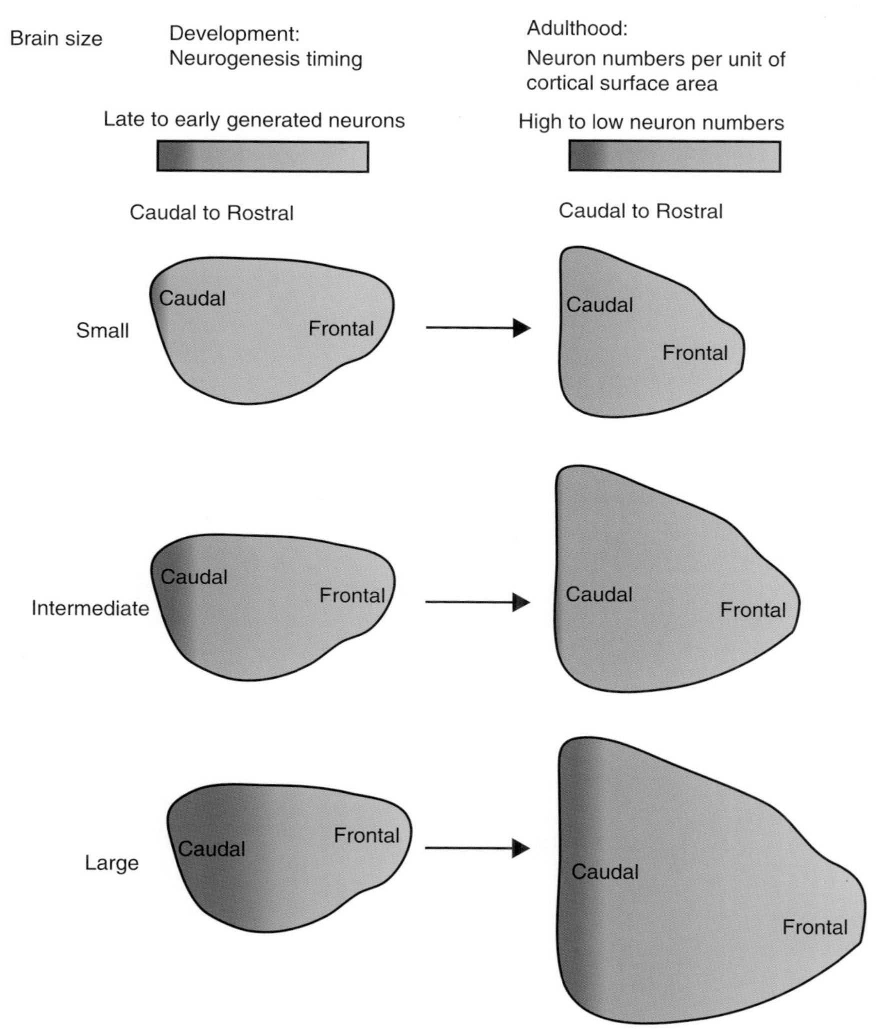

Figure 10.4. The diagram represents the gradient in neurogenesis timing in the isocortex during development and the gradient in neuron numbers under a unit of cortical surface area in adulthood. Gradients in neurogenesis timing vary across species, which would account for systematic variation in neuron numbers under a unit of cortical surface area in adulthood. Large-brained primate species would exhibit a greater disparity in neurogenetic schedules between the rostral and caudal poles leading to a greater disparity in neuron numbers per unit of cortical surface area across the isocortex in adulthood (Charvet *et al.* 2015).

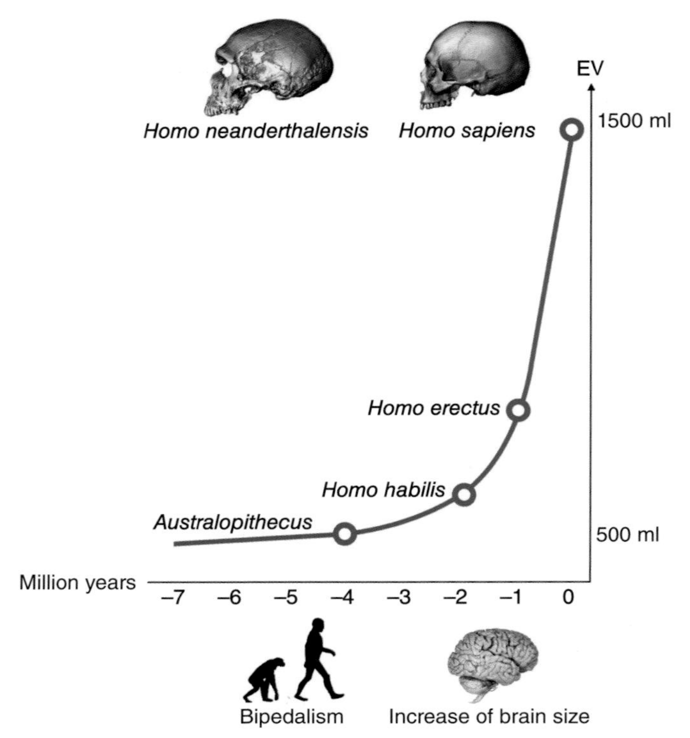

Figure 11.1. Brain-size evolution based on endocranial volumes. Evolutionary increases of brain size occurred within the genus *Homo*, particularly within the species *Homo erectus* and its descendants. Much of this dramatic increase, however, can be related to evolutionary increases in body size.

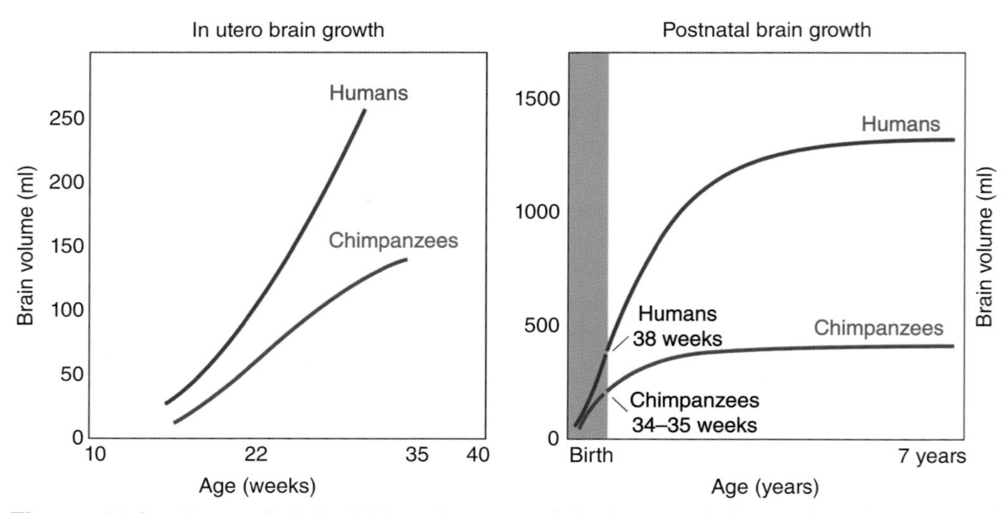

Figure 11.3. Prenatal (left side) and postnatal brain growth in modern humans and chimpanzees. Prenatal data are based on ultrasound (US) images of fetal brain development. The right panel shows endocranial volumes measured from computed tomographic scans; fetal US data (gray background) are the same as in left panel. Data based on Neubauer *et al.* (2012); Neubauer and Hublin (2012); Sakai *et al.* (2012).

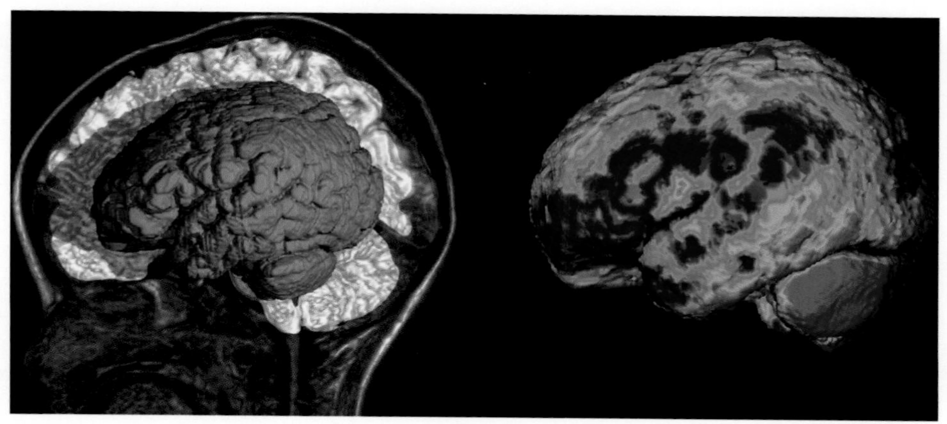

Figure 11.4. Globularization phase in modern humans. During early postnatal development, human brains do not only change their size, but also their shape. Magnetic resonance image scans of neonatal brain (blue) compared to a three year old. On the right hand side, the amount of shape change color coded from blue (small shape differences) to red (large shape differences). Developmental shape changes are most dramatic in the cerebellum, and the parietal lobe. Figure created using MRI data from ALBERT brain atlas (Gousias *et al.* 2012, 2013).

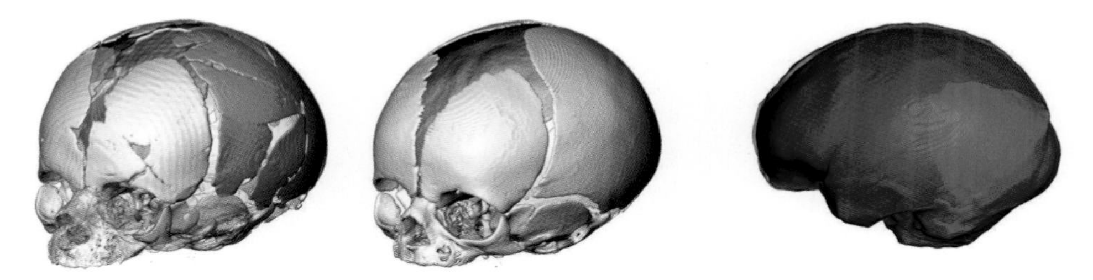

Figure 11.5. Virtual reconstruction of a Neanderthal neonate (left; based on Mezmaiskaya – Gunz *et al.* 2012) compared to the computed tomographic scan of a modern human baby (middle). Whereas Neanderthal facial characteristics like a large projecting face, large nose, and tall orbits are already visible, the braincases and endocasts (right panel) of Neanderthals (red) and modern humans (blue) are almost identical at the time of birth. The pronounced endocranial shape differences between *Homo sapiens* and *Homo neanderthalensis* develop in a postnatal globularization phase directly after birth.

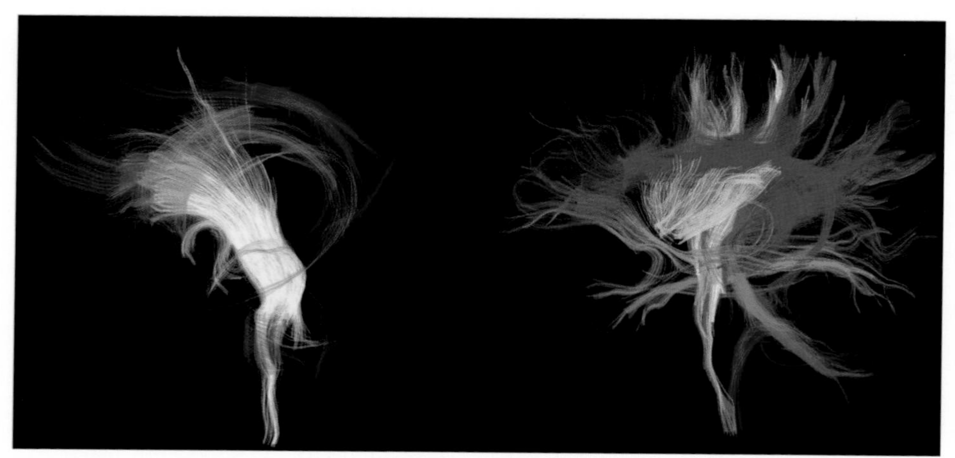

Figure 11.7. The brain's internal architecture – symbolized here by white matter connections – is the result of precisely timed sequences of synaptogenesis and the subsequent pruning of connections. However, the exact network topology is neither intrinsically predetermined nor static. Clinical data show that timing and rate of growth affect the wiring pattern of the brain. Fiber bundles (drawn in different colors) of white matter connections of a fetus in gestational week 17 (left panel), and a 3 year old (right panel). Figure created using MRI data are from the BrainSpan Atlas of the Developing Human Brain (Miller *et al.* 2014; http://brainspan.org).

Figure 12.3. The excavation protocol designed to obtain uncontaminated Neanderthal samples for DNA analysis at the El Sidrón site (Asturias, Northern Spain). The initial retrieval of the key amino acid substitutions at the *FOXP2* gene was done with two bone samples belonging to two different Neanderthal males from this site.

Chapter 8

Origin, Development, and Evolution of Primate Muscles, with Notes on Human Anatomical Variations and Anomalies

Rui Diogo[1] and Bernard Wood[2]
[1] *Department of Anatomy, Howard University College of Medicine, Washington, DC, USA*
[2] *Center for the Advanced Study of Hominid Paleobiology, Department of Anthropology, George Washington University, Washington, DC, USA*

Introduction

Most studies of the gross morphology of the soft tissues of primates are either in-depth investigations of a single structure or organ, or investigations that focus on a single taxon (reviewed in Gibbs *et al.* 2002 and Diogo and Wood 2012a). Recently, we undertook a detailed comparative study of the head, neck, pectoral, and forelimb myology of each of the major primate higher taxa (Diogo and Wood 2012a) including the 18 genera listed in Table 8.1, and we have now embarked on a second comparative study that focuses on the muscles of the trunk, pelvis, and lower limb. Our studies of primates are part of a long-term project to investigate the comparative anatomy, homologies, evolution, and development of the striated muscles of all of the major groups of vertebrates (e.g., Diogo 2004a, 2004b, 2005, 2007; Diogo and Abdala 2007, 2010; Diogo *et al.* 2008a, 2008b, 2009a, 2009b, 2012b). We have also presented the results of these analyses in the form of anatomical atlases of the apes (e.g., Diogo *et al.* 2010, 2012a, 2013a, 2013b), and reported the results of comprehensive parsimony and Bayesian cladistic analyses of the myology of these taxa (e.g., Diogo and Wood 2011, 2012b; Diogo *et al.* 2012b, 2013d). The most parsimonious tree obtained from the analysis of the 166 head, neck, pectoral, and upper limb myological characters in the 18 primate genera and in the outgroups (Rodentia: *Rattus*; Scandentia: *Tupaia*; and Dermoptera: *Cynocephalus*) was congruent with the molecular trees obtained recently by Fabre *et al.* (2009), Arnold *et al.* (2010), and Perelman *et al.* (2011). Most of the major primate clades were supported by high parsimony bootstrap support values (BSV) and/or Bayesian credibility support values (CSV) (e.g., six of the 15 non-hominoid primate clades have CSV and/or BSV≥98). Our analysis

Developmental Approaches to Human Evolution, First Edition. Edited by
Julia C. Boughner and Campbell Rolian.

Table 8.1. Summary of the total number of mandibular, hyoid (not including the small facial, extrinsic muscles of the ear), branchial, hypobranchial, pectoral, arm, forearm, and hand muscles in adults of some key primate genera. Data are from evidence provided by our own dissections and comparisons and from a review of the literature (Diogo and Wood 2011, 2012a); note that in some cases there are insufficient data to clarify whether a particular muscle is usually present, or not, in a taxon (e.g., the number of branchial muscles of *Gorilla* is given as 15 to 16 because it is not clear if the salpingopharyngeus is usually present, or not, as a distinct muscle in the members of this genus).

	Lemur	*Propithecus*	*Loris*	*Nycticebus*	*Tarsius*	*Pithecia*	*Aotus*	*Saimiri*	*Callithrix*	*Colobus*	*Cercopithecus*	*Papio*	*Macaca*	*Hylobates*	*Pongo*	*Gorilla*	*Pan*	*Homo*
Mandibular muscles	8	8	8	8	8	8	8	8	8	7–8	7–8	8	8	8	7	8	8	8
Hyoid muscles (not extrinsic ear)	25	24	24–26	26	24	22	23	21	22	24–25	26–27	25–26	26	26	26	26	26	27
Branchial muscles	14–16	14–16	15–17	14–17	16–17	14–16	14–16	15–16	14–16	13–14	16	14–15	16	17	14–15	15–16	15	16
Hypobranchial muscles	12	12	12–15	12–15	12	12–13	11–12	12	13	12	12	13	13	13	12–13	13	13	13
Pectoral muscles	17	15–16	16	16	17	15	16	16	17	16	17	17	17	14	15	14	14	14
Arm muscles	5	5	5	5	5	5	5	5	5	5	5	5	5	5	5	5	5	4
Forearm muscles	19	19	18	18	19	19	19	19	19	19	19	19	19	19	18	18	19	20
Hand muscles	30	30	30	34	32–36	22	22	22	21	27	27	27	27	27	20	20	26	21
Total number of muscles	**130–132**	**127–130**	**128–135**	**133–139**	**133–138**	**117–120**	**118–121**	**118–119**	**119–121**	**123–126**	**129–131**	**128–130**	**131**	**129**	**117–119**	**119–120**	**126**	**123**

was the first cladistic study based on a large morphology-based data matrix to provide compelling levels of support for the chimp-human clade (BSV 75, CSV 94). Apart from the utility of mapping myological characters on osteological or molecular phylogenies to address evolutionary questions, the results of our studies of primates, as well as of other vertebrates, suggest that myological characters can be used to generate phylogenies (see recent review of Diogo and Wood 2013).

The goal of this chapter is to draw together the broader evolutionary and developmental implications of our study of the musculature of primates. As well as increasing awareness of primate morphological evolution and evolutionary developmental biology in general, we hope that this synthesis will encourage others to analyze the myology, development, and evolution of other major mammalian and vertebrate groups.

As many of the issues discussed below concern the development of muscles and the spatial associations between them and bones/cartilages, we include a short introduction before discussing these issues. For example, with respect to limb development, fundamental ontogenetic characteristics are shared among distantly related tetrapod taxa (e.g., cartilage morphogenesis is remarkably similar across a wide range of tetrapod groups in that signaling for limb bone development usually precedes that for muscle development, cartilage in general is present prior to muscle formation (e.g., Fabrezi *et al.* 2007), and somatic limb muscle progenitor cells apparently do not carry intrinsic positional information (e.g., Duprez 2002)). Some studies suggest that the presence of the first tissues formed during limb development (condensations that will give rise to bones) may provide the positional signaling for the subsequent development of soft tissue (see Diogo and Abdala 2010). But muscles can also play an important role in at least some aspects of skeletal morphogenesis (e.g., muscle contraction might help regulate chondrocyte intercalation and skeletal elongation, facilitating coordination between muscular and skeletal development (e.g., Shwartz *et al.* 2012)) and muscle activity is required for survival of the tendon blastemas (undifferentiated cell masses) (Brand *et al.* 1985). However, it remains unknown whether tetrapods share a general, predictable spatial correlation between limb bones and muscles, and, if such a correlation exists, whether it reflects any link between the signaling involved in skeletal and muscle morphogenesis (e.g., Kardon 1998; DeLaurier *et al.* 2006; Blitz *et al.* 2013).

Concerning the head and neck muscles, studies using rhombomeric quail-to-chick grafts to investigate the influence of hindbrain segmentation on craniofacial patterning (e.g., Köntges and Lumsden 1996), showed that each rhombomeric neural crest cell population remains coherent throughout ontogeny, with rhombomere-specific matching of muscle connective tissue and their attachment sites for most head and neck muscles. One point not always well understood is that the specificity of muscle attachments referred to by Köntges and Lumsden (1996) relates to the connective tissue/fasciae associated with the muscles, and not with the ontogenetic and/or phylogenetic origin of the muscles. For example, the avian hyobranchialis ("branchiomandibularis" *sensu* Köntges and Lumsden 1996) is a branchial muscle, but it is attached anteriorly to hyoid (second arch) crest-derived skeletal domains (i.e., the retroarticular process of the mandible) (Diogo and Abdala 2010 and see next section).

This is because the anterior part of this muscle is associated with connective tissue/fasciae that is precisely derived from hyoid, and not mandibular, crest cells. The hyobranchialis was the *only* muscle studied by Köntges and Lumsden (1996) that derives its connective tissue from more than one branchial arch, because unlike its anterior component, the posterior portion is associated with connective tissue/fascia derived from the third and fourth arches. Accordingly, it inserts onto third and fourth arch crest-derived skeletal domains.

Other examples that illustrate the model proposed by Köntges and Lumsden (1996) concern the hypobranchial muscles hyoglossus, hypoglossus, and genioglossus. As explained by Diogo and Abdala (2010), previous mapping studies have shown that the myocytes and the innervation of these three muscles are derived from the much more posterior axial levels of the first somites (e.g., the fist six somites). However, as noted by Köntges and Lumsden (1996, pp.3240–3241) their "skeletal attachment fascia are derived from the more anterior axial levels of cranial neural crests." That is why the genioglossus and hypoglossus of birds, for instance, are attached to the paraglossals and the ventral basihyoid *sensu* Köntges and Lumsden (1996), which are derived from mandibular arch crest derived from the posterior midbrain. And that is why the hyoglossus ("ceratoglossus" *sensu* Köntges and Lumsden 1996), which is also ontogenetically and phylogenetically derived from the geniohyoideus (as are the genioglossus and hypoglossus: see Diogo and Abdala 2010), is attached to hyoid (second arch) crest-derived skeletal elements. Thus, the attachments of these three hypobranchial muscles are primarily determined by the origin of the connective tissues/fasciae with which they are associated. There are a few exceptions to the model proposed by Köntges and Lumsden (1996). For example, at least some facial muscles of mammals, which are derived from the second (hyoid) arch and are apparently associated with connective tissue/fascia also derived from this arch, move into midfacial and jaw territories populated only by frontonasal and first arch crest cells (e.g., Noden and Francis-West 2006). Also, Prunotto *et al.* (2004) have shown that the facial muscles behave, in terms of C-met mutations, as hypaxial migratory muscles. Thus, contrary to most other head muscles, with the exception of the hypobranchial muscles, the facial muscles are absent in organisms with C-met mutations. This suggests that these mammalian muscles migrate far away from their primary origin (see review of Diogo and Abdala 2010).

In the following sections we discuss in more detail the origin, ontogeny, and evolution of the head, neck, pectoral, and forelimb muscles.

The Developmental and Evolutionary Origins of the Head, Neck, Pectoral, and Forelimb Muscles

Head and Neck Muscles

According to the myological nomenclature used by Diogo and Abdala (2010) the main groups of head and neck muscles are: external ocular, mandibular, hyoid, branchial, epibranchial, and hypobranchial, corresponding to those proposed by Edgeworth

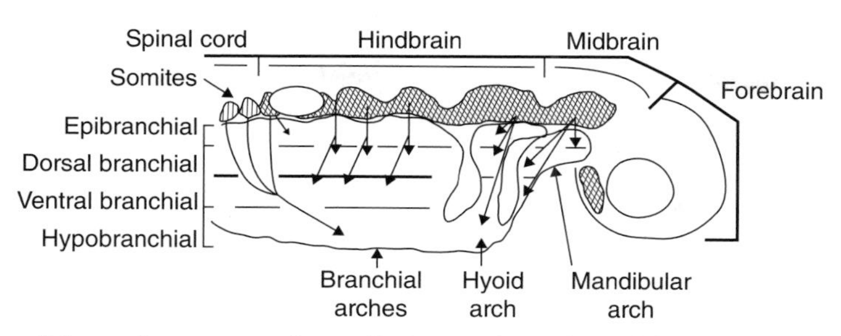

Figure 8.1. Schematic presentation of the embryonic origin of cranial muscles in gnathostomes based on Edgeworth's (e.g., 1935) analyses; premyogenic cells originate from the paraxial mesoderm (hatched areas) and several somites (areas with vertical bars); large arrows indicate a contribution of cells in segments of the mesoderm to muscle formation of different cranial arches; for more details, see text. Modified from Miyake *et al.* 1992; the nomenclature of the structures illustrated basically follows that of these authors.

(e.g., 1935). Edgeworth (1935) viewed the ontogeny of these muscles as discrete developmental pathways leading from presumptive premyogenic condensations to the different adult muscles in each cranial arch. The condensations of the first and second arches corresponded respectively to Edgeworth's "mandibular and hyoid muscle plates," and those of the more posterior, "branchial" arches corresponding to his "branchial muscle plates" (see Figure 8.1). According to Edgeworth these developmental pathways involve migration of premyogenic cells, differentiation of myofibers, directional growth of myofibers, and possibly interactions with surrounding structures. These events occur in specific locations (e.g., dorsal, medial, or ventral areas of each cranial arch, as shown in the scheme of Figure 8.1). For instance, the mandibular muscle plate gives rise dorsally to the premyogenic condensation of constrictor dorsalis, medially to the premyogenic condensation of adductor mandibulae, and ventrally to the intermandibularis (no description of a ventral mandibular premyogenic condensation was given by Edgeworth). The hyoid condensation usually gives rise to dorso-medial and ventral derivatives, and the hypobranchial condensation gives rise to the "genio-hyoideus" and to the "rectus cervicus" (as noted by Miyake *et al.* 1992, it is not clear if Edgeworth's "genio-hyoideus" and "rectus cervicus" represent separate premyogenic condensations or later states of muscle development). Although some authors use different nomenclatures, most authors and recent studies support the major divisions of head and neck muscles recognized by Edgeworth (1935) (i.e., mandibular, hyoid, branchial, and hypobranchial muscles).

Although exceptions may occur, the mandibular muscles are generally innervated by the Vth nerve. Modern humans have two ventral mandibular muscles (mylohyoideus and digastricus anterior) and six "adductor mandibulae" mandibular muscles (masseter, temporalis, pterygoideus lateralis, pterygoideus medialis, tensor tympani, and tensor veli palatini). The hyoid muscles are usually innervated by the VIIth nerve. Modern humans have three "dorso-medial" hyoid muscles (stylohyoideus, digastricus posterior, and stapedius) and several facial muscles that seemingly derive from the "dorso-medial" and particularly the "ventral" hyoid muscle groups (Diogo and Abdala 2010). The branchial muscles are usually innervated by the IXth and Xth nerves.

Diogo and Abdala (2010) divided the branchial muscles *sensu lato* into three main groups. The first, which comprises the "true" branchial muscles, are subdivided into: (A) the branchial muscles *sensu stricto* that are directly associated with the movements of the branchial arches and are usually innervated by the glossopharyngeal nerve (CNIX). Modern humans have a single muscle of this subgroup, the stylopharyngeus; (B) the protractor pectoralis and its derivatives, which are instead mainly associated with the pectoral girdle and are primarily innervated by the spinal accessory nerve (CNXI). Modern humans have two muscles belonging to this subgroup, the trapezius and sternocleidomastoideus. The second group consists of the pharyngeal muscles, which are only present as independent structures in extant mammals, and which are considered to be derived from arches 4–6, and are usually innervated by the vagus nerve (CNX). Modern humans have eight muscles belonging to this group (constrictor pharyngis medius, constrictor pharyngis inferior, cricothyroideus, constrictor pharyngis superior, palatopharyngeus, musculus uvulae, levator veli palatini, and salpingopharyngeus). The third group is made up of the laryngeal muscles, which are considered to be derived from arches 4–6 and are usually innervated by the vagus nerve (CNX). Modern humans have six muscles belonging to this group (thyroarytenoideus, vocalis, cricoarytenoideus lateralis, arytenoideus transversus, arytenoideus obliquus, and cricoarytenoideus posterior).

Regarding the epibranchial and hypobranchial muscles, according to Edgeworth (1935, p.189) these are "developed from the anterior myotomes of the body" and thus "are intrusive elements of the head"; they "retain a spinal innervation" and "do not receive any branches from the Vth, VIIth, IXth and Xth nerves." Mammals and thus primates have no epibranchial muscles. The "geniohyoideus" group of hypobranchial muscles is represented in modern humans by the geniohyoideus, genioglossus, hyoglossus, styloglossus, and palatoglossus and the instrinsic muscles of the tongue, and the "rectus cervicis" group by the sternohyoideus, omohyoideus, sternothyroideus, and thyrohyoideus. It is worth mentioning that apart from the mandibular, hyoid, branchial, hypobranchial, and epibranchial musculature, Edgeworth (1935, p.5) referred to a primitive "premandibular arch" in "which passed the IIIrd nerve". This IIIrd nerve, together with the IVth and VIth nerves – which according to Edgeworth (1935, p.5) are "not segmental nerves; they innervate muscles of varied segmental origin and are, phylogenetically, of later development than are the other cranial nerves" – innervate the external ocular muscles of most extant vertebrates. In modern humans, the group of external ocular muscles includes the superior, inferior, medial, and lateral rectus muscles, the inferior and superior oblique muscles, and the levator palpebrae superior.

Some of the hypotheses defended by Edgeworth have been contradicted by recent studies, but many of his conclusions have been corroborated by more recent developmental and genetic studies (see Diogo and Abdala 2010). For example, Miyake *et al.* (1992, p.214) noted that "Noden (1983, 1984, 1986) elegantly demonstrated with quail-chick chimeras that cranial muscles are embryologically of somitic origin, and not as commonly thought, of lateral plate origin, and in doing so corroborated the nearly forgotten work of Edgeworth." Miyake *et al.* (1992, p.214) also noted that developmental studies such as Hatta *et al.*

(1990, 1991) "have corroborated one of Edgeworth's findings: the existence of one premyogenic condensation (the constrictor dorsalis) in the cranial region of teleost fish." The existence of this and other condensations (e.g., the hyoid condensation) has received further support in developmental studies published in the last years (e.g., Knight *et al.* 2008; Kundrát *et al.* 2009). For example, in the zebrafish, immunoreactivity of the homeodomain transcription factor *engrailed* is only detected in the levator arcus palatini + dilatator operculi muscles (i.e., in the two muscles that are derived from the dorsal portion of the mandibular muscle plate – constrictor dorsalis *sensu* Edgeworth 1935). In mammals such as the mouse, *engrailed* immunoreactivity is detected in mandibular muscles that are likely derived from a more ventral ("adductor mandibulae") portion of that plate (i.e., in the masseter, temporalis, pterygoideus medialis, and/or pterygoideus lateralis (Knight *et al.* 2008)). Authors such as Tzahor (2009) have shown that even within a single species, muscles from the same arch (e.g., mandibular arch) may originate from different types of cells. For example, the mandibular "adductor mandibulae complex" and its derivatives (e.g., masseter) derive from cranial paraxial mesoderm, whereas the more ventral mandibular muscle intermandibularis and its derivatives (e.g., mylohyoideus) originate from medial splanchnic mesoderm.

Edgeworth's (1935) division of the head and neck muscles into external ocular, mandibular, hyoid, branchial, epibranchial, and hypobranchial muscles continues to be widely used by both comparative anatomists and developmental biologists. His scheme is similar to those used in numerous recent developmental and molecular works but, as expected, some researchers prefer to group the head and neck muscles in ways that do not always correspond to those proposed by Edgeworth. Thus Noden and Francis-West (2006) refer to three main types of head and neck muscles (Figure 8.2): (1) "extra-ocular" muscles, which correspond to Edgeworth's extra-ocular muscles; (2) "branchial" muscles, which correspond to the mandibular, the hyoid, and most of the branchial muscles *sensu* Edgeworth; and (3) "laryngoglossal" muscles, which include the hypobranchial muscles but also part of the branchial muscles *sensu* Edgeworth (namely the laryngeal muscles *sensu* Diogo and Abdala 2010). An advantage of recognizing these three main types is to stress that in vertebrate model taxa such as salamanders, chickens, and mice, the laryngeal muscles (for example the dilatator laryngis and constrictor laryngis) receive a contribution of somitic myogenic cells (e.g., Noden 1983; Piekarski and Olsson 2007) as do the hypobranchial muscles *sensu* Edgeworth (see above). That is, the main difference between the "branchial" and "laryngoglossal" groups *sensu* Noden and Francis-West (2006) is that, unlike the "branchial" group, the "laryngoglossal" receives a contribution of somitic cells. However, recent developmental studies (e.g., Piekarski and Olsson 2007) have shown that some of the "branchial" muscles *sensu* Noden and Francis-West (2006) (i.e., some "true," non-laryngeal, branchial muscles *sensu* the present work such as the protractor pectoralis and the levatores arcuum branchialium of salamanders and the trapezius of chickens and mice, and even some hyoid muscles such as the urodelan interhyoideus) also receive a contribution of somitic myogenic cells. Edgeworth (1935) included the protractor pectoralis and its derivatives – which

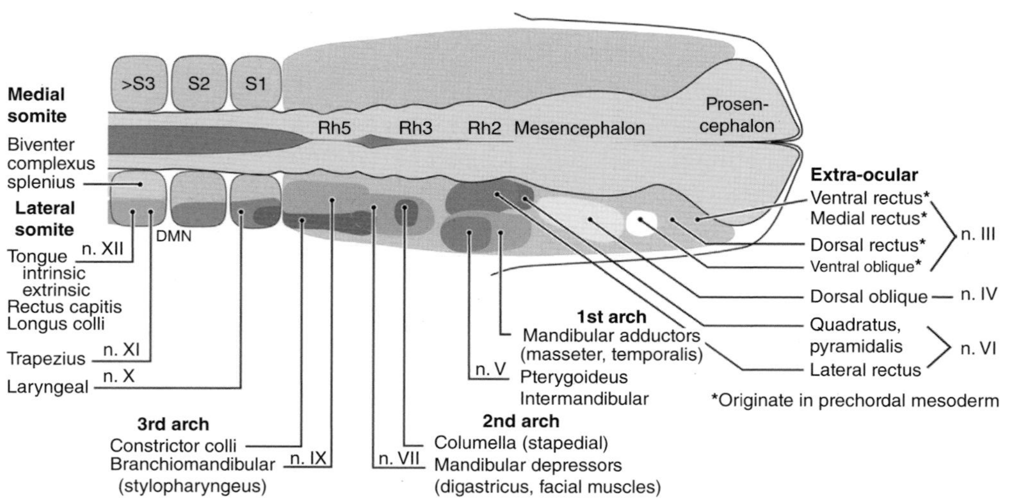

Figure 8.2. Noden and Francis-West's (2006) scheme showing the locations and main groups of muscle primordia within chick (Reptiles, Aves, *Gallus*) cephalic paraxial mesoderm, based on their interpretations on the results of recent developmental and molecular studies using techniques such as quail-chick transplants and retroviral injections; names in parentheses indicate some mammalian homologues. It is remarkable that the use of new developmental techniques has confirmed a great part of the Edgeworth's (e.g., 1935) hypotheses about the origin and homologies of the vertebrate head and neck muscles, for instance: that the "adductor mandibulae complex" ("mandibular adductors") derive from the first arch (mandibular muscles *sensu* Edgeworth) and that that the masseter and temporalis of mammals correspond to part of this "complex"; that the levator hyoideus ("columella") and the depressor mandibulae ("mandibular depressors") derive from the second arch (hyoid muscles *sensu* Edgeworth 1935); that the intrinsic and extrinsic tongue muscles are derived from somites and that they migrate anteriorly during the ontogeny in order to make part of the craniofacial musculature, that is, that they are hypobranchial muscles *sensu* Edgeworth. The main difference between Edgeworth's (1935) and Noden and Francis-West's (2006) schemes is that these latter authors include the laryngeal muscles and the trapezius in their "laryngoglossal" musculature, which also includes the hypobranchial muscles *sensu* Edgeworth; that is, they do not consider the trapezius and the laryngeal muscles as part of the "branchial musculature," as did Edgeworth, although more recent studies have shown that the trapezius is effectively mainly a branchial muscle, as originally proposed by Edgeworth (Theis *et al.* 2010) (NB: another difference between these schemes is that Noden and Francis-West 2006 consider (probably erroneously), that the "constrictor colli" (which is part of the interhyoideus *sensu* the present volume) is not a second arch muscle, that is, it is not a hyoid muscle *sensu* Edgeworth).

include the trapezius and sternocleidomastoideus of amniotes and thus of modern humans – in the branchial musculature, but he was aware that these muscles were at least partially originated from somites. However, it is important to note that recent studies have shown that although the trapezius might receive a somitic contribution, it is mainly derived from the branchial arches. Thus, it is essentially a branchial muscle, as originally proposed by Edgeworth (Theis *et al.* 2010).

Pectoral and Forelimb Muscles

Regarding the pectoral and forelimb musculature, we recognize five main groups of muscles (Diogo and Abdala 2010): the axial muscles of the pectoral girdle, the appendicular muscles of the pectoral girdle and arm, the appendicular muscles of the ventral forearm, the appendicular muscles of the hand, and the appendicular muscles of the dorsal forearm. The appendicular musculature of the pectoral girdle, arm, forearm, and hand mainly derives from the adductor and abductor muscle masses of the pectoral fin of sarcopterygian fish, and essentially corresponds to the "abaxial musculature" *sensu* Shearman and Burke (2009). The axial pectoral girdle musculature is derived from the postcranial axial musculature, and, together with most of the remaining epaxial and hypaxial muscles of the body (with the exception of, e.g., various muscles of the pectoral girdle and hind limb), form the "primaxial musculature" *sensu* Shearman and Burke (2009).

As explained by Shearman and Burke (2009), the muscles of the vertebrate body are classically described as epaxial or hypaxial according to their innervation from either the dorsal or ventral rami of the spinal nerves, respectively, whereas the terms "abaxial musculature" and "primaxial musculature" reflect embryonic criteria that are used to distinguish domains relative to embryonic patterning. The "primaxial" domain comprises somitic cells that develop within somite-derived connective tissue, and the "abaxial" domain includes muscle and bone that originates from somites but then mixes with, and develops within, lateral plate-derived connective tissue. Developmental studies carried out in recent years have shown there are important differences in the expression patterns associated with morphogenesis of the forelimb and of the hindlimb and between proximal and distal regions within each limb. For example, *Hox9* paralogs are active in the arm, but not in the thigh (e.g., Young and Hallgrímsson 2005) and studies have also shown that the formation of the muscles of the pectoral girdle occurs through mechanisms that are markedly different from those leading to the formation of the muscles of the arm, forearm, and hand that arise through the classic and well-studied migration from the somites to the limb bud. The superficial pectoral girdle muscles develop by an "In-Out" mechanism whereby myogenic cells migrate from the somites into the limb bud, followed by their extension from the proximal limb bud onto the thorax (Valasek *et al.* 2011 (NB: these superficial muscles correspond to the appendicular pectoral girdle muscles *sensu* the present work, which in modern humans include the pectoralis major and minor, infraspinatus, supraspinatus, deltoideus, teres minor and major, subscapularis, and latissimus dorsi). The deep pectoral girdle muscles are induced by the forelimb field (i.e., a localized domain of gene expression) that promotes myotomal extension (the migration of cells) directly from the somites (Valasek *et al.* 2011) (NB: these deep muscles correspond to the axial pectoral girdle muscles *sensu* the present work, which in modern humans include the serratus anterior, rhomboideus major, rhomboideus minor, levator scapulae, and subclavius).

According to Valasek *et al.* (2011) the appearance of the forelimb is followed by pectoral girdle development that attaches the forelimb to the axial skeleton. The developmental mechanisms involved in limb development are thus able to induce and recruit axial structures (e.g., the medial scapular border in mammals and the scapular blade in birds, as well as the deep girdle muscles and possibly even the cleithrum and sternum) for the anchorage of the limb (Valasek *et al.* 2011). Another important difference between the proximal portions of the limb and the distal portions of the limb (e.g., autopod) is that in the former, tendons are induced but do not segregate to form specific tendons in the absence of muscles, while in the latter, muscles are only present at later ontogenetic stages and tendon formation is initiated and segregation into individual tendons occurs in the absence of muscles. However, these distal tendons require contact with a muscle, for they have been shown to degenerate in a muscle-less limb environment (Hasson 2011).

Our myological cladistic analyses of primates (Diogo and Wood 2011, 2012a) allowed us to directly test the "In-Out" / "Myotomal Extension" morphogenetic hypothesis elaborated by Valasek *et al.* (2011) (Diogo and Wood 2013). According to this hypothesis, one would expect that the superficial (appendicular) pectoral girdle muscles would accumulate more anatomical changes at their proximal origin (because they extend from distal to proximal) and that the deep (axial) pectoral girdle muscles would accumulate more anatomical changes at their distal insertion (because they extend from proximal to distal). Within all of the evolutionary changes (steps) listed in that cladistic analysis there are 19 changes (of bony attachments) for a total of eight deep (axial) pectoral muscles; an average of 2.38 character state changes per muscle. Of these 19 changes in the eight muscles, two concern changes in proximal attachments (i.e., 0.25 per muscle), and five concern changes in distal attachments (i.e., 0.62 per muscle). Thus, within these deep (axial) pectoral muscles, there is a higher rate of change in the distal than in the proximal attachments, thereby supporting the hypothesis of Valasek *et al.* (2011). Regarding the superficial (appendicular) pectoral muscles, there are 26 changes for a total of 12 muscles (i.e., 2.17 character state changes per muscle). Within these 26 changes in the 12 muscles four concern the proximal attachments (i.e., 0.33 per muscle), and eight concern the distal attachments (i.e., 0.67 per muscle). So, within these superficial (appendicular) pectoral muscles, there is also a higher rate of change in the distal attachments than in the proximal attachments. Thus this evidence seems to contradict the hypothesis of Valasek *et al.* (2011), but the changes in proximal attachments relate to ribs and vertebrae. There is often intrageneric variation in the attachments of muscles in specific ribs/vertebrae so they are more difficult to polarize and code than changes in the distal attachments. Thus, the real rates of change per muscle of the proximal attachments of both the deep and the superficial pectoral muscles are therefore likely to be greater than the rate cited, therefore it would be unwise to use these data to contradict Valasek *et al.*'s (2011) hypothesis.

However, because coding bias concerns the proximal attachments of *both* the deep and the superficial muscles of all taxa, we can compare the rate of change per muscle for these two types of muscles, which, as noted above, are 2.38 and 2.17 changes per muscle, respectively. These rates are similar to the rate of 1.95 for the hand muscles (72 changes for a total of 37 muscles), but they are substantially lower than the rate of 6.4 for the arm muscles (32 changes for a total of five muscles) and higher than the rate of 1.24 for the forearm muscles (26 changes for a total of 21 muscles). Thus, our study of primate muscles does not support Weisbecker's (2011) suggestion that the structures of the distal limb, due to their later development and their being contingent on earlier-developing more proximal elements, are more morphologically variable than more proximally located ones.

To summarize this section, it can thus be said that by combining the results of phylogenetic and evolutionary studies of primates and other animals and developmental studies of key model organisms such as mice or chickens, we can now provide a much more comprehensive – although of course still imcomplete – picture of the origin, development, and evolution of the muscles of modern humans.

The Notions of Purpose and Progress in Evolution and the Parallelism Between Ontogeny and Phylogeny

Bakewell *et al.* (2007, p.7492) stated that their molecular (whole genome) studies show that "in sharp contrast to common belief, there were more adaptive genetic changes during chimp evolution than during human evolution" and they claim that their analysis "suggests more unidentified phenotypic adaptations in chimps than in humans." The results of our primate parsimony and Bayesian analyses using characters from the head, neck, pectoral, and forelimb muscles indicate that since the *Pan*/*Homo* split the clade Hominina has evolved faster than the clade Panina (about 2.3 times faster according to the lengths of the branches leading to modern humans (9) and to chimpanzees (4) in the parsimony tree obtained by Diogo and Wood 2011). In turn, since the split between *Gorilla* and the Hominini, gorillas have only accumulated two unambiguous muscular apomorphies, whereas there are eight (4 + 4) and 13 (4 + 9) unambiguous apomorphies leading to extant chimpanzees and to modern humans, respectively (Diogo and Wood 2011). In terms of their significance for our understanding of human evolution, our results could seem paradoxical. On the one hand, the cladistic analyses suggest there are more unambiguous evolutionary steps from the base of the tree to modern humans than to any other extant primate taxon included in Table 8.1. But, on the other hand, our comparative anatomical studies show that modern humans have fewer muscles than most other primates, including chimpanzees, and have many fewer muscles than do strepsirrhines and tarsiiforms, which are often seen as "anatomically plesiomorphic" primates (Table 8.1).

Such an apparent paradox is related to the fact that, as Gould noted in *The Structure of Evolutionary Theory* (2002), there is a general tendency to use "progressive trends" to tell evolutionary stories, particularly in paleontological publications, in which examples of stasis are often either unreported or underreported because stasis is interpreted as "no data." The results of our study support Gould's contention in the sense that there is no general trend to increase the number of muscles, or muscle bundles, at the nodes leading to hominoids and to modern humans. Within the context of our study and our myological comparisons of the taxa listed in Table 8.1, because each muscle is the result of parcellation, that is, of innovation trough differentiation leading to a morphogenetic semi-independence of the muscle (*sensu* Vermeij 1973), the simplest and most objective way to measure complexity was to compare the number of muscles of each taxon. This is because: (1) the number of muscles is an objective measure (e.g., the numerous researchers studying human gross anatomy basically agree on the number of muscles in the human body); (2) by using this way of measuring complexity, we combined a macroevolutionary definition of complexity similar to that of Bonner (1988; the only difference being that Bonner referred to the number of different types of cells instead of to the number of muscles) with a developmental definition of complexity that includes the notions of parcellation (e.g., Wagner and Altenberg, 1996) and morphogenetic semi-independence (e.g., Vermeij 1973). That is, with respect to the muscles in the regions we have investigated, although modern humans accumulated more evolutionary transitions than the other primates included in the cladistic study, these evolutionary transitions did not result in more muscles, or more muscle components (Table 8.1 and Diogo and Wood 2011, 2012a). For example, although some of the nine modern human apomorphies acquired since the *Pan*/*Homo* split involved the differentiation of new muscles (rhomboideus major and rhomboideus minor, extensor pollicis brevis and flexor pollicis longus), others involved the loss of muscles (levator claviculae and dorsoepitrochlearis) (Diogo and Wood 2011). As a result, more muscle changes were accumulated in our evolutionary history than in that of modern chimpanzees, but there are usually more head and neck and pectoral and upper limb muscles present in modern chimpanzees than in members our own species (Table 8.1). In fact, with respect to the number of head, neck, pectoral, and forelimb muscles and muscle bundles, one could make the case that modern humans are relatively simplified mammals (Diogo and Wood 2011, 2012a, 2013).

The notion of progress and purpose of evolution was closely related to the rise of developmental theories such as Haeckel's theory of recapitulation (e.g., Gould 1977). This theory was based on a "scalae naturae" view of nature in which white modern human males were seen as more complex and therefore placed at the top of the scale. This scheme had profound and unfortunate social and racial implications. Unfortunately, the notion of "scalae naturae" is still deeply embedded in many current textbooks and scientific papers (e.g., Diogo *et al.* 2015). As noted by Gould (1977), the theory of recapitulation is no longer accepted because the ontogeny of an animal does not recapitulate the *adult* stages of its ancestors. However, as also noted by Gould (1977), this does not mean that there is no parallel between ontogeny and phylogeny; in fact, such a parallel seems to occur often in animal ontogenies and our recent developmental studies of zebrafish, salamanders, and frogs support this view (Diogo *et al.* 2008a, 2008b; Ziermann and Diogo 2013, 2014; Diogo and Tanaka 2014; Diogo and Ziermann 2014).

As we now have detailed data about the ontogeny of the head and neck muscles of those taxa and about the phylogeny and evolution of these muscles within the vertebrates, it is possible to compare the order in which the muscles appear in ontogeny to the order in which the muscles evolved during phylogeny. Comparisons show that in the case of the zebrafish head muscles there is in general a parallel between ontogeny and phylogeny, but there are exceptions (e.g., the early ontogenetic appearance of muscles that evolved late in phylogeny and play a particularly important role in the feeding mechanisms not only of zebrafish adults but also of embryos (Diogo *et al.* 2008b)).

In the salamanders and frogs there is also a general parallelism between ontogeny and phylogeny, and our preliminary comparisons in modern humans also corroborate such a parallelism (Ziermann and Diogo 2013, 2014; Diogo and Tanaka 2014; Diogo and Ziermann 2014).

However, it is important to note that this does not mean that the commonly accepted view that during ontogeny the tendency is towards the differentiation (and not the dedifferentiation) of muscles is correct. That is, the order in which muscles appear in ontogeny is usually similar to the order in which they appear in phylogeny, but muscles are often also lost/reabsorbed later in ontogeny. For instance, in neotenic salamander species such as axolotls that lack a full metamorphosis, some muscles become indistinct during ontogeny (e.g., the pseudotemporalis profundus and the levator hyoideus become completely integrated in the pseudotemporalis superficialis and in the depressor mandibulae, respectively (Ziermann and Diogo 2013)). Our recent studies in modern humans have provided similar data about the loss of muscles during ontogeny (see below). Therefore, although the differentiation of muscles is more common during ontogeny than is the dedifferentiation, by absorption or by fusion, it is necessary to emphasize that dedifferentiation is much more common than previously assumed.

The Relationship between Trisomies, "Atavisms," Evolutionary Reversions, and Developmental Constraints

The most parsimonious tree obtained from the analysis of the complete dataset compiled by us about primate head, neck, pectoral, and upper limb muscles had a total length of 301 steps, of which 100 (33%) were non-homoplasic (i.e., they were not independently acquired or reverted elsewhere in the tree) evolutionary transitions (Diogo and Wood 2011). Of the 220 steps that were unambiguously optimized in the tree, 28 (i.e., 13%) were reversions to a plesiomorphic state (Table 8.2). Taking into account the total number of steps (301) within the tree, the number of characters (166), and the number of muscles (129) represented by the data in the cladistic analysis, there are about 1.8 evolutionary transitions per character and about 2.3 evolutionary transitions per muscle studied. These numbers stress the importance of homoplasy and of evolutionary reversions in morphological evolution. As stressed by Wiens (2011), less attention has been given historically to evolutionary reversions than to the two other types of homoplasic events, parallelism and convergence (e.g., Diogo 2005). Wiens listed several examples of violations of Dollo's Law (which states that once a complex structure is lost it is unlikely to be reacquired), including the loss of mandibular teeth in the ancestor of modern frogs >230 million years (MY) ago and their reappearance in the anuran genus *Gastrotheca* during the last 5–17 MY. In Diogo and Wood (2012b) we focused on the implications of our muscle studies for understanding the role played by reversions in primate and human evolutionary history and for developmental biology. We stressed that reversions played a substantial role in primate and human evolution because one in seven of the 220 evolutionary transitions unambiguously optimized in the most parsimonious tree obtained by

Table 8.2. List of the 28 evolutionary reversions obtained in Diogo and Wood's (2011) cladistic analysis of the primate taxa listed in Table 8.1 (Anat. region, anatomical region; Ch. state ch., character state change; Rev., reversion). For more details, see text and Diogo and Wood (2012b).

Description of reversion	Ch. state ch.	Anat. region
a) Rev. of "Biceps brachii has no bicipital aponeurosis"	[105:1-->0]	Arm
b) Rev. of "Digastricus anterior is not in contact with its counterpart for most of its length"	[3:1-->0]	Head/Neck
c) Rev. of "Spinotrapezius is not a distinct muscle"	[43:1-->0]	Head/Neck
d) Rev. of "Chondroglossus is present as a distinct bundle of the hyoglossus"	[58:1-->0]	Head/Neck
e) Rev. of "Chondroglossus is present as a distinct bundle of the hyoglossus"	[58:1-->0]	Head/Neck
f) Rev. of "Pterygoideus lateralis has well differentiated inferior and superior heads"	[9:1-->0]	Head/Neck
g) Rev. of "Depressor anguli oris is a distinct muscle"	[39:1-->0]	Head/Neck
h) Rev. of "Trapezius inserts onto the clavicle"	[45:1-->0]	Head/Neck
i) Rev. of "Rhomboideus major and rhomboideus minor are not distinct muscles"	[69:1-->0]	Pectoral
j) Rev. of "Opponens pollicis is a distinct muscle"	[143:1-->0]	Hand
k) Rev. of "Digastricus anterior is not in contact with its counterpart for most of its length"	[3:1-->0]	Head/Neck
l) Rev. of "Frontalis is a distinct muscle"	[32:1-->0]	Head/Neck
m) Rev. of "Sphincter colli profundus is not a distinct muscle" (either the muscle was lost in anthropoids and then reappeared in the Cebidae+Aotidae clade and in *Cercopithecus*, or was lost in *Pithecia* and catarrhines and then reappeared in *Cercopithecus*)	[24:1-->0]	Head/Neck
n) Rev. of "Digastricus anterior is not in contact with its counterpart for most of its length"	[3:1-->0]	Head/Neck
o) Rev. of "Cricoarytenoideus posterior does not meet its counterpart at the dorsal midline" (either the derived condition was acquired in anthropoids and then reverted in *Macaca*, *Hylobates*, and *Pongo*, or it was acquired in platyrrhines, in hominins, and in cercopithecids and then reverted in *Macaca*)	[56:1-->0]	Head/Neck
p) Rev. of "Geniohyoideus is fused to its counterpart in the midline"	[57:1-->0]	Head/Neck
q) Rev. of "Temporalis has a pars suprazygomatica" (either the derived condition was acquired in Euarchonta and then reverted in *Cynocephalus* and hominoids, or it was acquired in *Tupaia* and Primates then reverted in hominoids)	[8:1-->0]	Head/Neck
r) Rev. of "Pterygopharyngeus is not a distinct muscle" (either the derived condition was acquired in Euarchonta and then reverted in *Cynocephalus* and *Hylobates*, or it was acquired in *Tupaia* and Primates and then reverted in *Hylobates*)	[53:1-->0]	Head/Neck
s) Rev. of "Levator claviculae inserts onto a more medial portion of the clavicle" (derived condition considered to have probably arisen in catarrhines and then uniquely reverted in *Hylobates*, but see notes in text)	[76:1-->0]	Pectoral
t) Rev. of "Latissimus dorsi and teres major are fused"	[89:1-->0]	Pectoral
u) Rev. of "Pectoralis minor inserts onto the coracoid process"	[83:1-->0]	Pectoral

Table 8.2. (*Continued*).

	Description of reversion	Ch. state ch.	Anat. region
v)	Rev. of "Epitrochleoanconeus is not a distinct muscle"	[120:1-->0]	Forearm
w)	Rev. of "Contrahentes digitorum are missing"	[131:1-->0]	Hand
x)	Rev. of "Flexores breves profundi are fused with the intermetacarpales, forming the dorsal interossei" (either the derived condition was acquired in *Tupaia*, *Cynocephalus*, platyrrhines, and hominoids and then reverted in *Pan*, or it was acquired in Euarchonta and then reverted in strepsirrhines, *Tarsius*, cercopithecids, and in *Pan*)	[140:1-->0]	Hand
y)	Rev. of "Anterior portion of sternothyroideus extends anteriorly to the posterior portion of the thyrohyoideus"	[64:1-->0]	Head/Neck
z)	Rev. of "rhomboideus major and rhomboideus minor are not distinct muscles"	[69:1-->0]	Pectoral
α)	Rev. of "Tendon of flexor digitorum profundus to digit 1 is vestigial or absent"	[112:1-->0]	Forearm
β)	Rev. of "Flexor carpi radialis originates from the radius" (the muscle originates from the radius in gorillas, chimpanzees, and orangutans, but the usual condition for hylobatids is unclear and thus it is also unclear whether a radial origin constitutes a synapomorphy of hominoids or of hominids).	[123:1-->0]	Forearm

Diogo and Wood (2011) are reversions to a plesiomorphic state (N=28) (Table 8.2). Of those 28 reversions, six played a direct role in our own evolution in the sense that they occurred at the nodes that led to the origin of modern humans, and nine of the 28 reversions violate Dollo's law, with the average time between a structure's loss and reacquisition being about 50 MY (Table 8.3).

Our studies provide evidence to support the hypothesis that the reacquisition in adults of anatomical structures that were missing for long periods of time is possible because the developmental pathways responsible were maintained in the members of that clade. An example of this concerns the presence/absence of the contrahentes digitorum muscles in adult hominids. Chimpanzees display a reversion of a synapomorphy of the Hominidae (great apes and modern humans; acquired at least 15.4 MY ago: Tables 8.1 and Table 8.2) in which adult individuals have two contrahentes digitorum (in adults of other hominid taxa there are usually none) other than the muscle adductor pollicis; one going to digit 4 and the other to digit 5. Studies of the development of hand muscles (e.g., Cihak 1972) have shown that karyotypically normal modern human embryos *do have* contrahentes going to various fingers, but the muscles are usually lost during later embryonic development (Figure 8.3). Moreover, other studies (e.g., Dunlap *et al.* 1986) have shown that in karyotypically abnormal modern humans, such as individuals with trisomies 13, 18, or 21, the contrahentes muscles often persist (as "atavisms") until well after birth (Figure 8.3).

According to some authors, cases where complex structures are formed early in ontogeny just to become lost/indistinct in later developmental stages (the so called "hidden variation") may allow organisms to have a great ontogenetic potential early in development, so that if faced with external perturbations

Table 8.3. Time frame over which lost traits were regained within the euarchontan clade according to the cladistic analysis of Diogo and Wood (2011) and using the estimate times provided by Fabre *et al.* (2009). For more details, see text and Diogo and Wood (2012b).

Structure that was regained, violating Dollo's law (letters shown before the description of each structure correspond to letters shown in Table 8.2)	Number of evolutionary steps according to the hypothesis that violates Dollo's law (in bold) *versus* number of evolutionary steps that one would need to assume in order to not violate Dollo's law (in non-bold, following the description of these steps)	Date when feature was lost (*)	Date when feature was regained (**)	Time passed between (*) and (**)
a) Bicipital aponeurosis (NB: in modern humans the bicipital aponeurosis helps to reinforces the cubital fossa and to protect the branchial artery and the median nerve running underneath)	**2** *vs* 4 (loss in *Tupaia*, *Cynocephalus*, lorisiforms, and haplorrhines)	94.1 MY	63.2 MY	30.9 MY
c) Spinotrapezius (NB: the spinotrapezius is a separate muscle that corresponds to the descending part of the trapezius of modern humans, which mainly depresses the scapula)	**2** *vs* 4 (loss in *Tupaia*, *Cynocephalus*, strepsirrhines, and anthropoids)	94.1 MY	23.2 MY	70.9 MY
i) Rhomboideus minor (NB: in modern humans the rhomboideus minor acts together with the rhomboideus major to adduct the scapula and keep the scapula pressed against the thoracic wall)	**3** *vs* 10 (loss in *Tupaia*, *Cynocephalus*, tarsiers, *Pithecia*, *Aotus*, *Saimiri*, hylobatids, orangs, gorillas, and chimps: see text)	88.8 MY	3.6 MY	85.2 MY
m) Sphincter colli profundus (NB: the sphincter colli profundus is a thin facial muscle that lies deep to the platysma cervicale)	**3** *vs* 4 (loss in *Pithecia*, *Colobus*, Papionini and hominoidea)	39.8 MY or 25.0 MY	7.9 MY	31.9 MY or 17.1 MY (mean = 24.5 MY)
r) Pterygopharyngeus (NB: the pterygopharyngeus might corresponds to the pterygopharyngeal part of the superior pharyngeal constrictor of modern humans, which connects the medial pterygoid plate and its hamulus to the median raphe and mainly acts together with the other parts of the muscle to constrict the pharynx)	**3** *vs* 6 (loss in *Tupaia*, strepsirrhines, tarsiers, platyrrhines, cercopithecids, and hominoids)	94.1 MY or 82.2 MY	10.7 MY	83.4 MY or 71.5 MY (mean = 77.4MY)

v) Epitrochleoanconeus (NB: the epitrochleoanconeus mainly connects the medial epicondyle of the humerus to the olecranon process of the ulna, so it potentially helps to extend the forearm and/or stabilize the elbow joint)	**2** *vs* 4 (loss in hylobatids, orangs, gorillas, and Hominina)	19.5 MY	3.0 MY	16.5 MY
w) Contrahentes digitorum (NB: the contrahentes adduct the digits; see text)	**2** *vs* 3 (loss in orangs, gorillas, and Hominina)	15.4 MY	3.0 MY	12.4 MY
x) Intermetacarpales (NB: the intermetacarpales connect the metacarpals of adjacent digits; see text)	**5** *vs* 7 (loss in *Tupaia*, *Cynocephalus*, platyrrhines, hylobatids, orangs, gorillas, and Hominina)	94.1 MY or 19.5 MY	3.0 MY	91.1 MY or 16.5 MY (mean = 53.8 MY)
z) Rhomboideus minor (NB: in modern humans the rhomboideus minor acts together with the rhomboideus major to adduct the scapula and keep the scapula pressed against the thoracic wall)	**3** *vs* 10 (loss in *Cynocephalus*, strepsirrhines, tarsiers, *Pithecia*, *Aotus*, *Saimiri*, hylobatids, orangs, gorillas, and chimps: see text)	88.8 MY	2.4 MY	86.4 MY
				Total mean: 50.9 MY

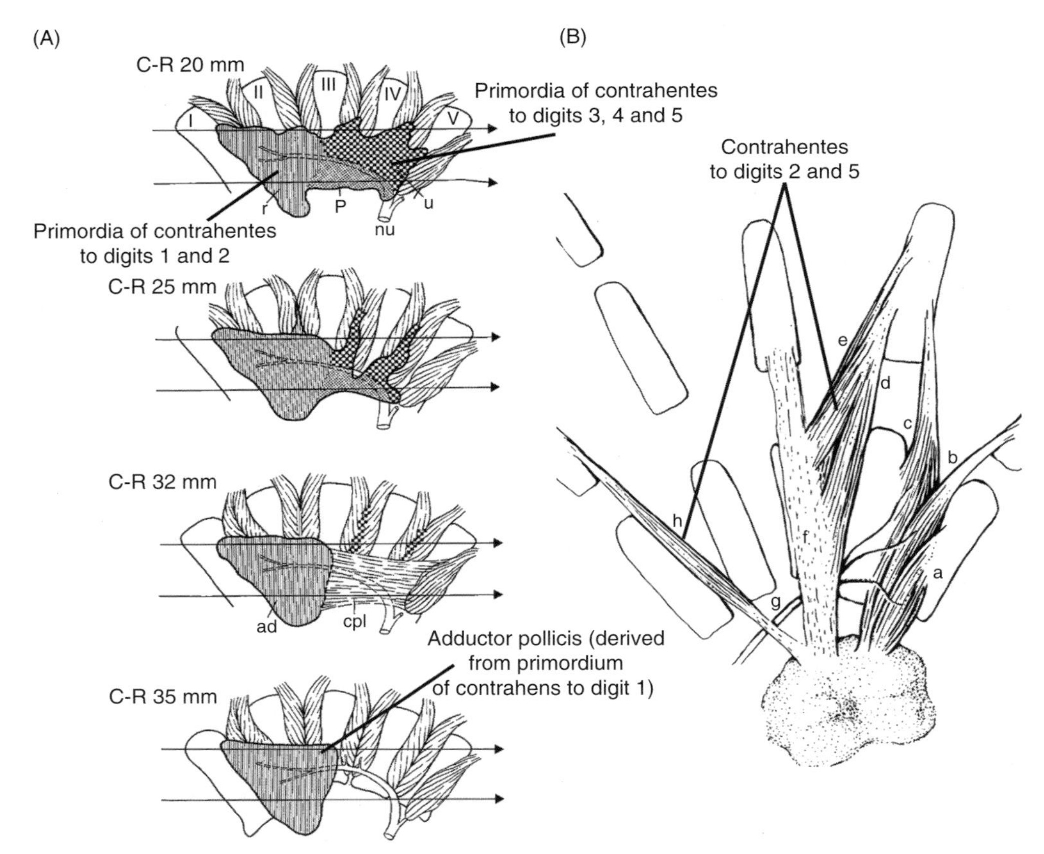

Figure 8.3. (A) Ontogeny of the contrahentes digitorum in the hand of karyotypically normal human embryos showing how the contrahentes to digits other than digit 1 are usually lost (reabsorbed) early in development (modified from Cihak 1972). Part of the interossei primordia (i.e., the flexores breves profundi layer) are shown between the metacarpals. r, u, p: radial, proximal, and ulnar parts of contrahentes layer; nu: ulnar nerve; ad: adductor pollicis; cpl: contrahens plate; I–V: metacarpals I–V; C-R: crown-rump length of the embryos. (B) Deep left-hand musculature of a trisomy 18 neonate (100 days old, female) showing the presence of contrahentes to digits 2 and 5 (the more superficial muscles, as well as the adductor pollicis, are not shown; modified from Dunlap *et al.* 1986). a: opponens pollicis; b: "interosseous palmaris of Henle"; c: interosseous dorsalis 1; d: contrahens to digit 2; e: interosseous palmaris 1; f: contrahens fascia/medial raphe; g: deep branch of ulnar nerve; h: contrahens to digit 5.

(e.g., climate change, habitat occupied by new species, etc.) evolution can use that potential (adaptive plasticity) (e.g, West-Eberhard 2003). However, authors such as Gould (1977) and Alberch (1989) have suggested that the occurrence of examples such as those cited above argue in favor of a "constrained" rather than an "adaptationist" view of evolution. This is in line with the view defended by the authors of more recent studies such as Galis and Metz (2007, pp.415–416), who stated that "without denying the evolutionary importance of phenotypic plasticity and genetic assimilation, we think that for the generation of macro-evolutionary novelties the evidence for the impact of hidden variation is limited." We are inclined to agree that hidden variation may have a limited role in the generation of evolutionary novelties. However, as explained above, hidden variation may have a more important role in the *reappearance* of some traits associated with these novelties, as in anatomical reversions that violate Dollo's law.

As explained above, in *Ontogeny and Phylogeny* Gould (1977) argues that although Haeckel's recapitulation hypothesis that the ontogeny of one organism recapitulates the adult stages of its ancestors (i.e., recapitulation) has been refuted, researchers often use it as a "straw man" to deny that there is often a parallel between ontogeny and phylogeny. According to Gould, such a parallel exists and it is probably driven more by phylogenetic/ontogenetic constraints than by adaptive plasticity, and studies of model organisms such as zebrafish or axolotls support this view (see above). The examples listed above about primate and modern human muscles, such as those involving the contrahentes and the intermetacarpales, also support Gould's arguments in the sense that in both cases there is a parallel between ontogeny and phylogeny. That is, in "normal" modern human ontogeny the muscles are evident and then became lost or indistinct later in ontogeny; during the recent evolutionary history of modern humans the muscles were plesiomorphically present and then became lost. This is not recapitulation in the Haeckelian sense, for the contrahentes digitorum and the intermetacarpales of karyotypically "normal" modern human embryos do not correspond to the muscles of *adult* primates such as chimpanzees and of other primate/mammalian adults, but instead to the muscles of the *embryos* of the latter taxa. That is, the developmental pathways that result in the presence of these muscles in the adults of the latter taxa have not been completely lost in modern humans, even after several million years. This is probably because these pathways are related to those involved in the development of other structures that *are* present and functional in modern human adults (Diogo and Wood 2012b). Regarding the occurrence of the evolutionary reversions that resulted in the presence of the contrahentes and of distinct intermetacarpales in extant chimpanzees, this is probably related to the occurrence of heterochronic, and specifically paedomorphic, events in the lineage leading to the genus *Pan* (Diogo and Wood 2012b). In fact, Cihak (1972) has shown that the intermetacarpales are also present as distinct muscles in early embryos of karyotypically normal modern humans, before they fuse with some flexor breves profundi muscles to form the muscles interossei dorsales. It is intuitively unlikely that the persistence of contrahentes in the later ontogenetic stages of karyotypically abnormal modern humans, such as individuals with trisomies 21, and particularly 13 and 18 (which usually die before or soon after birth), is the result of adaptive evolution and natural selection. What seems to be clear is that the presence of distinct contrahentes digitorum and intermetacarpales in adult chimpanzees is likely due to a prolonged or delayed development of the hand musculature of these apes, so in this respect extant chimpanzees are seemingly more neotenic than modern humans (Diogo and Wood 2012b). This is in line with recent studies that have pointed out that, although in the literature it is often stated that modern humans are in general more neotenic than other primates, both paedomorphic and peramorphic processes have likely been involved in the mosaic evolution of modern humans and of other hominoids (see Bufill *et al.* 2011 and references therein).

In order to investigate the developmental mechanisms related to the atypical development and abnormal phenotype of the striated muscles of modern human trisomic individuals, we are investigating the cadavers of fetal, neonatal, and adult modern humans with trisomies, as well as the cadavers of mouse models for Down Syndrome (DS) (e.g., Ts65Dn) (Diogo and Wood 2012b). The main hypothesis we want to test is that the disappearance of muscles such as the contrahentes and platysma cervicale during early developmental stages of karyotypically normal humans is related to apoptosis, and thus that the frequent persistence of these muscles until later ontogenetic stages in individuals with trisomies 13, 18, and 21 is associated with decreased muscle apoptosis. It has been suggested that modern human individuals with DS show an increase of apoptosis in neurons, granulocytes, and lymphocytes (e.g., Elsayed and Elsayed 2009). If our studies support the hypothesis that these individuals have decreased apoptosis, to the point of having additional muscles in later ontogenetic stages, this would suggest a more nuanced story with respect to apoptosis (i.e., a mosaic scenario where there is more apoptosis in some tissues and less in others). Moreover, by implying there is possibly a mismatch between the nervous (e.g., more apoptosis of cells of neurons) and muscular (i.e., less apoptosis and presence of extra muscles) systems, our hypothesis might also shed light on the etiology of the hypotonia (low muscle tone) that is present in almost all babies with DS.

From a developmental genetic perspective, it is now known that members of a small family of proteins, termed MCIP1 and MCIP2 (myocyte-enriched calcineurin interacting protein), are most abundantly expressed in striated and cardiac muscles and that such proteins form a physical complex with calcineurin A; MCIP1 is encoded by *DSCR1* (e.g., Gotlieb 2009). Expression of the MCIP family of proteins is upregulated during muscle differentiation, and its forced overexpression inhibits calcineurin signaling leading to a decrease of muscle apoptosis. When apoptosis is abnormally reduced in trisomic mice embryos, excess populations of myocytes can form in the atrioventricular region, where they may interfere with the normal migration of cells during cardiac development, leading to the occurrence of valvular abnormalities and atrioventricular or ventricular septal defects similar to the congenital heart defects typical of modern humans with DS (Gotlieb, 2009).

Tempo and Mode of Primate and Human Evolution, Modularity, and Ontogenetic Constraints

Recent studies (e.g., Chatterjee *et al.* 2009; Cooper and Purvis 2009; Perelman *et al.* 2011) suggest that rates of both morphological and molecular evolution vary among taxa. Differences in molecular rates are consistent with the concept of local molecular clocks. According to Tetushkin (2003, p.729), primates "provide the most interesting and striking example of such heterogeneity in the tempo of molecular evolution." However, most recent studies dealing with evolutionary rates within mammals and primates are based on molecular evidence, and the few non-molecular studies do not focus on detailed morphology but on global features of the phenotype such as body size (Diogo *et al.* 2013d). Recently, we calculated the rates of muscle evolutionary change within the primate clades generated from our myology-based cladistic analyses (Diogo *et al.* 2013d) and compared them with the molecular rates obtained by other authors, including the rates of molecular nucleotide substitution reported by Perelman *et al.* (2011). We addressed the following questions: (A) Are the rates of muscle evolution the same or different across different primate taxa and across different geological time periods?; (B) Are the muscle rates more in agreement with the gradualist model, with the punctuated equilibrium model, or with other models of evolutionary rate?; and (C) Are the muscle rates in general similar to the molecular rates provided in recent papers, or is there a mismatch between morphological and molecular rates as predicted by the neutral evolution model of Kimura (1968)?

Surprisingly, we found several examples where the muscle rates of evolution in various lineages of each of the major primate clades are strikingly similar. For instance, the rate of muscle evolution at the node leading to the clade including the platyrrhine families Cebidae plus Aotidae (0.32) is the same as that at the node leading to *Aotus* (0.32), a single aotid genus, and is similar to that at the node leading to the whole Platyrrhini clade (0.37). Moreover, the rate at the node leading to the Cercopithecidae (0.38) is the same as that at the node leading to the cercopithecid *Colobus* (0.38) and at the node leading to *Cercopithecus* (0.38). As these are the only rates that are similar across the whole primate tree obtained in our cladistic analysis (Diogo and Wood 2011), and as this happens inside two different large clades (New and Old World monkeys), it seems that for some reason the number of morphological evolutionary changes accumulated per period of time in at least some nodes within a clade is essentially constant. This would be the expectation for molecular evolutionary changes according to the neutral model of evolution, but this has not been reported previously for any type of morphological evolutionary changes, at least within the order Primates. Moreover, the muscle rates at the node leading to strepsirrhines (0.29) and then at the subsequent nodes leading to the lorisiforms (0.22) and to the lemuriforms (0.24) are also similar, particularly when one compares the differences with the range of different rates within all the primate nodes analyzed by us (0.00–2.72). The same can be said about the rates at the nodes leading to the Hominidae (1.47), Homininae (1.50), and Hominini (1.53). Thus these results lend support to the proposal by authors such as Gould (2002) that "internal" (e.g., ontogenetic) constraints play an important role in evolution. That is, in the last 25 MY, despite there having been major climate and environmental changes in Africa and Asia, the rates of muscle changes accumulated during that period at the nodes leading to the Cercopithecidae and then to *Colobus*, and also to *Cercopithecus*, are the same. Interestingly, these similarities in overall rates do not necessarily correspond to similarities in the rates for the head and neck and pectoral and forelimb regions. At the node leading to the Cercopithecidae the rate for the head and neck changes is 0.19, at the node leading to *Colobus* it is 0.00, and at the node leading to *Cercopithecus* it is 0.38: the respective rates for the pectoral and forelimb region are 0.19, 0.38, and 0.00.

From the point of view of the authors that support the paradigm of "internal constraints," the analysis of these partial rates would be viewed as evidence that ontogenetic constraints are so strong and interconnected that the potential for overall change accumulated in the different regions of the body is limited. This is in line with the results of recent studies showing that in early organogenesis, and particularly during the so-called "phylotypic stage," there is substantial interactivity among different body modules and thus there is low effective modularity (e.g., Galis and Metz 2007). It has also been argued that from a developmental perspective, if extensive somatic investment is made in one structure of one body module, this could limit investment dedicated to the formation of another structure, not only from that module but also from other body modules (e.g., Galis and Metz 2007). It is also possible that constructional trade-offs constrain investment in whole phenotypes because the structural space in organisms is limiting (e.g., Hulsey and Hollingsworth 2011). However, our study also provides examples of a shift towards a faster rate of muscle evolution that is then followed by a slow-down in the rate of muscle evolution (i.e., a punctuated equilibrium mode). For instance, the overall (head, neck, pectoral, and forelimb) muscle rate leading to the Cercopithecidae is 0.75, then that leading to the Papionini is 2.56, and that leading to *Papio* is 0.26, that is, 10 times smaller than that leading to Papionini (for more details see Diogo *et al.* 2013d).

With respect to the third question raised above, there are several examples of substantial differences between muscle and molecular evolutionary rates at the same primate nodes, as would be predicted by the neutral model of evolution. But at other places on the tree the muscle evolutionary rates obtained in our study are similar to the published rates of molecular evolution.

In summary, our study suggests that the tempo and mode of primate and human evolution is complex, and provides examples of different modes of evolution within the primate clade. It suggests that at the level of major clades, simplistic dichotomies such as "gradual versus punctuated" and "neutral versus non-neutral" (see review of Gould 2002) do not apply. Also, contrary to the proposal that there is a "general molecular slow-down" in hominoids (e.g., Goodman *et al.* 1985; Steiper *et al.* 2004) the muscle evolutionary rates at the nodes leading to, or within, the hominoids are among the highest within the primate clade.

The relationship between Modern Human Anomalies/Variations, Digit Loss/Gain, Muscle Changes and Homeotic Transformations

Limbs with digits first appeared in aquatic taxa with more than five digits, but the transitions to terrestrial locomotion and to pentadactyly, and then to digit reduction in some clades, are particularly poorly understood (e.g., Pierce *et al.* 2013). Information obtained from non-pentadactyl limbs is crucial to clarify how the functional and spatial associations between limb bones and muscles change during evolution and in the common limb birth defects found in modern humans and other species. However, anatomical and developmental studies usually focus on pentadactyl autopodia, and those dealing with non-pentadactyly tend to focus on hard tissues (e.g., Shubin and Alberch 1986; Young *et al.* 2009). The scarcity of information about soft tissues is paradoxical because non-pentadactyly and the specific spatial associations between limb bones and muscles are topics that have long attracted researchers' attention (e.g., Owen 1849), and particularly because non-pentadactyly is the most common human birth defect (Castilla *et al.* 1996). The study of non-pentadactyly is also of interest because it allows researchers to discuss broader evolutionary themes such as the occurrence of evolutionary trends, the frequency of anatomical convergence, and the existence of evolutionary reversals that violate Dollo's law (e.g., Diogo and Wood 2013).

Moreover, non-pentadactyly is often associated with one of the most important current topics in evolutionary developmental biology: homeotic transformations (i.e., the replacement of a body part by one that normally forms in another region of the body). For instance, in pre-axial polydactyly, one of the most common congenital anomalies of the human hand, duplication of the thumb leads to the two most radial digits having an homeotic identity of digit 1 (Castilla *et al.* 1996). Homeotic transformations have also played an important role in the evolution of normal phenotypes. For example, it is now commonly accepted that the digits of the adult bird wing derive from the second, third, and fourth developmental anlages (embryonic condensations), but that homeotically and morphologically these digits correspond to digits 1, 2, and 3 of other tetrapods; a similar homeotic transformation seemingly also occurred in the hand of the three-toed Italian skink *Chalcides chalcides* (Young *et al.* 2009).

In order to study the spatial associations between limb bones and muscles, we have compared the adult morphology, development, and regeneration of these structures in several wildtype and non-wildtype tetrapods (e.g., transgenic GFP – green fluorescent protein – amphibians; Diogo and Tanaka 2012, 2014; Diogo *et al.* 2013c; Diogo and Ziermann 2014). We also have investigated how human birth defects involving non-pentadactyl limbs influence muscle attachments and analyzed whether the anatomical patterns seen in those birth defects follow the patterns seen in wild-type non-human tetrapods with non-pentadactyl limbs, in order to infer whether the study of birth defects could reciprocally illuminate those patterns under extreme, unusual conditions. Interestingly, in both the non-pentadactyl limbs of wildtype taxa such as frogs, salamanders, crocodilians, and chickens and in the humans with birth defects, we found a surprisingly consistent pattern. This suggests that the identity and attachments of the fore and hindlimb muscles are mainly related to the physical (topological) position, and not to the number of the anlage or even to the homeotic identity of the digits to which the muscles are attached (Figure 8.4, color plate).

It is important to clarify what is meant here by "topological position," "number of anlage," and "homeotic identity" of the digits. Topological position refers to the adult relationship with other structures, and not to the position of the developmental anlages. For instance, the topological position of the adult avian digit derived from the second developmental anlage is digit 1, because this is the most radial digit in the

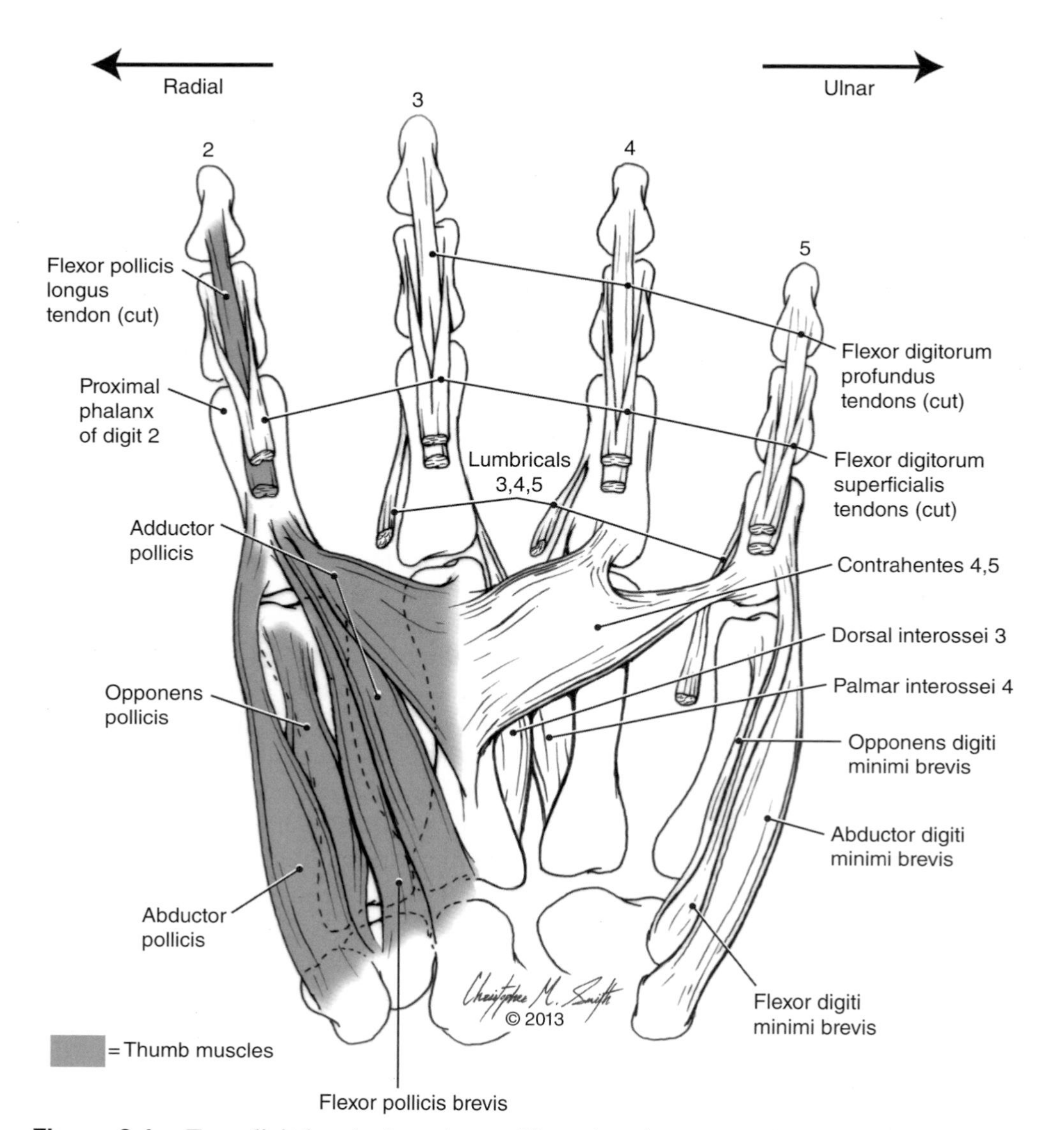

Figure 8.4. Four-digit hand of a trisomy 18 modern human newborn, showing that the muscles that usually attach to the thumb in karyotypically normal modern humans now attach to the digit that has a homeotic identity of digit 2 and develops from the anlage of digit 2, but that is now the most radial digit due to the loss of the thumb. Illustration by Christopher Smith (Johns Hopkins University, University School of Medicine Department of Art as Applied to Medicine).

adult; in this case the topological position (digit 1) and homeotic identity (digit 1) are the same and are different from the developmental anlage from which the digit develops (the second anlage: e.g., Young *et al.* 2009). Accordingly, and as predicted by our hypothesis, chickens do usually have an abductor pollicis brevis that inserts onto this most radial digit; in pentadactyl taxa this muscle is always inserted onto digit 1, which derives from the first, and not the second, anlage (Diogo and Abdala 2010).

This contrasts with the case of salamanders with four hand digits (e.g., axolotls), where digit 5 is missing and digit 4 develops from the anlage of digit 4 and has a homeotic identity of digit 4, but its topological position is similar to that of digit 5 in pentadactyl tetrapods because this is the most ulnar digit. The axolotl case illustrates and supports our hypothesis because although the homeotic identity of this most ulnar digit is that of digit 4, the digit is associated with muscles that normally go to digit 5 in pentadactyl tetrapods (e.g., abductor digiti minimi; Diogo and Tanaka 2012).

As explained above, our theory is also supported by our dissections of, and a review of the literature about, modern humans with birth defects leading to non-pentadactyly. This is illustrated, for instance, by the trisomy-18 newborn illustrated in Figure 8.4, which had six digits (with two thumbs) in one hand and four digits in the other hand (with no thumb). As predicted by our theory, a hand with no thumb has all the muscles that normally go to the thumb, but these muscles go to digit 2, which is the most radial digit (Figure 8.4, color plate). As also predicted, in a hand with two thumbs (preaxial polydactyly, which is one of the most common congenital anomalies of the modern human hand (Castilla *et al.* 1996)), the duplication of the thumb was not accompanied by a duplication of the muscles that normally go to the thumb. Instead, the muscles that normally go to the radial side of the humb (e.g., abductor pollicis brevis) were attached onto the radial side of the most radial thumb, while the muscles that normally go to the ulnar side of the thumb (e.g., adductor pollicis) were attached to the ulnar side of the most ulnar thumb. That is, the muscles are not simply duplicated like the thumb bones. Instead they go to each respective thumb according to the adult topological position of each of the duplicated digits; similar cases were also described by Light (1992).

There are, however, exceptions to our theory. For instance, Heiss (1957) described a case in which a modern human individual had two pentadactyl hands that had no thumbs, and in which, contrary to the cases referred to above, there were no major topological changes of the muscles (e.g., the attachment of normal thumb muscles to the most radial digit of that human subject). Instead, in both hands the normal thumb muscles were missing. In general, this configuration seems to be characteristic of the rare modern human disorder named "tri-phalangeal thumb," which is a malformation of digit 1 including a perfect homeotic transformation of the thumb into an index finger and in which the muscles normally associated with the thumb are absent (e.g., abductor opponens/adductor pollicis; see, e.g., Young and Wagner 2011). Therefore, the analysis of these cases seems to indicate that although the identity and attachments of the limb muscles are mainly related to the physical (topological) position of the digits to which they attach, and not to the number of the anlage or even the homeotic identity of these digits, this does not represent a strict rule. That is, the evolutionary/developmental factors that lead to this common anatomical pattern can be sometimes changed in a way that allows the occurrence of other patterns and thus of more evolutionary possibilities.

Importantly, the examples provided in the above paragraph and the generally predictable muscle changes associated with changes in the number/topological position of digits in both normal and abnormal individuals of different tetrapod taxa, support Alberch's (1989) "logic of monsters." According to this theory, which was also supported by a detailed skeletal study of digit reduction in amphibians (Alberch and Gale 1985), there is a parallel between the variation in normal and abnormal individuals of a certain taxon (e.g., modern humans) and the diversity observed in normal individuals of different taxa (e.g., species of lizards or amphibians). This parallel is achieved through regulation of a conserved developmental program (e.g., a set of genetic and/or epigenetic interactions) such that the structure of these internal interactions constrains the realm of possible variation upon which selection can operate. In theory, such a program can break down in the evolution of some clades, but within most clades this would lead to death of the embryos (Alberch 1989).

The Similarity of the Hind and Forelimb Structures of Modern Humans, Serial Homology and Homoplasy, and Developmental Biology

The idea that the structures of the forelimb (FL) and hindlimb (HL) are "serial homologues" was first proposed by authors such as Vicq d'Azyr (1774), Oken (1843), and Owen (1849). However, a careful examination of these and other original works of these authors reveals that their FL and HL comparisons were almost exclusively based on bones, and not on soft tissues such as muscles, nerves, and blood vessels. Moreover, in most cases the use of the term "limb serial homology" clearly referred to what is currently viewed as parallelism, that is, homoplasy and not true homology. For instance, in *On the Nature of Limbs* (Owen 1849) the examples of striking similarity between the FL and HL refer mainly to tetrapods with highly derived limbs (e.g., bats, horses, and plesiosaurs) and Owen uses the term parallelism more often than "serial homology." When he discusses phylogenetically plesiomorphic taxa (e.g., chondrichthyans) Owen states that those taxa "confuse" the notion of "archetype" and "serial homology," thus he was in fact referring to homoplasy and not to the concept of serial homology as we understand it today (Diogo *et al.* 2013c).

However, authors continued to cite these and other similar classical studies to "show" that the structures of the FL and HL, as well as of the fish pelvic and pectoral fins (PEL and PEC), were serial homologues. Instead of questioning this idea, authors thus preferred to focus on other details about the origin of the allegedly homologous paired appendages. For example, Gegenbaur proposed that fins evolved from the gill arches of the early limbless vertebrates; more recent studies suggest that fins evolved instead from continuous stripes of competency for appendage formation located ventrally and laterally along the embryonic flank (Shubin *et al.* 1997; Don *et al.* 2013). A continuation of this theory proposed that the paired appendages evolved with a shift in the zone of competency to the lateral plate mesoderm in conjunction with the establishment of the lateral somitic frontier, thus allowing for the formation of limb/fin buds with internal supporting skeletons (Don *et al.* 2013). The idea that paired appendages are serial homologues is generally associated with the notion that these appendages were originally similar to each other, and that there was a subsequent functional/anatomical divergence between them (Figure 8.5A, color plate). For instance, Don *et al.* (2013) explain that the ancestral Tbx4/5 cluster of vertebrates probably underwent a duplication event

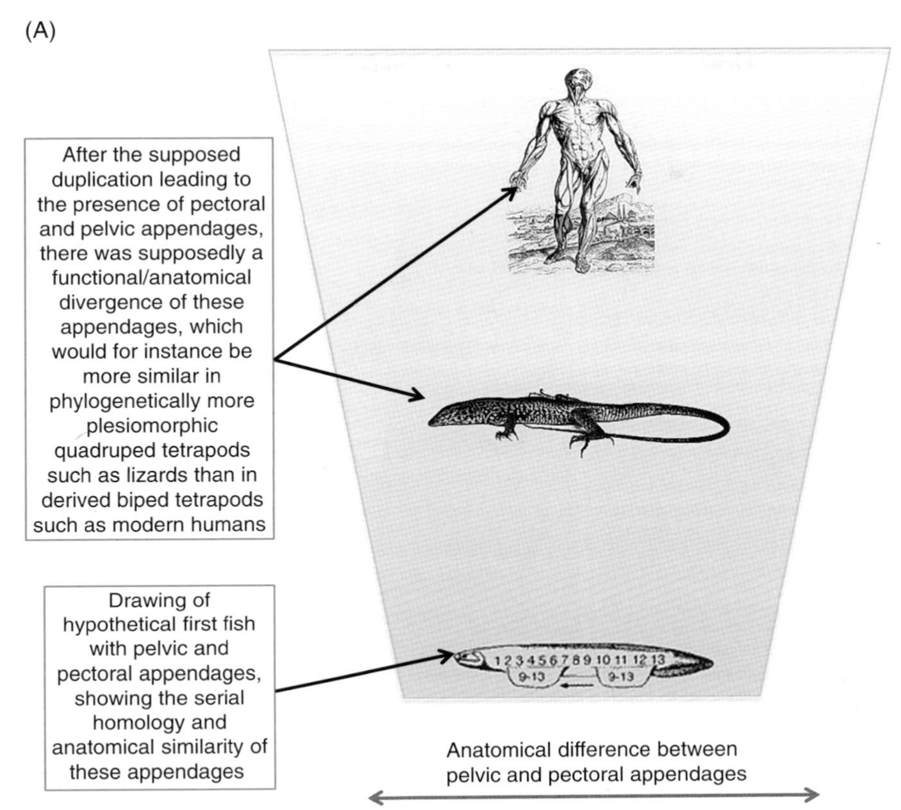

Figure 8.5. (A) Simplified scheme illustrating the "serial homology followed by functional/anatomical divergence" hypothesis often shown in textbooks and followed in more technical papers, particularly within the fields of developmental biology and evo-devo. The picture of the hypothetical fish is modified from Shubin *et al.*'s (1997) scheme showing the origin and evolution of paired appendages. According to that scheme, establishment of serially homologous appendages was proposed to result from gene cooption during the evolution of Paleozoic vertebrates. That is, *Hox* genes were initially involved in specifying regional identities along the primary body axis, particularly in caudal segments, and then during the origin of jawed fish there was a cooption of similar nested patterns of expression of *HoxD* genes in the development of both sets of paired appendages (numbers shown within the fish body). According to this scheme the cooption may have happened in both appendages simultaneously, or *Hox* expression could have been initially present in a pelvic appendage and been coopted in the development of an existing pectoral outgrowth (arrow below fish body; the pictures of the other taxa are modified from Diogo and Abdala 2010 and references therein). (B) The evolutionary history of the pelvic and pectoral appendages was more complex than the "serial homology followed by functional/anatomical divergence" hypothesis suggests. This is because it was more likely the result of a complex interplay between ontogenetic, functional, topological, and phylogenetic constraints leading to cases of anatomical divergence followed by cases of anatomical convergence ("similarity bottlenecks"). This is exemplified in this simplified scheme of the evolutionary muscle transitions leading to modern humans. Figure reused with permission from Diogo and Ziermann (2015).

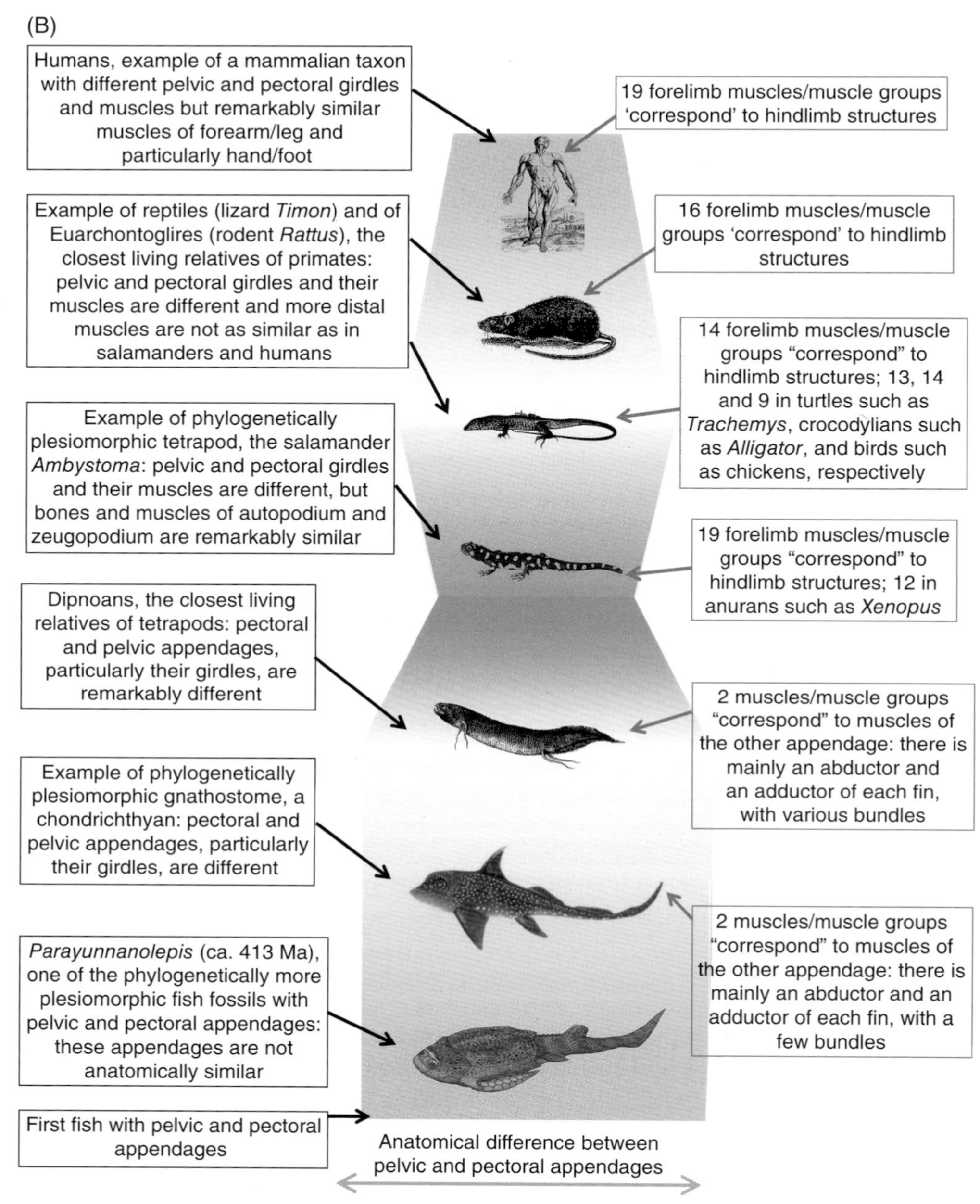

Figure 8.5. (*Continued*).

and that now Tb×4 is related to the HL and Tb×5 with the FL, and in their Figure 3 they state that "pectoral fins evolved first and then duplicated to form pelvic fins." This illustrates the confusion that is often seen in evo-devo studies about (A) a duplication of the Tb×4/5 cluster and subsequent cooption for the genetic pathways associated with the ontogeny of the different paired appendages, and (B) the morphological duplication of the appendages themselves and of their individual structures (i.e., muscles and bones). As we will argue below, A is true but B is not. Because a gene operates to facilitate an outgrowth that gives rise to different limbs in the same animal it does not mean that these limbs are serial homologues, but simply that genes and gene cascades/networks have been recruited homoplasically as organizers of limb development (e.g., Willmer 2003).

In our opinion, the failure to recognize that this idea of PEC and PEL and FL and HL serial homology is unfounded is due to four factors. The first is the recurrent citation of older authors such as Vicq d'Azyr (1774), Oken (1843), and Owen (1849) without a detailed, critical analysis of their arguments. The second is the almost exclusive focus on bones. The third factor, which is related to those above and to the decline of evolutionary vertebrate morphology in the second half of the 20th century (see Diogo and Abdala 2010 and Diogo *et al.* 2013c), is the scarcity of evidence from detailed and unbiased comparisons of both the hard and the soft tissues of the limbs across the major vertebrate clades. The fourth is the lack of a broader integration of the data obtained in such comparisons and in the fields of contemporary developmental biology, genetics, paleontology, functional morphology, and evolutionary biology, with the information provided in the older, classical texts.

As with the re-examination of the classic texts on the "serial homology" of FL and HL structures, detailed analysis of recent developmental, genetic, paleontological, and functional studies also raises serious questions about the serial homology concept. For example, a recent paper describing phylogenetically more plesiomorphic fossil fish with both pectoral and pelvic appendages (an antiarch placoderm) shows that these appendages are markedly different anatomically (Figure 8.5B, color plate; Zhu *et al.* 2012). The idea that these appendages were originally different and then later became more similar in fish such as osteichthyans has been defended by a few authors (Coates and Cohn 1998). Recent functional studies have also contradicted the old idea that the tetrapod FL and HL evolved mainly in a terrestrial environment and originally had similar functions (Pierce *et al.* 2012). Furthermore, recent developmental and genetic studies demonstrated a surprisingly distinct lag between the developmental modes of pectoral and pelvic appendage musculature not only in fish but also in tetrapods (e.g., Cole *et al.* 2011; Don *et al.* 2013). A review of the most detailed, and unfortunately often neglected, older developmental studies of the hard and soft tissues of the FL and HL of modern humans and other tetrapods also revealed major differences between FL and HL ontogeny (Bardeen 1906; Lewis 1910; Cihak 1972). A recent publication (Diogo *et al.* 2013c) has reviewed these and other lines of evidence that contradict the arguments used to support the FL-HL serial homology hypothesis. The final refutation of the "paired appendages serial homology followed by functional/anatomical divergence" hypothesis (Figure 8.5A, color plate) comes from the integrative analysis of lines of evidence obtained from various biological fields together with the results of a detailed examination and comparison of the soft tissues of the paired appendages of representatives of all major extant gnathostome clades. This is because the PEC and PEL of extant plesiomorphic gnathostome fish have basically undifferentiated adductor and abductor muscles masses (Figure 8.5B, color plate). All of the numerous and in many cases strikingly similar muscles (as discussed below), as well as various bones (e.g., Don *et al.* 2013) of the tetrapod FL and HL were therefore almost certainly acquired independently during the evolutionary transitions between early gnathostomes and tetrapods. That is, within a historical (phylogenetic: Wagner 1994) definition of homology these FL and HL structures cannot be considered serial homologues. One *could* argue that they may be homologues under the morphological or developmental definitions of homology (Wagner 1994), but this view is contradicted by the overall analysis of the evolution and homologies of the soft tissues, particularly the muscles, of the gnathostome paired appendages.

In fact, a hypothesis of morphological or developmental serial homology of these appendages implies that these appendages were originally similar and then diverged anatomically/functionally. But there is no good evidence that this is the case (Figure 8.5B, color plate). As noted above, in the more plesiomorphic fish found so far with PEC and PEL the hard tissues of these appendages are anatomically very different, and both the hard and soft tissues of the proximal region of the pectoral and pelvic appendages (particularly the girdles) remained markedly different throughout all gnathostome clades (Diogo *et al.* 2013c). For example, in all tetrapod clades listed in Figure 8.5B, including anatomically plesiomorphic taxa such as urodeles, there is no pelvic-thigh muscle that corresponds topologically to a pectoral-arm muscle. This may be due to phylogenetic constraints (e.g., the pelvic and pectoral girdles of fish and tetrapods were quite different anatomically). However, the more derived distal regions of the tetrapod FL and HL, particularly the autopodia (hand/foot), have a bony skeleton and a developmental plan that is quite different from those of the fish PEC and PEL (NB: even if tetrapod digits are considered to be derived from fish distal rays, as proposed for instance by Johanson *et al.* 2007, it is generally accepted that some tetrapod wrist/ankle bones are neomorphic structures: e.g., Don *et al.* 2013). That is, this case involves a major "evolutionary novelty" and thus less phylogenetic constraints, and the developmental constraints/factors resulting from further (derived) cooption of similar genes in the ontogeny of the FL and HL thus lead to a more marked similarity (a "similarity bottleneck": Figure 8.5B, color plate) between the distal regions of these limbs in plesiomorphic tetrapods. For example, in salamanders such as axolotls 19 muscles/muscle groups of the leg-foot clearly seem to "correspond" topologically to forearm-hand structures (Figure 8.5B, color plate). It should be emphasized that the developmental changes associated with the "transition fin-limb bottleneck" refer to a phylogenetically derived cooption of a few similar genes ("genetic piracy" *sensu* Roth 1994). It is now accepted that the cooption of similar ancestral genes to independently form complex structures such as eyes in vertebrate and non-vertebrate animals ("deep homology") is a case of evolutionary parallelism and thus a variety of homoplasy (Willmer 2003). It would therefore be very difficult to argue that the *derived* cooption of similar genes represents a true case of FL-HL serial homology under the developmental concept of homology.

Moreover, such ontogenetic factors/constraints are clearly not sufficient to explain the striking similarity between the leg-foot and forearm-hand muscles of (and thus, the other "similarity bottlenecks" leading to: Figure 8.5B, color plate) derived tetrapods such as horses or modern humans (Figure 8.6, color plate). The similarity bottlenecks shown in Figure 8.5B come from the empirical data obtained

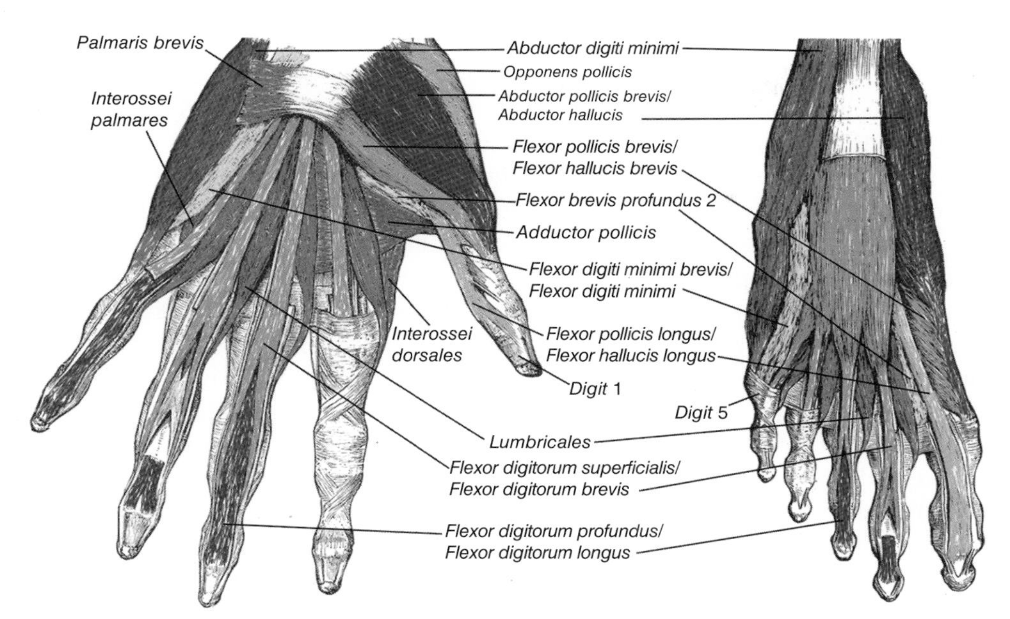

Figure 8.6. Superficial musculature of the modern human hand seen in palmar view (on the left) and of the modern human foot seen in plantar view (on the right). There are striking similarities between the muscles of the autopodium (hand/foot) of the forelimb and hindlimb of modern humans. Modified from Diogo *et al.* 2013c.

from our comparative studies and refer to cases in which there is a higher number of PEC muscles that are topologically similar to PEL muscles. For instance, in urodeles such as axolotls, and in modern humans, there are 19 muscles/muscle groups of the leg-foot that clearly seem to "correspond" topologically to forearm-hand structures. However, in anatomically more generalized quadrupedal mammals (e.g., rats) the number is 16 and in anatomically generalized reptiles such as lizards the number is only 14 (Figure 8.5B, color plate). As will be explained below, many of the human FL-HL muscles/muscle groups with "clear topological equivalents" were acquired independently during the evolutionary history of primates (Diogo and Wood 2012a; Diogo *et al.* 2013c). Such "similarity bottlenecks" have also led to various other tetrapod and even non-tetrapod derived clades (e.g., the PEL muscles of plesiomorphic teleosts were derived independently from, but are topologically very similar to, the PEC muscles of these fish: Winterbottom 1974). Apart from the influence of ontogenetic constraints, the "similarity bottlenecks" leading to derived fish and tetrapod clades (Figure 8.5B, color plate) are clearly also influenced by topological and functional constraints. For example, the only way to have a functional abductor hallucis or pollicis brevis (both muscles were homoplasically acquired during tetrapod evolution) is to have a muscle lying on the radial or tibial side and then having it insert onto the radial or tibial side of the first digit of the hand or foot (Figure 8.6, color plate).

A crucial point supporting the idea that ontogenetic constraints are not sufficient to explain the "similarity bottlenecks" leading to derived taxa such as modern humans

(Figure 8.5B, color plate) is the fact that various leg-foot muscles that are strikingly similar topologically to forearm-hand muscles in derived taxa do not develop from similar anlages. For example, the extensor pollicis longus and extensor hallucis longus of modern humans are remarkably similar topologically, but the former derives from the anlage of the short extensors of the hand, while the latter derives from the anlage of the long extensors of the leg (for more examples see Diogo *et al.* 2013c).

In summary, during the evolutionary history of the tetrapod FL and HL there are cases of evolutionary divergence leading to differences between the FL and HL musculature that are not seen in anatomically plesiomorphic tetrapods such as urodeles. There is also a substantial evolutionary parallelism/convergence leading to subsequent "similarity bottlenecks" between these limbs in derived taxa (Figure 8.5B, color plate). The strikingly similar FL and HL muscles seen for instance in groups such as modern humans are therefore the result of a complex interplay between ontogenetic, topological, functional, and even phylogenetic (Diogo *et al.* 2013c) constraints. A further critical point that shows the similarity of these muscles cannot be simply explained by a derived morphological/developmental integration of the FL and HL under a morphological or developmental serial homology concept (see above) is that several of these muscles, or their bundles, did not emerge at similar geological times/phylogenetic nodes. For example, the adductor hallucis and adductor pollicis of modern humans are similar because they have well-differentiated transverse and oblique heads. However, whereas the heads of the adductor hallucis are already well differentiated in phylogenetically plesiomorphic primates such as lemurs, those of the adductor pollicis only became well differentiated in the node leading to catarrhines (Old World monkeys + hominoids); for more examples see Diogo *et al.* (2013c).

We hope that the evidence and discussion above will at least make the readers aware of the serious problems faced by the fore-hindlimb serial homology hypothesis, and stimulate research on the puzzling and fascinating evolutionary history of the paired appendages using a new paradigm.

General Remarks and Future Directions

Muscle and other soft tissue evidence has been neglected in systematics and evolutionary and developmental biology. However, the few cladistic analyses based on soft tissues that have been published have shown that these tissues can be particularly useful for inferring phylogenetic relationships (e.g., Diogo and Wood 2012a). Moreover, the study of muscles also allows researchers to address evolutionary and developmental questions that are not tractable using other types of evidence. These include crucial questions about the evolution of the closest living relatives of modern humans and the evolution within our own clade. In the last few decades the emergence of evolutionary developmental biology has resulted in a resurgence of interest in comparative anatomy (e.g., Assis 2009). We suggest that the forthcoming decades will see a renaissance in the use of myology in evolutionary and developmental biology, including for taxa such as primates and modern humans.

Hopefully, by focusing on morphological and developmental evolution, this review will help with this renaissance by stimulating an interest in the study of muscles of vertebrates in general, and of primates and modern humans in particular. A line of future research that is particularly promising would be to marry the data obtained from comparative and developmental myological studies with paleontological and environmental data in order to better understand how, where, when, and even why changes to muscle morphology occurred during phylogeny.

References

Alberch, P. 1989. The logic of monsters: Evidence for internal constraint in development and evolution. *Geobios, Mémoire Special* 12:21–57.

Alberch, P., and Gale, E. A. 1985. A developmental analysis of an evolutionary trend: Digital reduction in amphibians. *Evolution* 39:8–23.

Arnold, C., Matthews, L. J., and Nunn, C. L. 2010. The 10kTrees website: A new online resource for primate phylogeny. *Evolutionary Anthropology* 19:114–118.

Assis, L. C. S. 2009. Coherence, correspondence, and the renaissance of morphology in phylogenetic systematics. *Cladistics* 25:528–544.

Bakewell, M. A., Shi, P., and Zhang, J. 2007. More genes underwent positive selection in chimpanzee evolution than in human evolution. *Proceedings of the National Academy of Sciences, USA* 104:7489–7494.

Bardeen, C. R. 1906. Development and variation of the nerves and the musculature of the inferior extremity and of the neighboring regions of the trunk in man. *American Journal of Anatomy* 6:259–390.

Blitz, E., Sharir, A., Akiyama, H., and Zelzer, E. 2013. Tendon-bone attachment unit is formed modularly by a distinct pool of *Scx*- and *Sox9*-positive progenitors. *Development* 140:2680–2690.

Bonner, J. T. 1988. *The evolution of complexity by means of natural selection*. Princeton: Princeton University Press.

Brand, B., Christ, B., and Jacob, H. J. 1985. An experimental analysis of the developmental capacities of distal parts of avian leg buds. *American Journal of Anatomy* 173:321–340.

Bufill, E., Agusti, J., and Blesam, R. 2011. Human neoteny revisited: The case of synaptic plasticity. *American Journal of Human Biology* 23:729–739.

Castilla, E. E., da Fonseca, R. L., da Graça Dutra, M. *et al.* 1996. Epidemiological analysis of rare polydactylies. *American Journal of Medical Genetics* 65:295–303.

Chatterjee, H. J., Ho, S. Y. M., Barnes, I., and Groves, C. 2009. Estimating the phylogeny and divergence times of primates using a supermatrix approach. *BMC Evolutionary Biology* 9:259.

Cihak, R. 1972. Ontogenesis of the skeleton and intrinsic muscles of the human hand and foot. *Advances in Anatomy, Embryology and Cell Biology* 46:1–194.

Coates, M. I., and Cohn, M. J. 1998. Fins, limbs, and tails: Outgrowths and axial patterning in vertebrate evolution. *BioEssays* 20:371–381.

Cole, N. J., Hall, T. E., Don, E. K. *et al.* 2011. Development and evolution of the muscles of the pelvic fin. *PLoS Biology* 9:e1001168.

Cooper, N., and Purvis, A. 2009. What factors shape rates of phenotypic evolution? A comparative study of cranial morphology of four mammalian clades. *Journal of Evolutionary Biology* 22:1024–1035.

DeLaurier, A., Schweitzer, R., and Logan, M. 2006. *Pitx1* determines the morphology of muscle, tendon, and bones of the hindlimb. *Developmental Biology* 299:22–34.

Diogo, R. 2004a. *Morphological evolution, aptations, homoplasies, constraints, and evolutionary trends: Catfishes as a case study on general phylogeny and macroevolution*. Enfield: Science Publishers.

Diogo, R. 2004b. Muscles versus bones: Catfishes as a case study for an analysis on the contribution of myological and osteological structures in phylogenetic reconstructions. *Animal Biology* 54:373–391.

Diogo, R. 2005. Evolutionary convergences and parallelisms: Their theoretical differences and the difficulty of discriminating them in a practical phylogenetic context. *Biology & Philosophy* 20:735–744.

Diogo, R. 2007. *On the origin and evolution of higher-clades: Osteology, myology, phylogeny and macroevolution of bony fishes and the rise of tetrapods*. Enfield: Science Publishers.

Diogo, R., and Abdala, V. 2007. Comparative anatomy, homologies and evolution of the pectoral muscles of bony fish and tetrapods: A new insight. *Journal of Morphology* 268:504–517.

Diogo, R., and Abdala, V. 2010. *Muscles of vertebrates: Comparative anatomy, evolution, homologies and development*. Oxford: Taylor & Francis.

Diogo, R., Abdala, V., Aziz, M. A. *et al.* 2009a. From fish to modern humans – comparative anatomy, homologies and evolution of the pectoral and forelimb musculature. *Journal of Anatomy* 214:694–716.

Diogo, R., Abdala, V., Lonergan, N. L., and Wood. B. 2008a. From fish to modern humans – comparative anatomy, homologies and evolution of the head and neck musculature. *Journal of Anatomy* 213:391–424.

Diogo, R., Hinits, Y., and Hughes, S. 2008b. Development of mandibular, hyoid and hypobranchial muscles in the zebrafish, with comments on the homologies and evolution of these muscles within bony fish and tetrapods. *BMC Developmental Biology* 8:24–46.

Diogo, R., Linde-Medina, M., Abdala, V., and Ashley-Ross, M. A. 2013a. New, puzzling insights from comparative myological studies on the old and unsolved forelimb/hindlimb enigma. *Biological Reviews* 88:196–214.

Diogo, R., Peng, Z., and Wood, B. 2013d. First comparative study of primate morphological and molecular evolutionary rates including muscle data: Implications for the tempo and mode of primate and human evolution. *Journal of Anatomy* 222:410–418.

Diogo, R., Potau, J. M., Pastor, J. F. *et al.* 2010. *Photographic and descriptive musculoskeletal atlas of gorilla*. Oxford: Taylor & Francis.

Diogo, R., Potau, J. M., Pastor, J. F. *et al.* 2012a. *Photographic and descriptive musculoskeletal atlas of gibbons and siamangs (Hylobates)*. Oxford: Taylor & Francis.

Diogo, R., Potau, J. M., Pastor, J. F. *et al.* 2013c. *Photographic and descriptive musculoskeletal atlas of chimpanzees (Pan)*. Oxford: Taylor & Francis.

Diogo, R., Potau, J. M., Pastor, J. F. *et al.* 2013d. *Photographic and descriptive musculoskeletal atlas of orangutans (Pongo)*. Oxford: Taylor & Francis.

Diogo, R., Richmond, B. G., and Wood, B. 2012b. Evolution and homologies of primate and modern human hand and forearm muscles, with notes on thumb movements and tool use. *Journal of Human Evolution* 63:64–78.

Diogo, R., and Tanaka, E. M. 2012. Anatomy of the pectoral and forelimb muscles of wildtype and green fluorescent protein-transgenic axolotls and comparison with other tetrapods including humans: A basis for regenerative, evolutionary and developmental studies. *Journal of Anatomy* 221:622–635.

Diogo, R., and Tanaka, E. M. 2014. Development of fore- and hindlimb muscles in GFP-transgenic axolotls: Morphogenesis, the tetrapod bauplan, and new insights on the forelimb-hindlimb enigma. *Journal of Experimental Zoology, Part B: Molecular and Developmental Evolution* 322:106–127.

Diogo, R., and Wood, B. 2011. Soft-tissue anatomy of the primates: Phylogenetic analyses based on the muscles of the head, neck, pectoral region and upper limb, with notes on the evolution of these muscles. *Journal of Anatomy* 219:273–359.

Diogo, R., and Wood, B. 2012a. *Comparative anatomy and phylogeny of primate muscles and human evolution*. Oxford: Taylor & Francis.

Diogo, R., and Wood, B. 2012b. Violation of Dollo's Law: Evidence of muscle reversions in primate phylogeny and their implications for the understanding of the ontogeny, evolution and anatomical variations of modern humans. *Evolution* 66:3267–3276.

Diogo, R., and Wood, B. 2013. The broader evolutionary lessons to be learned from a comparative and phylogenetic analysis of primate muscle morphology. *Biological Reviews* 88:988–1001.

Diogo, R., Wood, B., Aziz, M. A., and Burrows, A. 2009b. On the origin, homologies and evolution of primate facial muscles, with a particular focus on hominoids and a suggested unifying nomenclature for the facial muscles of the Mammalia. *Journal of Anatomy* 215:300–319.

Diogo, R., and Ziermann, J. M. 2014. Development of fore- and hindlimb muscles in frogs: Morphogenesis, homeotic transformations, digit reduction, and the forelimb-hindlimb enigma. *Journal of Experimental Zoology, Part B: Molecular and Developmental Evolution* 322:86–105.

Diogo, R., Ziermann, J. M., and Linde-Medina, M. 2015. Is evolutionary biology becoming too politically correct? A reflection on the scalae naturae, phylogenetically basal clades, anatomically plesiomorphic taxa, and "lower" animals. *Biological Reviews* 90:502–521.

Don, E. K., Currie, P. D., and Cole, N. J. 2013. The evolutionary history of the development of the pelvic fin/hindlimb. *Journal of Anatomy* 222:114–133.

Dunlap, S. S., Aziz, M. A., and Rosenbaum, K. N. 1986. Comparative anatomical analysis of human trisomies 13, 18 and 21: I, the forelimb. *Teratology* 33:159–186.

Duprez, D. 2002. Signals regulating muscle formation in the limb during embryonic development. *International Journal of Developmental Biology* 46:915–925.

Edgeworth, F. H. 1935. *The cranial muscles of vertebrates*. Cambridge: Cambridge University Press.

Elsayed, S. M., and Elsayed, G. M. 2009. Phenotype of apoptopic lymphocytes in children with Down syndrome. *Immunity & Ageing* 6:2.

Fabre, P.-H., Rodrigues, A., and Douzery, E. J. P. 2009. Patterns of macroevolution among Primates inferred from a supermatrix of mitochondrial and nuclear DNA. *Molecular Phylogenetics and Evolution* 53:808–825.

Fabrezi, M., Abdala, V., and Oliver, M. I. M. 2007. Developmental basis of limb homology in lizards. *Anatomical Record* 290:900–912.

Galis, F., and Metz, J. A. J. 2007. Evolutionary novelties: The making and breaking of pleiotropic constraints. *Integrative and Comparative Biology* 47:409–419.

Gibbs, S., Collard, M., and Wood, B. 2002. Soft-tissue anatomy of the extant hominoids: A review and phylogenetic analysis. *Journal of Anatomy* 200:3–49.

Goodman, M., Czelusniak, J., and Beeber, J. E. 1985. Phylogeny of primates and other eutherian orders: A cladistic analysis using amino acid and nucleotide sequence data. *Cladistics* 1:171–185.

Gotlieb, N. R. 2009. *Quantification and analysis of apoptosis in embryonic atrioventricular endocardial cushions of the Ts65Dn mouse model for Down syndrome.* Graduate thesis, Franklin and Marshall College.

Gould, S. J. 1977. *Ontogeny and phylogeny*. Cambridge, MA: Harvard University Press.

Gould, S. J. 2002. *The structure of evolutionary theory*. Cambridge, MA: Belknap.

Hasson, P. 2011. "Soft" tissue patterning: Muscles and tendons of the limb take their form. *Developmental Dynamics* 240:1100–1107.

Hatta, K., Bremiller, R., Westerfield, M., and Kimmel, C. B. 1991. Diversity of expression of engrailed-like antigens in zebrafish. *Development* 112:821–832.

Hatta, K., Schilling, T. F., Bremiller, R., and Kimmel, C. B. 1990. Specification of jaw muscle identity in zebrafish: Correlation with engrailed-homeoprotein expression. *Science* 250:802–805.

Heiss, H. 1957. Beidseitige kongenitale daumenlose Fünffingerhand bei Mutter und Kind. *Zeitschrift für Anatomie und Entwicklungsgeschichte* 120:226–231.

Hulsey, C. D., and Hollingsworth, P. R. 2011. Do constructional constraints influence cyprinid (Cyprinidae: Leuciscinae) craniofacial evolution? *Biological Journal of the Linnean Society* 103:136–146.

Johanson, Z., Joss, J., Boisvert, C. A. *et al.* 2007. Fish fingers: Digit homologues in sarcopterygian fish fins. *Journal of Experimental Zoology, Part B: Molecular and Developmental Evolution* 15B:757–768.

Kardon, G. 1998. Muscle and tendon morphogenesis in the avian hind limb. *Development* 125:4019–4032.

Kimura, M. 1968. Evolutionary rate at the molecular level. *Nature* 5129:624–626.

Knight, R. D., Mebus, K., and Roehl, H. H. 2008. Mandibular arch muscle identity is regulated by a conserved molecular process during vertebrate development. *Journal of Experimental Zoology, Part B: Molecular and Developmental Evolution* 310B:355–369.

Köntges, G., and Lumsden, A. 1996. Rhombencephalic neural crest segmentation is preserved throughout craniofacial ontogeny. *Development* 122:3229–3242.

Kundrát, M., Janacek, J., and Martin, M. 2009. Development of transient head cavities during early organogenesis of the Nile Crocodile (*Crocodylus niloticus*). *Journal of Morphology* 270:1069–1083.

Lewis, W. H. 1910. The development of the muscular system. *In:* F. Keibel and F. P. Mall (eds) *Manual of embryology, vol.* 2. Philadelphia: J. B. Lippincott. pp.455–522.

Light, T. R. 1992. Thumb recostruction. *Hand Clinics* 8:161–175.

Miyake, T., McEachran, J. D., and Hall, B. K. 1992. Edgeworth's legacy of cranial muscle development with an analysis of muscles in the ventral gill arch region of batoid fishes (Chondrichthyes: Batoidea). *Journal of Morphology* 212:213–256.

Noden, D. M. 1983. The embryonic origins of avian cephalic and cervical muscles and associated connective tissues. *American Journal of Anatomy* 168:257–276.

Noden, D. M. 1984. Craniofacial development: New views on old problems. *Anatomical Record* 208:1–13.

Noden, D. M. 1986. Patterning of avian craniofacial muscles. *Developmental Biology* 116:347–356.

Noden, D. M., and Francis-West, P. 2006. The differentiation and morphogenesis of craniofacial muscles. *Developmental Dynamics* 235:1194–1218.

Oken, L. 1843. *Lehrbuch der Naturphilosophie*. Zürich: Friedrich Schulthess.

Owen, R. 1849. *On the nature of limbs*. London: John Van Voorst.

Perelman, P., Johnson, W. E., Roos, C. *et al.* 2011. A molecular phylogeny of living primates. *PLoS Genetics* 7:e1001342.

Piekarski, N., and Olsson, L. 2007. Muscular derivatives of the cranialmost somites revealed by long-term fate mapping in the Mexican axolotl (*Ambystoma mexicanum*). *Evolution & Development* 9:566–578.

Pierce, S. E., Clack, J. A., and Hutchinson, J. R. 2012. Three-dimensional limb joint mobility in the early tetrapod *Ichthyostega*. *Nature* 486:523–626.

Pierce, S. E., Hutchinson, J. R., and Clack, J. A. 2013. Historical perspectives on the evolution of tetrapodomorph movement. *Integrative and Comparative Biology* 53:1–15.

Prunotto, C., Crepaldi, T., Forni, P. E. *et al.* 2004. Analysis of *Mlc-lacZ* Met mutants highlights the essential function of Met for migratory precursors of hypaxial muscles and reveals a role for Met in the development of hyoid arch-derived facial muscles. *Developmental Dynamics* 231:582–591.

Roth, L. V. 1994. Within and between organisms: Replicators, lineages, and homologues. *In:* B. K. Hall (ed.) *Homology: The hierarchical basis of comparative biology*. San Diego: Academic Press. pp.301–337.

Shearman, R. M., and Burke, A. C. 2009. The lateral somitic frontier in ontogeny and phylogeny. *Journal of Experimental Zoology, Part B: Molecular and Developmental Evolution* 312B:602–613.

Shubin, N. H., and Alberch, P. 1986. A morphogenetic approach to the origin and basic organization of the tetrapod limb. *Evolutionary Biology* 20:319–387.

Shubin, N., Tabin, C., and Carrol, S. 1997. Fossils, genes and the evolution of animal limbs. *Nature* 388:639–648.

Shwartz, Y., Farkas, Z., Stern, T. *et al.* 2012. Muscle contraction controls skeletal morphogenesis through regulation of chondrocyte convergent extension. *Developmental Biology* 370:154–163.

Steiper, M. E., Young, N. M., and Sukarna, T. Y. 2004. Genomic data support the hominoid slowdown and an Early Oligocene estimate for the hominoid-cercopithecoid divergence. *Proceedings of the National Academy of Sciences, USA* 101:17021–17026.

Tetushkin, E. Y. 2003. Rates of molecular evolution of primates. *Russian Journal of Genetics* 39:721–736.

Theis, S., Patel, K., Valasek, P. *et al.* 2010. The occipital lateral plate mesoderm is a novel source for vertebrate neck musculature. *Development* 137:2961–2971.

Tzahor, E. 2009. Heart and craniofacial muscle development: A new developmental theme of distinct myogenic fields. *Developmental Biology* 327:273–279.

Young, N. M., and Hallgrímsson, B. 2005. Serial homology and the evolution of mammalian limb covariation structure. *Evolution* 59:2691–2704.

Young, R. L., Caputo, V., Giovannotti, M. *et al.* 2009. Evolution of digit identity in the three-toed Italian skink *Chalcides chalcides*: A new case of digit identity frame shift. *Evolution & Development* 11:647–658.

Young, R. L., and Wagner, G. P. 2011. Why ontogenetic homology criteria can be misleading: Lessons from digit identity transformations. *Journal of Experimental Zoology, Part B: Molecular and Developmental Evolution* 316B:165–170.

Valasek, P., Theis, S., DeLaurier, A. *et al.* 2011. Cellular and molecular investigations into the development of the pectoral girdle. *Developmental Biology* 357:108–116.

Vermeij, G. J. 1973. Biological versatility and earth history. *PNAS* 70:1936–1938.

Vicq-d'Azyr, F. 1774. Parallèle des os qui composent les extrémités. *Mémoires de l'Académie Royale des Sciences* 1774:519–557.

Wagner, G. P. 1994. Homology and the mechanisms of development. *In:* B. K. Hall (ed.) *Homology: The hierarchical basis of comparative biology*. San Diego: Academic Press. pp.273–299.

Wagner, G. P., and Altenberg, L. 1996. Perspective: Complex adaptations and the evolution of evolvability. *Evolution* 50:967–976.

Weisbecker, V. 2011. Monotreme ossification sequences and the riddle of mammalian skeletal development. *Evolution* 65:1323–1335.

West-Eberhard, M. J. 2003. *Developmental plasticity and evolution.* Oxford: Oxford University Press.

Wiens, J. J. 2011. Re-evolution of lost mandibular teeth in frogs after more than 200 million years, and re-evaluating Dollo's Law. *Evolution* 65:1283–1296.

Willmer, P. 2003. Convergence and homoplasy in the evolution of organismal form. *In:* G. Muller and S. Newman (eds) *Origination of organismal form: Beyond the gene in developmental and evolutionary biology*. Cambridge, MA: MIT Press. pp.40–49.

Winterbottom, R. 1974. A descriptive synonymy of the striated muscles of the Teleostei. *Proceedings of the Academy of Natural Sciences of Philadelphia* 125:225–317.

Zhu, M., Yu, X., Choo, B. *et al.* 2012. An antiarch placoderm shows that pelvic girdles arose at the root of jawed vertebrates. *Biology Letters* 23:453–456.

Ziermann, J. M., and Diogo, R. 2013. Cranial muscle development in the model organism *Ambystoma mexicanum*: Implications for tetrapod and vertebrate comparative and evolutionary morphology and notes on ontogeny and phylogeny. *Anatomical Record* 296:1031–1048.

Ziermann, J. M., and Diogo, R. 2014. Cranial muscle development in frogs with different developmental modes: Direct development vs. biphasic development. *Journal of Morphology* 275:398–413.

Chapter 9

The Evolutionary Biology of Human Neurodevelopment: Evo-Neuro-Devo Comes of Age

Bernard Crespi and Emma Leach
Department of Biological Sciences, Simon Fraser University, Burnaby, BC, Canada

Introduction

The field of evolutionary developmental biology arose from the joining of two research traditions: century-old conceptualizations of how embryonic development has evolved, and recent discoveries of how genes orchestrate changes and variation in development. The goals of this field are manyfold, and famously include the roles of pleiotropy and "constraints" in evolutionary change compared to unconstrained polygenic inheritance, the extent and control of modularity in developmental-genetic phenotypes, and the importance of heterochrony in evolutionary-developmental trajectories.

Most applications of evolutionary-developmental research have centered on morphological or life-historical timing phenotypes, probably due to ease of quantification. However, some of the most interesting traits that have evolved through changes in development are behavioral, and ultimately in a mechanistic sense, neurological. Such phenotypes are more challenging to measure due to the complexities of brain development and function. Recent, accelerating progress in the study of human neurodevelopment, in the contexts of both typical and atypical brain development, are making possible the first tripartite connections between evolutionary biology, developmental biology, and neurological phenotypes. These connections are especially important because they allow the joint study of how humans evolved, how the human brain develops, and how variation among individuals in human brain development can manifest as neurodevelopmental conditions, most prominently autism and schizophrenia. In turn, such studies can dovetail with anthropological, paleontological, and comparative-primatological data on human evolution, to uncover convergent lines of evidence that lend rigor to the intrinsically challenging goal of inferring how modern human brain development and cognition have evolved.

Developmental Approaches to Human Evolution, First Edition. Edited by
Julia C. Boughner and Campbell Rolian.

In this chapter, we seek to inaugurate the field of evo-neuro-devo, the study of how neurodevelopment evolves. We focus on humans, the species for whom most salient data are available.

Our general approach is straightforward: segregating genetic variation, and *de novo* (new germline) mutations, provide novel insights into human neurodevelopmental gene functions, including effects from pleiotropy, polygenic inheritance, and developmental-genetic convergence. Phenotypes characterizing psychiatric conditions mediated by neurodevelopment are generated via gene-environment interactions, which have been more or less highly canalized by effects of selection whose impacts are expected to decrease with evolutionary time. Current phenotypic effects from genetic variation thus allow direct insights into psychiatric conditions, typical neurogenetic architecture (the genetic basis of neurological variation), and evolutionary histories of neurodevelopment: how modularity, connectivity, timing, and information-processing trajectories have evolved. This methodology is analogous to using experimental alterations of genes, proteins, or pathways to infer functions and tradeoffs, in the context of phenotypic, evolutionary trajectories inferred from independent sources of information.

We address two main questions. First, how are the causes and phenotypes of the primary human neurodevelopmental conditions, especially autism and schizophrenia, related to recent neurodevelopmental and cognitive changes in human evolutionary history? In this context, how have risks for particular neurodevelopmental psychiatric conditions, as constellations of associated phenotypes, evolved, as extremes of normal variation? We refer to these psychiatric reifications as "conditions," more than "disorders," to emphasize that their psychological and neurological phenotypic spectra grade continuously into so-called normality, and to pre-empt consideration of psychiatric conditions as "diseases" that solely involve dysfunction.

Second, to what degree does variation in specific sets of evo-neuro-devo phenotypes underlie the causes and phenotypes of neurodevelopmental conditions? In particular, we focus on how genetic factors that cause changes in the rate and timing patterns of neurodevelopmental events, and associated psychological phenotypes, are involved in autism spectrum conditions and schizophrenia. We thus put forward a "developmental heterochronic" model for helping to explain the genetic bases of these two conditions, and connect the model with heterochronic change in human ancestry. This question generates a new perspective on how typical development is related to atypical development in autism and schizophrenia, in the general framework of how recently-evolved human neurological and psychological traits have generated risks for particular extremes of variation.

The Genetical Evolution of Neurodevelopment

The evolutionary genetics of human neurodevelopment can be studied from two perspectives, which we call "genes-up" and "phenotypes-down." The genes-up approach involves analyses of specific genes with documented functional roles in both human evolution and human neurodevelopment. The best examples of such genes are those that have mediated the evolutionary tripling of human brain size since our divergence from the chimpanzee lineage. Such genes were originally designated as "microcephaly genes" by medical geneticists because their mutational losses of function lead to brain sizes about one-third of normal.

Molecular-biological studies indicate that microcephaly genes (which include *ASPM*, *CDK5RAP2*, *CENPJ*, *CEP152*, *DUF1220*, *KCTD13*, *MCPH1*, *STIL*, and *WDR62*, among others) (Mahmood *et al.* 2011; Dumas *et al.* 2012; Golzio *et al.* 2012) exhibit convergent functions in cell-cycle dynamics during early brain development, which increase numbers of neural progenitor cells and thus increase the overall size of the brain (Megraw *et al.* 2011). Most importantly, molecular-evolutionary studies have shown that the evolution of microcephaly genes is characterized by episodes of functional amino-acid evolution – so-called positive selection – along the human lineage as well as among our primate relatives (e.g., Ponting and Jackson 2005; Montgomery *et al.* 2011). Such findings indicate that, from an evolutionary perspective, these are actually genes "for" increased brain size along the human lineage.

Based on these independent lines of molecular, developmental, and evolutionary evidence, we can describe a "microcephaly paradigm" for evo-neuro-devo change: humans undergo a series of naturally selected allelic substitutions that led to the evolution of human-specific phenotypes via changes in development. In turn, genetic alterations to the resultant human-evolved developmental pathways (for microcephaly, losses of function in key regulators) generate phenotypes with evolutionary structure – that is, architecture – that links the human disorder with human evolutionary change and with development. Other sorts of large alterations to microcephaly genes, such as duplications (involving *de novo* mutations), may result in opposite-direction phenotypes: larger brain size, apparently through opposite alterations (gains of function) in the relevant developmental pathway (e.g., Golzio *et al.* 2012). In turn, small-scale variation in brain size has been linked with allelic variation in single nucleotide polymorphisms for several microcephaly genes (Rimol *et al.* 2010). This polygenic, small-scale variation indicates that small genetic effects on brain size development may also reflect phylogenetic history. To the extent that such segregating allelic variation is maintained in human populations due to tradeoffs (as opposed to genetic drift, or as opposed to alleles changing in frequency due to selection; Crespi 2011a), there should be both costs and benefits to the alternative alleles at a locus; for example, larger brains are energetically much more costly, but have been associated with increased scores on tests of "intelligence" (Schoenemann 2006).

The "microcephaly paradigm" (Figure 9.1) involves a simple phenotype (overall brain size), and relatively simple mechanisms (such as extensions of the proliferation stages of early brain growth). A second, more specialized human-evolved phenotype, speech and language, also appears to fit the paradigm. Thus, the *FOXP2* gene has evolved under positive selection in humans, loss of function in the gene results in a reduced human-specific phenotype (here, in speech, language, and articulation skills), and segregating allelic variation in the gene is associated with speech and language–related phenotypes (in autism spectrum conditions and in schizophrenia). Here, the molecular-developmental mechanism involves localized neural-growth effects on

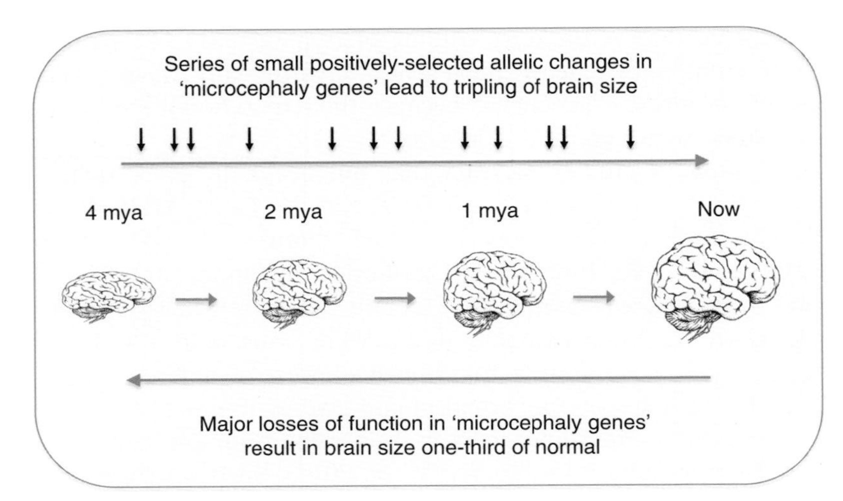

Figure 9.1. The "microcephaly paradigm" posits that human-derived and human-elaborated developmentally based phenotypes, such as brain size, evolve through a series of small, positively selected changes, and can be reduced or lost (or, in some cases, increased) due to large-scale mutational or environmental alterations. This paradigm connects human evolution with human development, in the contexts of their joint genetic underpinnings and risks of particular forms of disease. The microcephaly paradigm is illustrated here with a morphed human brain.

neurodevelopment, in speech-related regions of the brain (Vernes *et al.* 2011). The cognitive effects of alternative segregating alleles on brain development and language remain to be investigated in detail, although for two single nucleotide polymorphisms (SNPs), one allele has been reported to increase risk of autism spectrum disorders, whereas the alternative allele increases risk of schizophrenia with auditory hallucinations (Gong *et al.* 2004; Sanjuan *et al.* 2005; Casey *et al.* 2012). In theory, alternative alleles at SNPs of *FOXP2* may thus involve tradeoffs in language-related cognitive functions among non-clinical populations, as described in more detail below.

Can the microcephaly paradigm be expanded to encompass not just brain size and structure but also major patterns of human cognitive variation underlain by neurodevelopment? To do so, we must adopt a "phenotypes-down" perspective, as the genetic, genomic, and developmental underpinnings of complex human cognitive phenotypes are insufficiently known to start usefully and comprehensively from genes.

The Evolution of Human Cognition and its Disorders

Humans evolved large brains in conjunction with a spectacular constellation of new abilities for complex cognition. Let us therefore expand the microcephaly paradigm to a large suite of human-evolved (uniquely human) and human-elaborated (enhanced and more complex in humans) neurodevelopmental, cognitive phenotypes (Figure 9.2). Each of these phenotypes has evolved in the human lineage, each has a polygenic basis (so far as known), and each undergoes some trajectory of development

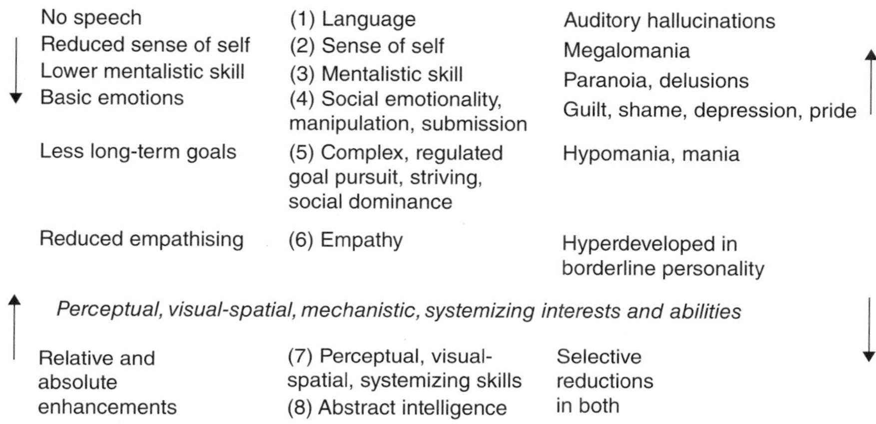

Figure 9.2. The diametric brain hypothesis represents an application of the microcephaly paradigm to human perceptual, cognitive, and affective architecture. By this hypothesis, humans have a suite of social brain adaptations, and a suite of non-social brain adaptations, which tend to exhibit tradeoffs with one another with or without statistical adjustment for general intelligence. Psychotic-affective conditions involve maladaptively over-developed social-brain phenotypes in conjunction, to some degree, with maladaptive under-development of non-social brain phenotypes, and autism spectrum conditions involve the converse. For evidence regarding these alterations, in psychiatric conditions, see Kravariti *et al.* (2006); Mottron *et al.* (2006); Crespi and Badcock (2008); Soulières *et al.* (2009, 2011); Baron-Cohen (2010, 2012); Crespi (2011b); Johnson *et al.* (2012). "Mentalistic" refers here to psychological traits with social or self-referential content.

as humans mature throughout infancy, childhood, adolescence, and young adulthood. Thus, for example, (1) language is unique to humans, (2) humans exhibit highly developed social cognition, social emotionality. and social behavior (the so-called social brain), which is commonly manifest in complex, highly regulated goal pursuit, (3) human technical, mechanical, and systematic thinking skills far exceed those of our close primate relatives, and (4) we show much better developed abstract, so-called "fluid" intelligence, defined as pure problem-solving ability independent of learned, culturally based knowledge (Reader and Laland 2002; Saxe 2006; Reader *et al.* 2011; Suddendorf *et al.* 2011; Frith and Frith 2012; Nisbett *et al.* 2012).

These human-evolved and human-elaborated phenotypes are certainly associated with, and potentiated by, large absolute brain size, but they also involve specializations, as indicated, for example, by the notable degree of neuroanatomical and functional modularity in social-brain (Frith and Frith 2012) and non-social-brain (Stout *et al.* 2008) regions. Moreover, although large suites of human cognitive

Table 9.1. Evidence of negative correlations between social skills and non-social skills, which are indicative of tradeoffs.

Social skill	Non-social skill	Citations
Verbal skills	Spatial-imagery	Johnson and Bouchard 2007
Empathizing	Systemizing	Nettle 2007
Empathizing	Mental rotation	Cook and Saucier 2010
Theory of mind	Embedded figures	Jarrold *et al.* 2000
Social skills	Embedded figures	Pellicano *et al.* 2006 Russell-Smith *et al.* 2012
Social skills	Raven's matrices	Fugard *et al.* 2011
Social skills	Visual search	Keehn *et al.* 2013 Joseph *et al.* 2002, 2009
Reading mind in the eyes	Embedded figures	Baron-Cohen and Hammer 1997 Baron-Cohen *et al.* 1997
Social interest	Mental rotation	Dinsdale *et al.* (2013)

abilities that depend on general intelligence show strong pairwise positive correlations (embodied by the latent factor "g", which appears to reflect some measure of general information-processing skills; Johnson and Bouchard 2005), some sets of human cognitive abilities show negative correlations with one another, with or without statistical adjustments for general intelligence (Table 9.1). In particular, extensive evidence has accumulated, especially in the literature on autism spectrum conditions, for negative associations of social, verbal phenotypes with visual-spatial, perceptual, and mechanistic phenotypes. These findings are indicative of tradeoffs between different sets of human cognitive traits, such that abilities in one large domain, verbal and social skills, tend to negatively covary with abilities in another domain, non-social skills. Different skills also, of course, tend to recruit different sets of regions of the brain, involving social-brain areas including midline areas, the temporoparietal junction, language processing in Broca's and Wernicke's areas, and the orbitofrontal cortex (e.g., Saxe 2006; Frith and Frith 2012), in contrast to parietal and occipital regions, and some additional areas, for non-social skills (e.g. Schoenemann 2006; Stout *et al.* 2008; Hoppe *et al.* 2012).

A central role for cognitive tradeoffs in human brain development and functioning is also concordant with the current best-supported model for the architecture of human intelligence, the verbal-perceptual-rotational model (Major *et al.* 2012), which involves a negative correlation between verbal skills and visual-spatial abilities, when the general factor "g" is parceled out (Johnson and Bouchard 2009).

Why are cognitive tradeoffs important? In Figure 9.2, each of the human-evolved and human-elaborated phenotypes can vary in either of two directions, towards a lower or higher level of development and expression. Reduced expression of social and verbal phenotypes, with concomitant enhanced expression of visual-spatial, mechanistic, technical, and perceptual abilities, characterize autism spectrum conditions, especially among individuals with a higher general level of intellectual functioning (Caron *et al.* 2006; Mottron *et al.* 2006; Crespi and Badcock 2008). By contrast, relatively higher expression of social and verbal phenotypes, concomitant to reduced abilities in visual-spatial, mechanistic, technical, and perceptual domains, characterizes the psychotic-affective spectrum, which includes the related, overlapping conditions schizophrenia, bipolar disorder, major depression, borderline personality disorder, and schizotypal personality disorder (e.g., Crespi and Badcock 2008; Crespi *et al.* 2009; Crespi 2011b). This diametric difference between the autistic spectrum and the psychotic-affective spectrum is notably demonstrated by the enhanced empathic skills, compared to non-clinical individuals, reported among individuals with borderline personality disorder (Dinsdale and Crespi 2013); by contrast, reduced empathic skills and interests are specifically characteristic of autism (Baron-Cohen 2010).

Are cognitive tradeoffs genetically based? A study by Kravariti *et al.* (2006) documented that pedigree-based genetic risk of schizophrenia (e.g., risk based on degree of genetic relatedness to a family member with schizophrenia) was highly significantly correlated, among non-clinical individuals, with high verbal skills relative to visual-spatial skills; having more alleles predisposing to schizophrenia was thus associated with a higher disparity between verbal and visual-spatial abilities. Similarly, Leach *et al.* (2013) showed that higher genetic risk of schizophrenia, as determined from genotyping 32 of the best-supported schizophrenia risk loci (single nucleotide polymorphisms), was associated with lower scores on a test of mental rotation; by contrast, several autism-risk alleles were associated with higher scores on the same test (Leach and Crespi, unpublished data). And comparably, mice knocked out for Shank1, a putative autism-risk gene, have shown enhanced abilities at spatial learning relative to control mice (Hung *et al.* 2008).

A strong prediction that follows from these considerations is that tradeoffs between verbal-social and visual-spatial skills should be mediated by allelic variation at loci that also underlie risk of autism, and risk of psychotic-affective conditions such as schizophrenia. We have evaluated this prediction in a non-clinical population by genotyping a well-documented schizophrenia risk locus, the SNP rs3916971 in the DAOA (D-amino acid oxidase activator) gene, and testing for associations of genotype with scores on a test of verbal skills (vocabulary), and a test of visual-spatial skills (mental rotation) (Leach *et al.* 2013). The DAOA gene is of particular interest because: (1) it has apparently evolved recently in primates, with its full-length protein product found only among humans (Chumakov *et al.*, 2002); (2) it is one of the best-documented risk genes for schizophrenia, as well as for bipolar disorder (Detera-Wadleigh and McMahon 2006); (3) its functional roles include modulation of the NMDA receptor, which mediates symptoms of schizophrenia; and (4) SNPs in the gene have been associated with better performance on several cognitive tasks, including verbal skills (Goldberg *et al.* 2006; Opgen-Rhein *et al.* 2008; Jansen *et al.* 2009a, 2009b).

We found that males bearing two risk alleles for the SNP rs3916971 in the DAOA gene showed significantly better vocabulary performance, but significantly lower mental rotation performance, than males with one or no risk alleles (Figure 9.3). These results require replication, but they suggest that, as suggested by the findings of Kravariti *et al.* (2006), some schizophrenia risk genes mediate tradeoffs between higher verbal skills and lower visual-spatial skills. Documenting additional loci that

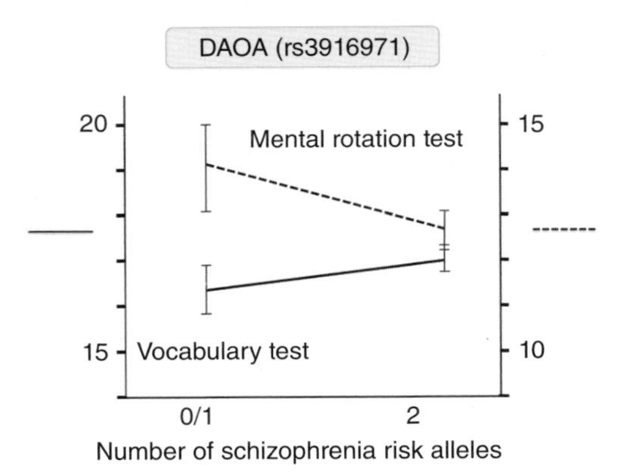

Figure 9.3. In a non-clinical population of males, having more schizophrenia risk alleles for the single nucleotide polymorphism rs3916971, in the gene DAOA (D-amino acid oxidase activator) is associated with better vocabulary skills, but worse mental rotation skills ($p<0.05$ for each, t-tests, SEs shown on plot; data from Leach *et al.* 2013). This locus thus provides evidence of a genetically based tradeoff between verbal skills and visual-spatial skills.

show such effects would help to close the loop between phenotypes-down and genes-up approaches, initially with single threads, but progressively strengthening as more and more data accumulate on neurodevelopment and neurocognitive function.

The upshot of what we can call the "evo-neuro" model in Figure 9.1 and Figure 9.2 is that each of the two major sets of so-called disorders of human sociality and cognition, the autism spectrum and the psychotic-affective spectrum, can be considered to involve extremes of tradeoffs, specifically for traits that have become elaborated in recent human evolutionary history. Given that each of these two spectra grades continuously, in its constituent phenotypes, into populations of typically developing individuals (Crespi and Badcock 2008; Crespi 2011b), this model for the architecture of human social-cognitive disorders also represents a simple, testable model for understanding cognitive variation among all human populations (Figure 9.4). The primary usefulness of this model is that it can direct research along promising paths by suggesting specific data to collect, and provide a unifying framework for existing results, such as familial associations of autism spectrum conditions with technical interests and abilities, and psychotic-affective conditions with literature and the humanities (Baron-Cohen 2012; Campbell and Wang 2012; Wei *et al.* 2013.

Most broadly, the evo-neuro model can be conceptualized as a generalization of the "microcephaly paradigm" described above, whereby recent human-evolutionary trajectories have structured the variation among humans in cognitive phenotypes, and in doing so, potentiated risk for psychiatric conditions with particular sets of phenotypes. The causes and correlates of psychiatric conditions may thus provide direct insights into normal human cognitive variation, as well as its evolutionary history. How well, then, is the model supported by the available data that connects genetic variation and perturbations, in two opposite directions, with diametric variation in cognitive phenotypes?

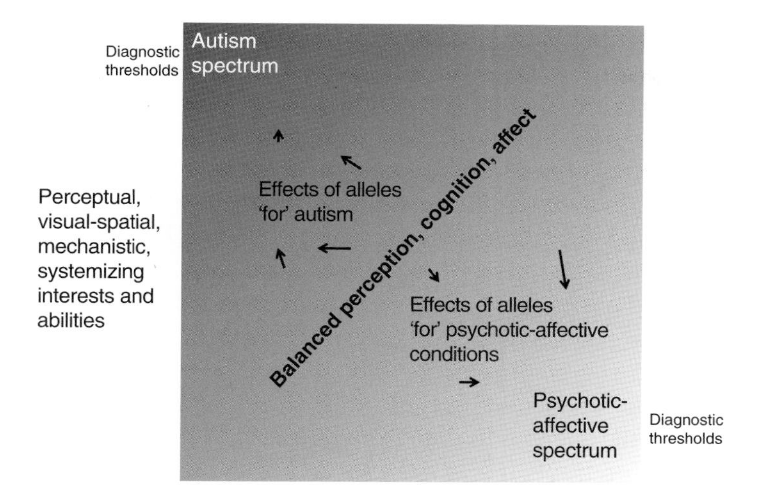

Figure 9.4. Autism spectrum and psychotic-affective spectrum conditions, both of which are strongly modulated by neurodevelopment, may be conceptualized as representing extremes of joint variation in social (and mentalistic), relative to non-social (and mechanistic), interests and abilities. Under this framework, alleles "for" either set of conditions are represented by arrows, of variable magnitude, that point towards the conditions. These alleles result in either adaptively increased interests and abilities in one domain (nearer the diagonal balance line) with moderate decreases in the other domain, or maladaptively over-developed interests and abilities in one domain, and maladaptively reduced interest and abilities in the other domain (near the upper left, or lower right, corners). Intensities of shading represent degrees of expression of perceptual, cognitive, and affective phenotypes characteristic of the autism spectrum and psychotic-affective spectrum.

Diametric Disorders of Neurodevelopment

Autism spectrum and psychotic-affective conditions as discussed above are usually "idiopathic" in causation, which means simply that the causes – genetic, environmental, or both – are unknown. For the best-studied conditions, autism and schizophrenia, a polygenic basis has recently been well established for "explaining" some subset of genetic risk, which is consistent with the high heritabilities of both sets of conditions (Corvin *et al.* 2012). Thus, for both, segregating allelic variation from hundreds or thousands of single nucleotide polymorphisms, each of which influences risk to a tiny yet estimable degree, contributes to distinguishing case from control populations (International Schizophrenia Consortium 2009; Vorstman *et al.* 2013; Skafidas *et al.* 2014). These are not genes "for" autism or schizophrenia at all, *per se* – they are genes with different alleles that cause slight differences in neurodevelopment, with differential impacts on social compared to non-social cognition (Kendler 2005). The effects of such genes interact with variation in social and non-social environments – "nurture" – in complex ways; for example, some alleles increase sensitivity to social-environmental effects, some alleles elevate risks of psychiatric conditions only in specific environments such as childhood maltreatment, and some alleles are expected to alleviate risks in specific environments, including early-life behavioral therapies.

In our evo-neuro model, effects of an "autism risk" allele could be represented by a small vector pointed left and up at some angle (Figure 9.4). This vector can be conceptualized as the direction and magnitude of change in position on the plot caused by "replacing" one allele with the other – essentially the same as Fisher's "additive effects." The alternative allele at this locus would, of course, be protective against autism – and also move brain development towards the psychotic-affective zone, usually ever so slightly. The position of any individual on this plot, (which ignores effects from the environment, and interactive effects of alleles), can be imagined as a summation of hundreds or thousands of small genetic vectors, which include effects from the alleles inherited from mother and father, plus any new mutations. The existence of tradeoffs between social and non-social cognition suggests that some notable proportion of vectors orient between upper left and lower right. Autism spectrum conditions may thus result, in part, from harboring "too many" alleles for non-social, compared to social, cognitive functions, and psychotic-affective conditions result, in part, from the converse. A notable feature of this model is that it is fully compatible with the other major psychological models for autism, including Baron-Cohen's model of high systemizing relative to empathizing (Baron-Cohen 2010), Happé and Frith's (2006) model based on relatively weak central coherence, and Mottron's model of enhanced perceptual functioning (Mottron *et al.* 2006).

The genetic framework that we have described is simple, polygenic, and incomplete, because we also know that some subset of autism cases, and schizophrenia cases, are mediated by genetic alterations of larger effect. These large vectors are extremely useful for evaluating the evo-neuro model, because they commonly involve large genetic or genomic changes in two opposite directions from normality. Thus, to the extent that the model corresponds with our cognitive nature, if changes in one direction predispose to autism spectrum conditions, then changes in the other direction should predispose to psychotic-affective conditions.

Three main forms of large, diametric genetic perturbation have been associated with risk of autism and schizophrenia: (1) genomic copy number variants, (2) imprinted gene effects, and (3) X chromosome gains versus losses (Figure 9.5). Each form of perturbation involves *de novo* (not inherited) gains or losses of gene expression, genetic composition, or gene activity, involving the same regions of DNA. As such, each represents a sort of natural "experiment" in changing gene dosages, for some stretch of DNA, from 2 to 1 or 3, or from 1 to 0 or 2. The primary drawback of such "experiments" is that in humans they are, of course, uncontrolled, and usually engender effects from multiple genes, such that determining causation from the gene to phenotype levels is especially challenging; large changes also tend to be relatively pathological, variable, and syndromic (exhibiting a suite of characteristic phenotypes) in their effects, which can obscure relative cognitive deficits and enhancements. By contrast, mice can readily be engineered

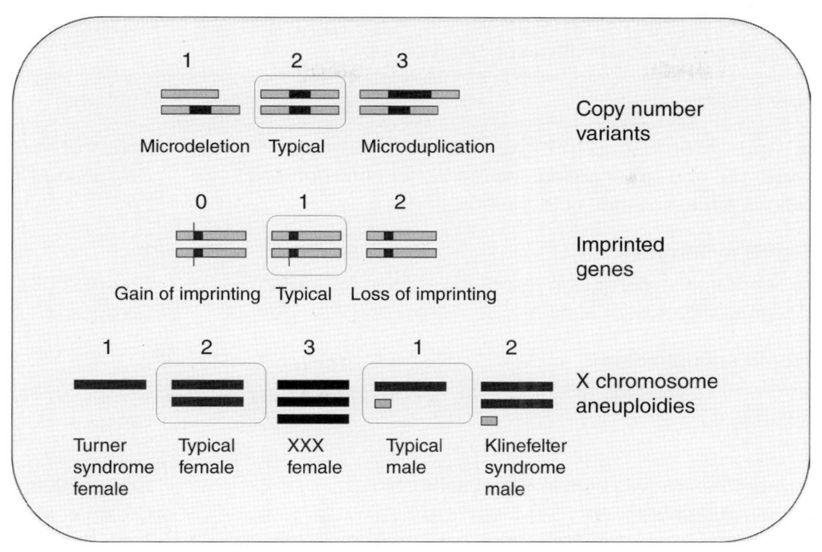

Figure 9.5. Three forms of genomic and epigenetic alteration can result in large alterations to gene expression in two opposite directions, towards increases or decreases from the typical range (which is circled, in each case). Copy number variants may include one to dozens of genes with higher or lower expression, imprinted genes involve losses or gains of expression due to losses or gains of epigenetic marks, and X chromosome aneuploidies involve partial or complete loss of an X in females, or gain of an X in males or females. To the extent that these alterations change expression of genes that mediate social and non-social brain development, they provide tests of the diametric brain hypothesis, as described in Table 9.2.

with higher or lower gene copy numbers, or doses of specified gene products, commonly resulting in opposite effects on neurological phenotypes and behavior (Crespi 2013a).

Table 9.2 summarizes currently available information on the prevalence of autism and schizophrenia in association with opposite alterations in genomic copy number variants, imprinted gene regions, and X chromosome numbers. These data demonstrate clear, overall support for the diametric model, in that opposite genetic or genomic alterations are commonly associated with autism spectrum conditions versus psychotic-affective conditions. Moreover, some sets of syndromes, such as Angelman syndrome, Rett syndrome, and Pitt-Hopkins syndrome (all associated with the autism spectrum), and Prader-Willi syndrome, Smith-Magenis syndrome, and Kleine-Levin syndrome (all associated with the psychotic-affective spectrum), involve phenotypic canalization, such that for each set a range of different genetic alterations gives rise to notably overlapping physical, behavioral, and psychiatric phenotypes (Table 3 in Crespi 2008a). Such canalization reflects the patterns observed in idiopathic autism and schizophrenia (myriad highly diverse genetic alterations, but similar endpoint phenotypes) (Happé 1994), and demonstrates that human neurodevelopment is structured, at least in part, along a canalized axis from autism spectrum to psychotic-affective cognition.

Table 9.2. Effects of large diametric genomic or epigenetic alterations on risk of autism spectrum conditions versus psychotic-affective spectrum conditions.

Copy-number variants	Citations
At four loci, 1q21.1, 16p11.2, 22q11.2, and 22q13.3, deletions are associated with one set of conditions (autism or schizophrenia), and duplications are associated with the other	Crespi *et al.* 2010 Crespi and Crofts 2012
Imprinted-gene syndromes	
Prader-Willi syndrome involves high risk of psychosis, Angelman syndrome involves high risk of autism	Crespi 2008a Crespi *et al.* 2009
X chromosome aneuploidies	
Turner syndrome involves high risk of autism, XXX involves high risk of schizophrenia	Crespi 2008a, 2008b Crespi *et al.* 2009
Klinefelter syndrome involves high levels of schizotypy, high risk of schizophrenia	Crespi 2008b Crespi *et al.* 2009

The Dawn of Evo-Neuro-Devo

Our evo-neuro model conceptually links the study of human cognitive evolution with the analysis of human psychiatric conditions. As such, it generates a non-subjective, non-arbitrary medical model for organizing the semi-chaotic plethora of named "disorders" reified by the *Diagnostic and Statistical Manual of Mental Disorders* (American Psychiatric Association 2013). Thus, psychiatric conditions, like all other medical conditions, can be understood strictly in terms of what evolved, adaptive biological system has become dysfunctional, and how (Johnson *et al.* 2012; Nesse and Stein 2012). The only problem with our model is that it lacks an essential component – perhaps *the* essential component – development.

Human brain and cognitive development have been studied in two largely disparate domains: (1) neuroscience (including neurogenetics), which focuses bottom-up on neural mechanisms that orchestrate neurodevelopment, and (2) developmental psychology, which typically involves the top-down application of sequential-stage models for how cognitive development proceeds from birth to adulthood. Both of these domains have been applied, albeit separately, to the study of psychiatric, neurodevelopmental conditions. However, neither has been used, systematically, to connect processes and patterns of human brain and cognitive development with trajectories of recent human evolution.

Thus far, we have discussed how the human brain has evolved to be large overall, and to specialize in aspects of social as well as non-social cognition. From a developmental perspective, however, the evolution of the human brain and cognition is most strikingly characterized by heterochronic extension: temporal lengthening of all of the fundamental neural processes that brains undergo from early growth to adulthood. This extension, sometimes associated with slowing of developmental times, has been reported for humans in neurological traits ranging from brain gene expression (Somel *et al.* 2009; Liu *et al.* 2012), to synaptic plasticity (Bufill *et al.* 2011), synaptic spine generation and development (Huttenlocher and Dabholkar 1997; Petanjek *et al.* 2011; Charrier *et al.* 2012), expansion of neocortical surface area (Rakic 2009; Lui *et al.* 2011), and myelinization (Miller *et al.* 2012). These manifestations of neurodevelopmental extension are, apparently, integral to the life-history shift, along the human lineage, towards extension of the period of pre-adult development (Bogin and Smith 1996; Bjorklund *et al.* 2009; Zollikofer and de Leon 2010; Bogin 2012), though with an earlier time of weaning. Evolutionary expansion of the human childhood stage, with resultant more child-like human adult phenotypes, has, of course, usually been described in terms of neoteny – the retention of juvenile characteristics in adults due to evolutionary changes in rates and timing of development (Godfrey and Sutherland 1996; Brüne 2000).

The concept of neoteny has a long history in the study of human development and evolution (Brüne 2000), yet has seldom been subject to systematic study using data from different disciplines. We will use neurodevelopmental conditions, especially autism and schizophrenia (the conditions for which most data are available), as windows to analyze patterns of change in rates and timing of human neurological and cognitive development. This approach is predicated on the assumption that sets of genes and pathways that underlie neurodevelopment should be expected to overlap between: (1) human evolutionary-developmental changes (and evolutionary changes earlier in primate and mammalian development), and (2) genetic alterations, as well as segregating variation, that distinguish typically developing individuals of different ages from individuals with autism, or with schizophrenia. This approach represents the microcephaly paradigm writ neurodevelopmental, over the largest neurological scales from perception and cognition to behavior. The primary previous deployment of this perspective comes from work by Crow (1997), Horrobin (2001), and Burns (2007), who have described extensive evidence regarding the hypothesis that schizophrenia represents "the illness that made us human," because it centrally involves the suite of human neurodevelopmental and cognitive phenotypes that are most highly elaborated, or unique (such as language), along the human lineage. This perspective, of course, is exemplified in our Figure 9.2, for a set of broad psychological phenotypes. As such, we are now adding a neurodevelopmental, heterochronic dimension to the variation in Figure 9.2 (and its neurological and psychological underpinnings), as each of the phenotypes shown must be a product of some time-dependent trajectory as ontogeny proceeds.

Developmental Heterochrony and Variation in Human Neurodevelopment

We will evaluate perhaps the simplest possible heterochronic model for the links of typical human neurodevelopment with human neurodevelopment in autism or in schizophrenia: shifts in the timing and rates of typical development, towards either slowing or lengthening (or both) (which may include non-completion of a typical trajectory), or acceleration or shortening (or both) (which may involve early differentiation). An initial treatment of this question, which focused mainly on autism, was described in Crespi (2013b); here, we expand on the model and link it more directly to human evolutionary history.

Two general types of perceptual, cognitive, behavioral, psychological, and neurological developmental phenotypes will be considered. First, some phenotypes involve progressive, sequential, largely quantitative increases in complexity and maturity as development progresses from infancy through adolescence or some period therein. Bjorklund *et al.* (2009) refer to these as "deferred adaptations," and language skills represent a clear example. Second, some childhood phenotypes instead represent stage-specific adaptations (from Bjorklund *et al.* 2009, "ontogenetic adaptations") that change qualitatively over development, such that children at each stage express more or less different phenotypes, each adapted specifically for that period (Bjorklund 1997; Bjorklund *et al.* 2009; Thompson-Schill *et al.* 2009). Examples would include private speech, pretend play, and a lack of cognitive control, all of which have been postulated as traits specific to early childhood that promote social and cultural learning (Bjorklund *et al.* 2009; Thompson-Schill *et al.* 2009).

The primary hypothesis that we address is that autism involves, in part, simply delay or non-completion of typical neurodevelopmental and psychological trajectories. By contrast, schizophrenia, and to some extent related psychotic-affective conditions such as bipolar disorder and depression, involve the opposite: acceleration, early differentiation, and maladaptive "over"-development. This hypothesis can be addressed by comparing phenotypes characteristic of autism and schizophrenia with phenotypes of younger versus older typically developing individuals (Crespi 2013b), and by extrapolating the trajectories of typical development beyond their usual bounds.

Progressive, sequentially developing, and increasingly complex human phenotypes include, most notably, theory of mind, conception and coherence of self, linguistic abilities, abstract thought, imagination including mental time travel and hypothetical scenario-building, controlled and long-term goal-seeking, social emotionality including pride, guilt, embarrassment, and shame, and, most generally, what we can call mentalistic cognition (e.g., Badcock 2004; Crespi and Badcock 2008). As noted above, each of these phenotypes shows evidence of under-development in autism spectrum conditions, whereby typical trajectories are not completed (Crespi and Badcock 2008; Badcock 2009). Woodard and Van Reet (2011) characterize such psychological trajectories as involving a series of four stages in "object identification," which can be defined as assimilation of external stimuli into the maturing structure of the internal, developing self:

1. *Part-object/inanimate object identification;*
2. *Part-object/initial or emerging human as part-object identification*;
3. *Non-complete or non-integrated whole-object human identification*;
4. *Whole-object, complete human identification* [the endpoint for typical development].

This conceptualization of stage-specific typical development is of particular interest because the intermediate stages so strikingly characterize several major, otherwise-disparate aspects of autism, including fascination with parts of objects, perceptions focused on local more than global and integrated features of the environment, treating people as inanimate "things," a reduced concept of self and self-agency (Crespi and Badcock 2008; Gray *et al.* 2011; Uddin 2011), and lower levels of empathy towards other humans. In the developmental heterochronic lexicon, autism may thus involve some combination of a slower rate (leading to so-called neoteny) and an earlier endpoint (referred to as progenesis). Comparable stage-specific analyses can be derived based on developmental-psychological theory from Vygotsky (Fernyhough 1996), or based on ideas from metarepresentation and theory of mind (Lombardo and Baron-Cohen 2011).

What phenotypes do we expect if such sequences of development are accelerated, or continue for a longer than usual time ("hypermorphosis"), along the psychological dimensions described above? As depicted in Figure 9.2, hyper-developed theory of mind descends into paranoia and delusions of persecution; over-developed sense of self extends to delusions of grandeur; linguistic abilities chaotically expand and fracture in thought disorder; imagination runs amok in hallucinations and delusions that manifest elaborate psychotic scenarios with no basis in reality (Nettle 2001); uncontrolled, risky goal seeking manifests as manic episodes; and social emotions, especially guilt, shame, and embarrassment, hyper-express in depression (e.g., Zahn-Waxler *et al.* 2006, 2008). With regard to object identification, we continue in psychotic-affective conditions to a novel step (5): ascribing animacy, agency, and human attributes to non-human objects (Gray *et al.* 2011), and sometimes (6) identification with the spiritual external world as a whole, in association with magical ideation, paranormal experiences (Leonhard and Brugger 1998), hyper-developed empathy in exaggerated mirror neuron system activation (McCormick *et al.* 2012), hyper-agency in megalomania, and neurological alterations including temporal lobe epilepsy underlining hyper-religiosity and symptoms of psychosis (Trimble 2007).

These considerations indicate how even apparently bizarre and inexplicable manifestations of psychotic-affective conditions can be understood as extremes of typical cognitive phenomena. From our neurodevelopmental perspective, proximate causes of such phenotypes can be considered to include loss of negative regulation of social, linguistic, and emotional phenotypes, and loss of homeostatic control over positively and negatively valenced emotional and behavioral feedbacks, in schizoaffective disorder, bipolar disorder, and major depression. In each case, the alterations result from certain forms of change to brain structure and neurochemistry, as described in more detail below. By contrast, ultimate causes of such symptoms derive from the recent human evolutionary trajectory towards increasingly elaborated social, emotional, and linguistically based perception, cognition, and behavior. This framework can be evaluated empirically by determining the degree to which the genes and alleles underlying psychotic-affective conditions, and their subclinical psychological manifestations in diverse aspects of human personality, have undergone adaptive molecular-evolutionary change along the human and great ape lineages (e.g. Crespi *et al.* 2007; Torri *et al.* 2010). Such tests can indeed determine the degree to which risk of schizophrenia (and the psychotic-affective spectrum more generally) represents a direct, pleiotropic by-product of human evolution.

Above, we have discussed "deferred adaptations," which involve developmental ratcheting of better and better skills in some specific domain as development proceeds. "Ontogenetic adaptations," by contrast, entail qualitative differences between younger and older children, which allow direct tests of our developmental-heterochronic model. Thus, to the degree that the model is correct, perceptual, cognitive, and behavioral differences between younger and older typically developing individuals should reflect the differences between individuals with autism, typically developing individuals, and individuals with schizophrenia.

Crespi (2013b) reviews data across four major research areas involving ontogenetic adaptations: (1) restricted interests and repetitive behavior (in autism), (2) local and parts-focused compared to global and gestalt processing of visual information, (3) absolute compared to relative pitch processing for auditory information, and (4) relatively short-range compared to long-range neural connectivity, considered in terms of distributions of path lengths between interacting neurons or brain regions. For each of these four domains, evidence from a broad swath of literature supports the idea that autism involves retention of phenotypes typical of relatively early childhood: high levels of restricted interests and repetitive behavior, a local bias to visual processing, auditory processing that differentially involves absolute pitch, and relatively short-range connectivity (Crespi 2013b). These inferences are, moreover, fully compatible with two of the most influential theories for understanding autistic traits, weak central coherence (Happé 1994) and enhanced perceptual functioning (Mottron *et al.* 2006).

For visual and auditory processing phenotypes, and neural connectivity, schizophrenia shows notable evidence of patterns opposite to those for autism, although direct comparisons of schizophrenia with autism and typical development, based on the same psychological or neurophysiological tests, are seldom available (Crespi 2013b). The most telling comparison, though, is the best-documented: lower relatively long-range neural connectivity in autism (and higher short-range connectivity than in typical individuals), but lower relatively short-range connectivity in schizophrenia, due in part to high rates of neuronal loss differentially affecting short-range connections. In principle, this connectivity difference may explain many of the diametric perceptual, cognitive, and behavioral differences between autism and schizophrenia, including such well-documented patterns as: (1) larger versus smaller head size; (2) superior visual and auditory perception and processing in autism but degraded sensory functions in schizophrenia, which apparently contribute to hallucinations (Waters *et al.* 2012; Pynn and Desouza 2013); (3) the local versus global visual-processing biases discussed above; and (4) absent, highly literal, or hyperlexic language skills in autism but, in schizophrenia, dyslexic reading profiles, auditory hallucinations, loose associations in speech, chaotic and metaphorical language use including neologisms (coining of new words) and clanging (focus on rhyming sounds of words), and, among poets, high levels of schizotypal cognition (Nettle 2001; Crespi 2008b; Crespi and Badcock 2008; Poirel *et al.* 2011). Most notably, typically developing individuals undergo a trajectory of neuronal, synaptic, and dendritic pruning that starts in early childhood and continues through early adulthood and beyond (e.g., Shaw *et al.* 2006; Paus *et al.* 2008). In schizophrenia, this process apparently accelerates, or continues for a relatively long time, indicating that heterochronic alterations – failures of neoteny, as it were (Brüne 2000; Burns 2004) – underpin a substantial component of neurodevelopmental changes in schizophrenia. Diametric variation in neural short-range relative to long-range connectivity patterns may also provide, at least in part, a simple basis for the cognitive tradeoffs described above between social and non-social psychological functions.

Postulating and testing diametric variation in developmentally based connectivity patterns in autism compared to schizophrenia is well and good – but not evolutionary *per se*, and so not evo-neuro-devo. In principle, brain connectivity has evolved along the human lineage, in the general direction of psychotic-affective (mentalistic) cognition, and also towards increases in both mechanistic and mentalistic cognitive abilities with tradeoffs between them. A simple scenario of such a possible evolutionary trajectory is shown in Figure 9.6, in the context of variation in brain connectivity among extant humans, and in autism and schizophrenia. But what empirical data

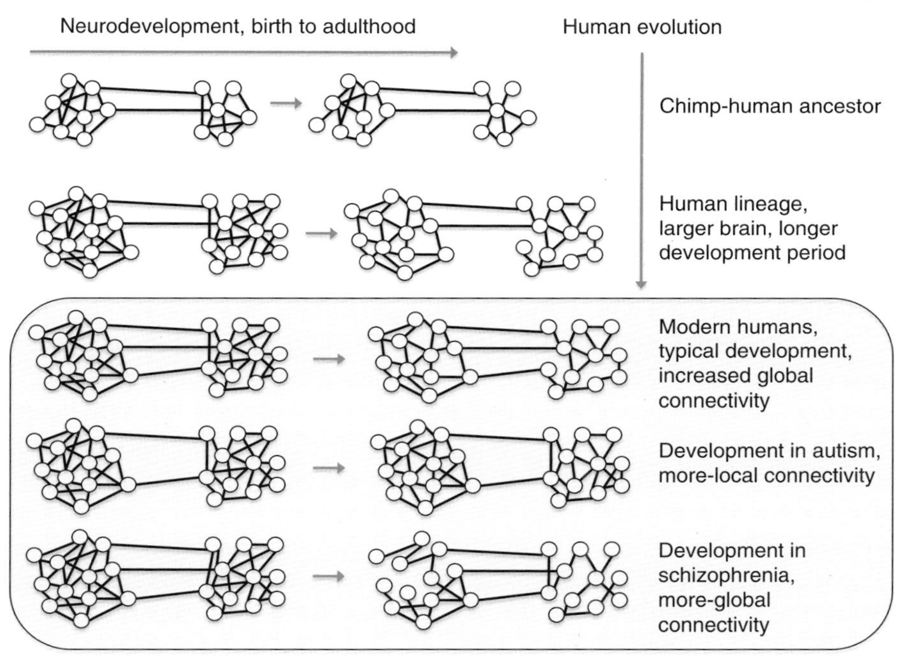

Figure 9.6. A highly simplified "evo-neuro-devo" model for how human local and global brain connectivity may have evolved, and how their development is altered in autism and in schizophrenia. Two connected brain regions are shown, with each region showing "small-world" connectivity patterns with highly connected hubs, and decreased connectivity with age due to (primarily short-range) synaptic and neuronal pruning. By this model, increases in brain size along the human lineage involved strong selection to maintain and increase long-range connectivity, which otherwise decreases with absolute size. Examples of brain structures manifesting such increases in connectivity along the human lineage include the arcuate fasciculus (connecting Broca's area with Wernicke's area) (Rilling *et al.* 2008), and the corpus callosum itself (connecting the left and right hemispheres) (Rilling and Insel 1999). Among modern humans, autism involves relatively reduced long-range connectivity and reduced pruning, both of which lead to relatively more-local connectivity. By contrast, schizophrenia involves increased local pruning (and gray matter loss), which can lead differentially to connectivity that is less local and more global.

can be, or has been, collected to evaluate such hypotheses, or other hypotheses that seek to connect evolution with human neurodevelopment?

The Origin of Modern Human Neurodevelopment

A voluminous literature has accumulated on the so-called "genes that make us human," which are inferred from studies of molecular evolution driven by statistically inferred natural selection. Such studies have produced large lists of ranked genes, about which we know virtually nothing substantial about function in body and brain, or roles in human evolution (Hughes 2007), except in a few aforementioned cases involving microcephaly and speech. This unfortunate situation has arisen because deciphering molecular-developmental functions for particular amino acids or haplotypes is challenging, and because research in human molecular-evolutionary genetics proceeds largely independently from research on human neurodevelopment and its disorders.

Three simple and feasible methods can, however, be deployed to close the loops between human evolution, neurodevelopment, and neurodevelopmental conditions such as autism and schizophrenia.

First, studies of the "genes that make us human" can be complemented by studies that systematically determine what cognitive and psychological phenotypic variation can be ascribed to ancestral versus derived amino acids, haplotypes, or other genetic polymorphisms. Such analyses simply require genotyping large sets of normal, non-clinical individuals (to determine if each individual has the derived or ancestral allele, for a given site inferred from previous work to be subject to positive selection), and phenotyping them for a broad set of cognitive and psychological traits (to determine what traits, if any, differ between individuals with derived and ancestral alleles). Cognitive and psychological traits can, of course, be chosen based on previous knowledge of what particular genes appear to do, where they are expressed, and if they are known to be associated with particular phenotypic domains or diseases. In one of the few examples of this approach, Wong *et al.* (2012) found that having more derived alleles at a specific amino acid site in the *ASPM* gene was associated with better performance in perception of auditory tone. The primary limitation of this approach is that it can be used only for analyzing polymorphic genetic variation, not alleles that have gone to fixation along the human lineage. Its primary advantage is that, given data from enough genes and phenotypes, we should be able to infer the trajectories of ongoing evolution of human cognition and neurodevelopment.

Second, DNA sequencing of recent human fossils, from the last tens of thousands of years (i.e., along the recent human lineage), can be used to directly determine what series of genetic changes has taken place in conjunction with the origin of modern humans. Such studies are technologically feasible (Meyer *et al.* 2012), though they would require ancient DNA samples from at least several tens of humans for robust inferences to be drawn. Such analyses could be complemented by phylogenetic analyses of DNA from extant humans (to infer ancestral states and time frames), and they would also dovetail nicely with the first method described above.

Third, the cognitive correlates of well-validated risk alleles for autism and for schizophrenia can be analyzed in non-clinical populations, to determine just what these alleles actually "do" in normal neurodevelopment and neurological-psychological function. How many such alleles mediate tradeoffs between social and non-social cognition, or between other sets of domains? How do they do so, in terms of brain function? Studies such as these are especially useful because they will generate insights into both the cognitive architecture of typically developing individuals, and the causes of psychiatric conditions, with novel implications for therapy.

For each of the three approaches above, relevance of the alleles to neurodevelopment *per se*, and to short-range compared to long-range connectivity patterns, can be evaluated using brain imaging studies of typically developing individuals that differ in genotype, studies of individuals with psychiatric conditions, and studies of mice engineered to express human alleles such as the human-derived allele of *FOXP2* (Enard *et al.* 2009). Convergence of results from independent evolutionary, neurological, developmental, and genetic approaches should lead to robust insights into human evo-neuro-devo, and bring this nascent field fully to life.

Conclusions

In this chapter we have constructed a rudimentary scaffold for the field of evo-neuro-devo, an emerging discipline that focuses on how human neurodevelopment, and risk for psychiatric conditions mediated by neurodevelopment, have evolved together. The field is challenging due to the complexity of human brain development and the paucity of hard evidence regarding the evolutionary history of human neurological and cognitive traits. However, the connections of genetic data to both human neurodevelopment and evolution can be used to leverage robust, empirically based inferences, given the causal primacy of genes in neurodevelopment and their evolutionary encoding of population-genetic forces past.

The primary pillars of our scaffold are twofold. First, the major features of autism and psychotic-affective spectrum conditions can be derived by hypo-development, or hyper-development, of human-unique or human-elaborated phenotypes (Figure 9.2). This explanatory framework explicitly connects human evolution with psychiatric conditions that involve variation in human social and non-social brain systems, indicating that both can be analyzed together, most notably in the context of genetically based cognitive tradeoffs. Second, we have presented evidence that developmental heterochrony characterizes, at least to some degree, the neurological bases of autism and schizophrenia (Figure 9.5), such that simple shifts in the rates and timing of neurodevelopmental processes can help to explain a broad suite of diverse phenotypes that characterize the two sets of conditions. The question then becomes, both for therapy and for understanding human brain evolution, what processes orchestrate temporal sequences and shifts in how the brain changes through childhood, adolescence, and beyond? Answers to this question should raise evo-neuro-devo to maturity, and, eventually, into the forefront of evolutionary biology, anthropology, neuroscience, and psychology.

Acknowledgements

We thank NSERC for financial support, and S. Read for technical assistance.

References

American Psychiatric Association. 2013. *Diagnostic and statistical manual of mental disorders* (5th edn). Washington, DC: APA.

Badcock, C. 2004. Mentalism and mechanism: The twin modes of human cognition. *In*: C. B. Crawford and C. A. Salmon (eds) *Evolutionary psychology, public policy and personal decisions*. Mahwah, NJ: Lawrence Erlbaum Associates. pp.99–116.

Badcock, C. 2009. *The imprinted brain: How genes set the balance between autism and psychosis*. Philadelphia, PA: Jessica Kingsley Publishers.

Baron-Cohen, S. 2010. Empathizing, systemizing, and the extreme male brain theory of autism. *Progress in Brain Research* 186:167–175.

Baron-Cohen, S. 2012. Autism and the technical mind: Children of scientists and engineers may inherit genes that not only confer intellectual talents but also predispose them to autism. *Scientific American* 307(5):72–75.

Baron-Cohen, S., and Hammer, J. 1997. Parents of children with Asperger syndrome: What is the cognitive phenotype? *Journal of Cognitive Neuroscience* 9:548–554.

Baron-Cohen, S., Wheelwright, S., and Jolliffe, T. 1997. Is there a "language of the eyes"? Evidence from normal adults and adults with autism or Asperger syndrome. *Visual Cognition* 4:311–331.

Bjorklund, D. F. 1997. The role of immaturity in human development. *Psychological Bulletin* 122:153–169.

Bjorklund, D. F., Periss, V., and Causey, K. 2009. The benefits of youth. *European Journal of Developmental Psychology* 6:120–137.

Bogin, B. 2012. The Evolution of Human Growth. *In*: N. Cameron and B. Bogin (eds) *Human Growth and Development*, 2nd edn. New York: Academic Press. pp.287–324.

Bogin, B., and Smith, B. H. 1996. Evolution of the human life cycle. *American Journal of Human Biology* 8:703–716.

Brüne, M. 2000. Neoteny, psychiatric disorders and the social brain: Hypotheses on heterochrony and the modularity of the mind. *Anthropology and Medicine* 7:301–318.

Bufill, E., Agustí, J., and Blesa, R. 2011. Human neoteny revisited: The case of synaptic plasticity. *American Journal of Human Biology* 23:729–739.

Burns, J. K. 2004. An evolutionary theory of schizophrenia: Cortical connectivity, metarepresentation, and the social brain. *Behavioral and Brain Sciences* 27:831–885.

Burns, J. K. 2007. *The descent of madness. Evolutionary origins of psychosis*. London: Routledge.

Campbell, B. C., and Wang, S. S. 2012. Familial linkage between neuropsychiatric disorders and intellectual interests. *PLoS One* 7(1):e30405.

Caron, M. J., Mottron, L., Berthiaume, C., and Dawson, M. 2006. Cognitive mechanisms, specificity and neural underpinnings of visuospatial peaks in autism. *Brain* 129(Pt 7): 1789–1802.

Casey, J. P., Magalhaes, T., Conroy, J. M. *et al.* 2012. A novel approach of homozygous haplotype sharing identifies candidate genes in autism spectrum disorder. *Human Genetics* 131(4):565–579.

Charrier, C., Joshi, K., Coutinho-Budd, J. *et al.* 2012. Inhibition of SRGAP2 function by its human-specific paralogs induces neoteny during spine maturation. *Cell* 149(4):923–935.

Chumakov, I., Blumenfeld, M., Guerassimenko, O. *et al.* 2002. Genetic and physiological data implicating the new human gene G72 and the gene for D-amino acid oxidase in schizophrenia. *Proceedings of the National Academy of Sciences, USA* 99(21):13675–13680.

Cook, C. M., and Saucier, D. M. 2010. Mental rotation, targeting ability and Baron-Cohen's empathizing-systemizing theory of sex differences. *Personality and Individual Differences* 49(7):712–716.

Corvin, A., Donohoe, G., Hargreaves, A. *et al.* 2012. The cognitive genetics of neuropsychiatric disorders. *Current Topics in Behavioral Neurosciences* 12:579–613.

Crespi, B. J. 2008a. Genomic imprinting in the development and evolution of psychotic spectrum conditions. *Biological Reviews of the Cambridge Philosophical Society* 83(4):441–493.

Crespi, B. J. 2008b. Language unbound: Genomic conflict and psychosis in the origin of modern humans. *In*: D. Hughes and P. D'Ettorre (eds) *Sociobiology of communication: An interdisciplinary perspective*. Oxford: Oxford University Press. pp.225–248.

Crespi, B. J. 2011a. The emergence of human-evolutionary medical genomics. *Evolutionary Applications* 4(2):292–314.

Crespi, B. J. 2011b. One hundred years of insanity: Genomic, psychological and evolutionary models of autism in relation to schizophrenia. *In*: M. S. Ritsner (ed.) *Textbook of schizophrenia-spectrum disorders*. New York: Springer. pp.163–186.

Crespi, B. J. 2013a. Diametric gene-dosage effects as windows into neurogenetic architecture. *Current Opinion in Neurobiology* 23(1):143–151.

Crespi, B. J. 2013b. Developmental heterochrony and the evolution of autistic perception, cognition and behavior. *BMC Medicine* 11:119.

Crespi, B. J., and Badcock, C. 2008. Psychosis and autism as diametrical disorders of the social brain. *Behavioral and Brain Sciences* 31(3):241–320.

Crespi, B. J., and Crofts, H. J. 2012. Association testing of copy number variants in schizophrenia and autism spectrum disorders. *Journal of Neurodevelopmental Disorders* 4(1):15.

Crespi, B., Stead, P., and Elliot, M. 2010. Comparative genomics of autism and schizophrenia. *Proceedings of the National Academy of Sciences, USA* 107:1736–1741.

Crespi, B. J., Summers, K., and Dorus, S. 2007. Adaptive evolution of genes underlying schizophrenia. *Proceedings of the Royal Society of London, Series B: Biological Sciences* 274(1627):2801–2810.

Crespi, B. J., Summers, K., and Dorus, S. 2009. Genomic sister-disorders of neurodevelopment: An evolutionary approach. *Evolutionary Applications* 2:81–100.

Crow, T. J. 1997. Is schizophrenia the price that *Homo sapiens* pays for language? *Schizophrenia Research* 28(2–3):127–141.

Detera-Wadleigh, S. D., and McMahon, F. J. 2006. G72/G30 in schizophrenia and bipolar disorder: Review and meta-analysis. *Biological Psychiatry* 60(2):106–114.

Dinsdale, N. L., Hurd, P. L., Wakabayashi, A. *et al.* 2013. How are autism and schizotypy related? Evidence from a non-clinical population. *PLoS ONE* e63316.

Dinsdale, N., and Crespi, B. J. 2013. The Borderline Empathy Paradox: Evidence and Conceptual Models for Empathic Enhancements in Borderline Personality Disorder. *Journal of Personality Disorders* 27(2):172–195.

Dumas, L. J., O'Bleness, M. S., Davis, J. M. *et al.* 2012. *DUF1220*-domain copy number implicated in human brain-size pathology and evolution. *American Journal of Human Genetics* 91(3):444–454.

Enard, W., Gehre, S., Hammerschmidt, K. *et al.* 2009. A humanized version of *FOXP2* affects cortico-basal ganglia circuits in mice. *Cell* 137(5):961–971.

Fernyhough, C. 1996. The dialogic mind: A dialogic approach to the higher mental functions. *New Ideas in Psychology* 14(1):47–62.

Frith, C. D., and Frith, U. 2012. Mechanisms of social cognition. *Annual Review of Psychology* 63:287–313.

Fugard, A. J. B., Stewart, M. E., and Stenning, K. 2011. Visual/verbal-analytic reasoning bias as a function of self-reported autistic-like traits: A study of typically developing individuals solving Raven's Advanced Progressive Matrices. *Autism* 15(3):327–340.

Godfrey, L. R., and Sutherland, M. R. 1996. Paradox of peramorphic paedomorphosis: Heterochrony and human evolution. *American Journal of Physical Anthropology* 99:17–42.

Goldberg, T. E., Straub, R. E., Callicott, J. H. *et al.* 2006. The G72/G30 gene complex and cognitive abnormalities in schizophrenia. *Neuropsychopharmacology* 31:2022–2032.

Golzio, C., Willer, J., Talkowski, M. E. *et al.* 2012. *KCTD13* is a major driver of mirrored neuroanatomical phenotypes of the 16p11.2 copy number variant. *Nature* 485:363–367.

Gong, X., Jia, M., Ruan, Y. *et al.* 2004. Association between the *FOXP2* gene and autistic disorder in Chinese population. *American Journal of Medical Genetics, Part B: Neuropsychiatric Genetics* 127B(1):113–116.

Gray, K., Jenkins, A. C., Heberlein, A. S., and Wegner, D. M. 2011. Distortions of mind perception in psychopathology. *Proceedings of the National Academy of Sciences, USA* 108(2):477–479.

Happé, F. 1994. *Autism: An introduction to psychological theory*. London: UCL Press.

Happé, F., and Frith, U. 2006. The weak coherence account: Detail-focused cognitive style in autism spectrum disorders. *Journal of Autism and Developmental Disorders* 36(1):5–25.

Hoppe, C., Fliessbach, K., Stausberg, S. *et al.* 2012. A key role for experimental task performance: Effects of math talent, gender and performance on the neural correlates of mental rotation. *Brain and Cognition* 78(1):14–27.

Horrobin, D. 2001. *The madness of Adam and Eve: How schizophrenia shaped humanity*. London: Bantam Press/Transworld Publishers.

Hughes, A. L. 2007. Looking for Darwin in all the wrong places: The misguided quest for positive selection at the nucleotide sequence level. *Heredity* 99(4):364–373.

Hungm, A. Y., Futai, K., Sala, C. *et al.* 2008. Smaller dendritic spines, weaker synaptic transmission, but enhanced spatial learning in mice lacking Shank1. *Journal of Neuroscience* 28(7):1697–1708.

Huttenlocher, P. R., and Dabholkar, A. S. 1997. Regional differences in synaptogenesis in human cerebral cortex. *Journal of Comparative Neurology* 387(2):167–178.

International Schizophrenia Consortium, Purcell, S. M., Wray, N. R. *et al.* 2009. Common polygenic variation contributes to risk of schizophrenia and bipolar disorder. *Nature* 460(7256):748–752.

Jansen, A., Krach, S., Krug, A. *et al.* 2009a. Effect of the G72 (DAOA) putative risk haplotype on cognitive functions in healthy subjects. *BioMed Central Psychiatry* 9:60.

Jansen, A., Krach, S., Krug, A. *et al.* 2009b. A putative high risk diplotype of the G72 gene is in healthy individuals associated with better performance in working memory functions and altered brain activity in the medial temporal lobe. *NeuroImage* 45(3):1002–1008.

Jarrold, C., Butler, D. W., Cottington, E. M., and Jimenez, F. 2000. Linking theory of mind and central coherence bias in autism and in the general population. *Developmental Psychology* 36(1):126–138.

Johnson, W., and Bouchard T. J., Jr. 2005. The structure of human intelligence: It is verbal, perceptual, and image rotation (VPR), not fluid and crystallized. *Intelligence* 33(4):393–416.

Johnson, W. and Bouchard T. J., Jr. 2007. Sex differences in mental abilities: *g* masks the dimensions on which they lie. *Intelligence* 35(1):23–39.

Johnson, W., and Bouchard T. J., Jr. 2009. Linking abilities, interests, and sex via latent class analysis. *Journal of Career Assessment* 17(1):3–38.

Johnson, S. L., Leedom, L. J., and Muhtadie, L. 2012. The dominance behavioral system and psychopathology: Evidence from self-report, observational, and biological studies. *Psychological Bulletin* 138(4):692–743.

Joseph, R. M., Keehn, B., Connolly, C. *et al.* 2009. Why is visual search superior in autism spectrum disorder? *Developmental Science* 12:1083–1096.

Joseph, R. M., Tager-Flusberg, H., and Lord, C. 2002. Cognitive profiles and social-communicative functioning in children with autism spectrum disorder. *Journal of Child Psychology and Psychiatry and Allied Disciplines* 43(6):807–821.

Keehn, B., Shih, P., Brenner, L. A. *et al.* 2013. Functional connectivity for an "island of sparing" in autism spectrum disorder: An fMRI study of visual search. *Human Brain Mapping* 34(10):2524–2537.

Kendler, K. S. 2005. "A gene for...": The nature of gene action in psychiatric disorders. *American Society for Psychiatry* 162:1243–1252.

Kravariti, E., Toulopoulou, T., Mapua-Filbey, F. *et al.* 2006. Intellectual asymmetry and genetic liability in first-degree relatives of probands with schizophrenia. *British Journal of Psychiatry* 188:186–187.

Leach, E., Hurd, P., and Crespi, B. 2013. Schizotypy, cognitive performance, and genetic risk for schizophrenia in a non-clinical population. *Personality and Individual Differences* 55(3):334–338.

Leonhard, D., and Brugger, P. 1998. Creative, paranormal, and delusional thought: A consequence of right hemisphere semantic activation? *Neuropsychiatry, Neuropsychology, and Behavioral Neurology* 11(4):177–183.

Liu, X., Somel, M., Tang, L. *et al.* 2012. Extension of cortical synaptic development distinguishes humans from chimpanzees and macaques. *Genome Research* 22:611–622.

Lombardo, M. V., and Baron-Cohen, S. 2011. The role of the self in mindblindness in autism. *Consciousness and Cognition* 20(1):130–140.

Lui, J. H., Hansen, D. V., and Kriegstein, A. R. 2011. Development and evolution of the human neocortex. *Cell* 146(1):18–36.

Mahmood, S., Ahmad, W., and Hassan, M. J. 2011. Autosomal Recessive Primary Microcephaly (MCPH): Clinical manifestations, genetic heterogeneity and mutation continuum. *Orphanet Journal of Rare Diseases* 6:39.

Major, J. T., Johnson, W., and Deary, I. J. 2012. Comparing models of intelligence in Project TALENT: The VPR model fits better than the CHC and extended Gf-Gc models. *Intelligence* 40(6):543–559.

McCormick, L. M., Brumm, M. C., Beadle, J. N. *et al.* 2012. Mirror neuron function, psychosis, and empathy in schizophrenia. *Psychiatry Research* 201(3):233–239.

Megraw, T. L., Sharkey, J. T., and Nowakowski, R. S. 2011. *Cdk5rap2* exposes the centrosomal root of microcephaly syndromes. *Trends in Cell Biology* 21(8):470–480.

Meyer, M., Kircher, M., Gansauge, M. T. *et al.* 2012. A high-coverage genome sequence from an archaic Denisovan individual. *Science* 338(6104):222–226.

Miller, D. J., Duka, T., Stimpson, C. D. *et al.* 2012. Prolonged myelination in human neocortical evolution. *Proceedings of the National Academy of Sciences, USA* 109:16480–16485.

Montgomery, S. H., Capellini, I., Venditti, C. *et al.* 2011. Adaptive evolution of four microcephaly genes and the evolution of brain size in anthropoid primates. *Molecular Biology and Evolution* 28(1):625–638.

Mottron, L., Dawson, M., Soulières, I. *et al.* 2006. Enhanced perceptual functioning in autism: An update, and eight principles of autistic perception. *Journal of Autism and Developmental Disorders* 36(1):27–43.

Nesse, R. M., and Stein, D. J. 2012. Towards a genuinely medical model for psychiatric nosology. *BMC Medicine* 10:5.

Nettle, D. 2001. *Strong imagination: Madness, creativity and human nature*. Oxford: Oxford University Press.

Nettle, D. 2007. Empathizing and systemizing: What are they, and what do they contribute to our understanding of psychological sex differences? *British Journal of Psychology* 98:237–255.

Nisbett, R. E., Aronson, J., Blair, C. *et al.* 2012. Intelligence: New findings and theoretical developments. *American Psychologist* 67(2):130–159.

Opgen-Rhein, C., Lencz, T., Burdick, K. E. *et al.* 2008. Genetic variation in the DAOA gene complex: Impact on susceptibility for schizophrenia and on cognitive performance. *Schizophrenia Research* 103:169–177.

Paus, T., Keshavan, M., and Giedd, J. N. 2008. Why do many psychiatric disorders emerge during adolescence? *Nature Reviews Neuroscience* 9(12):947–957.

Pellicano, E., Maybery, M., Durkin, K., and Maley, A. 2006. Multiple cognitive capabilities/deficits in children with an autism spectrum disorder: "Weak" central coherence and its relationship to theory of mind and executive control. *Development and Psychopathology* 18(1):77–98.

Petanjek, Z., Judaš, M., Šimic, G. *et al.* 2011. Extraordinary neoteny of synaptic spines in the human prefrontal cortex. *Proceedings of the National Academy of Sciences, USA* 108:13281–13286.

Poirel, N., Simon, G., Cassotti, M. *et al.* 2011. The shift from local to global visual processing in 6-year-old children is associated with grey matter loss. *PLoS One* 6:e20879.

Ponting, C., and Jackson, A. P. 2005. Evolution of primary microcephaly genes and the enlargement of primate brains. *Current Opinion in Genetics & Development* 15(3):241–248.

Pynn, L. K., and Desouza, J. F. 2013. The function of efference copy signals: Implications for symptoms of schizophrenia. *Vision Research* 76:124–133.

Rakic, P., Ayoub, A. E., Breunig, J. J., and Dominguez, M. H. 2009. Decision by division: Making cortical maps. *Trends in Neurosciences* 32(5):291–301.

Reader, S. M., Hager, Y., and Laland, K. N. 2011. The evolution of primate general and cultural intelligence. *Philosophical Transactions of the Royal Society of London, Series B: Biological Sciences* 366(1567):1017–1027.

Reader, S. M., and Laland, K. N. 2002. Social intelligence, innovation, and enhanced brain size in primates. *Proceedings of the National Academy of Sciences, USA* 99(7):4436–4441.

Rilling, J. K., Glasser, M. F., Preuss, T. M. et al. 2008. The evolution of the arcuate fasciculus revealed with comparative DTI. *Nature Neuroscience* 11(4):426–428.

Rilling, J. K., and Insel, T. R. 1999. Differential expansion of neural projection systems in primate brain evolution. *Neuroreport* 10(7):1453–1459.

Rimol, L. M., Agartz, I., Djurovic, S. *et al.* 2010. Sex-dependent association of common variants of microcephaly genes with brain structure. *Proceedings of the National Academy of Sciences, USA* 107(1):384–388.

Russell-Smith, S. N., Maybery, M. T., Bayliss, D. M., and Sng, A. A. 2012. Support for a link between the local processing bias and social deficits in autism: An investigation of embedded figures test performance in non-clinical individuals. *Journal of Autism and Developmental Disorders* 42(11):2420–2430.

Sanjuan, J., Tolosa, A., González, J. C. *et al.* 2005. *FOXP2* polymorphisms in patients with schizophrenia. *Schizophrenia Research* 73(2–3):253–256.

Saxe, R. 2006. Uniquely human social cognition. *Current Opinion in Neurobiology* 16(2):235–239.

Schoenemann, P. T. 2006. Evolution of the size and functional areas of the human brain. *Annual Review of Anthropology* 35: 379–406.

Shaw, P., Greenstein, D., Lerch, J. *et al.* 2006. Intellectual ability and cortical development in children and adolescents. *Nature* 440(7084):676–679.

Skafidas, E., Testa, R., Zantomio, D. *et al.* 2014. Predicting the diagnosis of autism spectrum disorder using gene pathway analysis. *Molecular Psychiatry* 19:504–510.

Somel, M., Franz, H., Yan, Z. *et al.* 2009. Transcriptional neoteny in the human brain. *Proceedings of the National Academy of Sciences, USA* 106:5743–5748.

Soulières, I., Dawson, M., Gernsbacher, M. A., and Mottron, L. 2011. The level and nature of autistic intelligence II: What about Asperger syndrome? *PLoS One* 6(9):e25372.

Soulières, I., Dawson, M., Samson, F. *et al.* 2009. Enhanced visual processing contributes to matrix reasoning in autism. *Human Brain Mapping* 30(12):4082–4107.

Stout, D., Toth, N., Schick, K., and Chaminade, T. 2008. Neural correlates of Early Stone Age toolmaking: Technology, language and cognition in human evolution. *Philosophical Transactions of the Royal Society of London, Series B: Biological Sciences* 363(1499): 1939–1949.

Suddendorf, T., Addis, D. R., and Corballis, M. C. 2009. Mental time travel and the shaping of the human mind. *Philosophical Transactions of the Royal Society of London, Series B: Biological Sciences* 364(1521):1317–1324.

Thompson-Schill, S. L., Ramscar, M., and Chrysikou, E. G. 2009. Cognition without control: When a little frontal lobe goes a long way. *Current Directions in Psychological Science* 18:259–263.

Torri, F., Akelai, A., Lupoli, S. *et al.* 2010. Fine mapping of AHI1 as a schizophrenia susceptibility gene: From association to evolutionary evidence. *FASEB Journal: Official Publication of the Federation of American Societies for Experimental Biology* 24(8):3066–3082.

Trimble, M. R. 2007. *The soul in the brain: The cerebral basis of language, art, and belief.* Baltimore: Johns Hopkins University Press.

Uddin, L. Q. 2011. The self in autism: An emerging view from neuroimaging. *Neurocase* 17(3):201–208.

Vernes, S. C., Oliver, P. L., Spiteri, E. *et al.* 2011. *FOXP2* regulates gene networks implicated in neurite outgrowth in the developing brain. *PLoS Genetics* 7(7):e1002145.

Vorstman, J. A., Anney, R. J., Derks, E. M. *et al.* 2013. No evidence that common genetic risk variation is shared between schizophrenia and autism. *American Journal of Medical Genetics, Part B: Neuropsychiatric Genetics* 162(1):55–60.

Waters, F., Allen, P., Aleman, A. *et al.* 2012. Auditory hallucinations in schizophrenia and nonschizophrenia populations: A review and integrated model of cognitive mechanisms. *Schizophrenia Bulletin* 38(4):683–693.

Wei, X., Yu, J. W., Shattuck, P. *et al.* 2013. Science, technology, engineering, and mathematics (STEM) participation among college students with an autism spectrum disorder. *Journal of Autism and Developmental Disorders* 43(7):1539–1546.

Wong, P. C., Chandrasekaran, B., and Zheng, J. 2012. The derived allele of *ASPM* is associated with lexical tone perception. *PLoS One* 7(4):e34243.

Woodard, C. R., and Van Reet, J. 2011. Object identification and imagination: An alternative to the meta-representational explanation of autism. *Journal of Autism and Developmental Disorders* 41:213–226.

Zahn-Waxler, C. Crick, N. F, Shirtcliff, E. A., and Woods, K. E. 2006. The origins and development of psychopathology in females and males. *In*: D. Cicchetti and D. J. Cohen (eds) *Developmental psychopathology, vol. 1: Theory and method*. Hoboken: John Wiley & Sons, Inc. pp.76–138.

Zahn-Waxler, C., Shirtcliff, E. A., and Marceau, K. 2008. Disorders of childhood and adolescence: Gender and psychopathology. *Annual Review of Clinical Psychology* 4:275–303.

Zollikofer, C. P. E., and Ponce de León, M. S. 2010. The evolution of hominin ontogenies. *Seminars in Cell & Developmental Biology* 21(4)441–452.

Chapter 10

Evolving the Developing Cortex: Conserved Gradients of Neurogenesis Scale and Channel New Functions in Primates

Christine J. Charvet and Barbara L. Finlay
Behavioral and Evolutionary Neuroscience Group, Department of Psychology, Cornell University, Ithaca, NY, USA

Introduction: Evo-Devo and the Evolution of Computation

Graceful unfolding of the capacities of computational devices in larger scales has come to be an expectation of users of computers with ever-increasing power in the context of the mushrooming internet. The word "graceful" is often used in computer and computational sciences to describe the desirable behavior of systems in response to addition or loss of elements, such as memory resources or transmission bandwidth. A good application maintains its essential functions and performs at maximum efficiency at all scales, employing all its available resources, and minimally, does not crash! Comparing the evolution of computer devices to the evolution of brains in a general way is an entertaining exercise, and we will argue, an instructive one (Nagarajan and Stevens 2008). We can identify taxonomic differences in computers, competition and selection in sales, and evolutionary "grade shifts" over time as new operating systems are introduced. We can also find an analogy to biological cultural selection, where information can be offloaded from the genetic load of individuals to culture. In the cultural evolution of computers, all the information for functions on individual computers no longer needs to be stored on it when programs and data can be kept instead in the "cloud."

The level of the evolution of circuitry in the evolving-computer versus evolving-brain comparison is a more difficult yet potentially very useful way to generate new descriptions and ideas about how brains have changed, and can change (Moriarty and Miikkulainen 1997; Stanley and Miikkulainen 2003; Kaplan and Oudeyer 2007; Nagarajan and Stevens 2008; Lehman and Stanley 2013). We may ask what kinds of architectures in each tend to be scalable and evolvable, able to efficiently employ new information yet not derailed by every minor environmental change (Finlay and

Developmental Approaches to Human Evolution, First Edition. Edited by Julia C. Boughner and Campbell Rolian.

Darlington 1995; Striedter 2005; Lee and Stevens 2007; Dyer *et al.* 2009; Charvet *et al.* 2011; Finlay *et al.* 2011; Snider *et al.* 2010). The fundamental charge, of course, is not to mistake changing computing systems for the brain, but to use them as sources of insight and analogy. Anthropologists have not typically been concerned with the scalability and flexibility of computation, turning their attention instead to brain size as a morphological index of behavioral complexity and intelligence, and sometimes life history (Dunbar 1993; Dunbar and Shultz 2007; Hofman 2014). This computer analogy is intended to bring the problems of scalability and evolvability of computational devices to the forefront before delving into the neuro-embryological details of the conserved pattern of cortical generation. We will return to this analogy periodically.

Conservation versus Constraint

Metaphors drawn from devices at hand have a long history in discussions of development and evolution. The original illustration of how a "developmental constraint" might channel evolution, the QWERTY keyboard (named after the first six letters of the top row of typewriter keyboards), comes from Stephen J. Gould (1987). The QWERTY keyboard reduced key jamming in manual, ribbon-and-key typewriters by deliberately spacing commonly used letters apart from each other. Although the QWERTY pattern does not permit optimal speed once the key-and-ribbon bug is gone, conserved versions of this keyboard with only minor alterations are probably in the immediate view of any readers of this article. Gould was interested equally in explaining the persistence of traits as well as understanding adaptation of new traits, but he gave conserved mechanisms and structures a generally negative cast by terming them "constraints" (Gould 1980). He argued that because of deep embedding of developmental mechanisms inside each other and pleiotropic effects of genes, selection on a single trait might not be possible without changing or damaging other ones (Gould 1977; Gould and Lewontin 1979; Gould 1989). Thus, stability in organization over evolutionary time came to be viewed as an essentially negative feature showing potential adaptations reined in by unwanted side effects called "developmental constraint" (Finlay *et al.* 2001; Linde-Medina and Diogo 2014).

Beginning with work in conserved body and brain plans in both invertebrates and vertebrates, evo-devo researchers profoundly changed the view of conserved structure from constraint by examining what organizational benefits conserved developmental programs might provide, and introducing the concept of "evolvability" (Raff 2000; Raff and Raff 2000; West-Eberhard 2003; Wagner 2005; Kirschner and Gerhart 2006; Gerhart and Kirschner 2007; Draghi and Wagner 2008; Masel and Trotter, 2010; Charvet *et al.* 2013; Lehman and Stanley 2013). These evo-devo researchers discovered that myriad developmental programs had been maintained in detail over hundreds of millions of years, simultaneously producing the most remarkable variation in adult morphology and behavior. The current research focus of evo-devo investigates the mechanisms by which conserved developmental programs (usually with a genomic focus) can produce diverse structures (Striedter 2005; Carroll 2009; Dyer *et al.* 2009; Charvet *et al.* 2011). We will apply this approach to the mammalian brain and the well-described suite of neuro-embryological mechanisms for its own constructions, with the focus on the computing consequences of these mechanisms described at the cell biological and brain architecture level. In addition, we will consider how computers and the internet can vary over many orders of magnitude in number of elements, volume, and speed, as a source of new insights about how computational devices can change progressively. Both neurodevelopmental research and the consideration of how our own computational devices have progressively enlarged lead us away from the modular, "add-on" account, the original guess about the nature of brain evolution.

Understanding the Brain Correlates of Behavioral and Cognitive Adaptations

At the most general level of natural selection, brain organization must ultimately depend on functionality and fitness. Organisms must recognize their energetic needs; traverse their environments without mortal accident; recognize and counter parasites and diseases; successfully forage; and recognize, compete, and mate with other. When anthropologists discuss these abilities and how to relate them to the brain, they rarely explicitly discuss more general computational parameters such as "processing speed" or "hierarchical reinforcement learning capacity," but rather proceed directly to how each specific behavior might be realized in the brain, usually attributing new or improved competence to an alteration in the size or organization of a pre-existing brain region. Depending on academic discipline, this view of brain evolution is termed "mosaic brain evolution" in evolutionary neurobiology and anthropology (Holloway 1968; Barton and Harvey 2000; Iwaniuk *et al.* 2004) or "massive modularity" in cognitive science (Fodor 1983; Sperber 2001; see Anderson and Finlay, 2014, for an extended discussion). In this tradition, researchers attempt to locate the telltale marks in endocasts for gyri sub-serving language; or the location of the increased neuromass for the "social brain," trichromacy, tool use, or kin calculations (Geschwind 1974; Dunbar 1993; Barton 1998; Dunbar and Shultz 2007; Stout *et al.* 2008).

Similarly (but superficially), a competence list for computers will involve specific abilities, "applications," such as word processing, number crunching over multiple orders of magnitude, game-play of all sorts, and speech and face recognition. The way we have learned to employ computers and observe their multiple uses and changes over time, however, makes it unlikely that any experienced computer user will assume that an ability like "word processing" or "face recognition" implies a specialized component or module has been introduced into the computer's hardware. In addition, improvements in generic computer features, such as the replacement of floppy disks with hard-disk drives, and then those with solid-state drives, can improve performance wholesale across multiple applications, and allow entirely new applications. Why are we convinced that the way we should understand the process of mapping function to structure should be so different in the cases of brains and silicon computing devices?

Extending Evo-Devo Work on Body Plan to the Brain Plan

The growth of evo-devo as a field has highlighted the deep conservation of developmental programs across species in features such as the basic body plan, and physiological pathways (Gerhart and Kirschner 1997; Carroll 2009). The conservation of developmental mechanisms can be hard to appreciate, as it is often hidden by the diversity of its morphological and physiological outcomes. Evo-devo researchers interested in morphology have actively sought the key organizational features, which produce robustness and evolvability of the basic body plan (Heffer and Pick 2013). For example, the pattern of *Hox* genes, which set up the anterior-to-posterior polarity of the body and subsequent segmentation along that axis, is conserved in detail (with various numbers of duplications of the whole *Hox* group) across vertebrates and invertebrates (Lewis 1978). The prominent segmental organization of this pattern promotes evolvability by allowing changes in the control of particular gene products to be confined to a particular location, allowing a leg to be produced in one location and gills in another (Akam 1995; Lemons and McGinnis 2006). For another example at the physiological level, apparent strict constraint on or conservation of the types of signaling pathways between the cell body and nucleus of any cell allow peripheral divergence to be translated to a common intracellular language (Rutledge *et al.* 1974; Burke *et al.* 1995; Raff and Raff 2000). To continue our computer metaphor, this constraint in cell biology resembles the requirement that all computers must use the same protocols to communicate with the internet.

More recently, the brain plan embedded in the body plan has been the subject of similar study, looking at the same features of overall establishment of axes, segmentation, determination of the "identity" of neurons, and so forth (Lumsden 1990; Rubenstein *et al.* 1994; Finlay *et al.* 1998; Sylvester *et al.* 2011; Puelles and Ferran 2012). Here we will discuss some features of mammalian brain development that similarly confer computational, and hence behavioral, robustness and evolvability at multiple scales with a focus on the isocortex. We will focus particularly on "standing" gradients of neuronal proliferation that cause the isocortex to scale over six orders of magnitude in volume, yet reliably produce species-specific capacities like whisker grooming or tool-making, nocturnal or diurnal vision, echolocation or language. The intent here is to begin to specify the computational organization entailed by the well-known association of brain size with behavioral complexity, so that we might begin to reify the nature of "processing speed" and "computational capacity."

The Special Role of "Simple" Brain Size (Absolute or Relative to Body Size)

Proponents of modular accounts of human capacities have reasonably balked at the idea that the simple increases in neuron numbers could be adequate to account for such complex capacities as language, social calculations, or visual discrimination. In fact, computational studies have shown that simply adding more and more new neurons to neural nets designed to solve specific problems does not improve them, and in fact, typically harms them (e.g., Jacobs 1997). We will first argue that neurons are not simply "piled on." Rather, the development of the cortex across mammals embodies a conserved processing strategy, which has an intriguing scaling with size (Gould 1977; Finlay and Darlington 1995; Finlay *et al.* 2001; Stevens 2002; Lee and Stevens 2007). This processing device of the cortex in combination with selected changes in peripheral "devices" and inherited and environmental changes in motivation can account for how new complex capacities develop. This framework is better than positing more and more *ad hoc* computational modules installed in the cortex by natural selection, whose numbers have become excessive (Finlay 2007; Penn *et al.* 2008).

The size of the brain and a few of its larger components has long been the staple of anthropological analyses, though the crude nature of brain size as a measurement of behavior has also been ruefully acknowledged (Jerison 1973; Stephan *et al.* 1981; Striedter 2005; Bauernfeind *et al.* 2013). Relative brain size may not be so bad after all, however, as a first step in understanding human computational advantages. The work of Louis Lefebvre and colleagues, over a number of years, has found that across birds, mammals generally, and primates particularly, useful types of behavioral plasticity (innovation in food acquisition, tool use, and complexity of social organization) strongly correlate with each other, rather than appearing in isolation (Lefebvre *et al.* 2004; Lefebvre 2013a). Additionally, these same behaviors also correlate strongly with two unequivocal measures of evolutionary fitness: the ability to successfully invade new environments, and relative yearly mortality. Finally, these measures covary with relative brain size and to a small extent with laboratory "IQ" tests (Wickett *et al.* 1994). Quoting Lefebvre (2013b):

> This correspondence suggested that birds and primates show similar, convergent relationships between innovation rate and relative forebrain volume, innovation rate and tool use, and innovation rate and individual learning. In turn, these positive relationships suggested that a common, general factor might underlie interspecific differences in animal cognition, an idea that is gaining ground [Deaner *et al.* 2006; van Schaik *et al.* 2012] after a period dominated by the concept of massive modularity.

This repeated covariation of food-source innovation, tool use, and niche invasion across taxa is evidence against the selection of each as an independent brain module. Understanding their shared computational resources or each separately would greatly inform our understanding.

We would like to understand Lefebvre's "general factor" as a specific, selectable feature of brain organization, and not the "residual brain mass" left when all specifically named behavioral adaptations and their presumed brain modules have been accounted for (Charvet and Finlay 2012). Plasticity to adapt developmental strategies to conform to stress and scarcity versus stability and plenty, to vary social structures to the size and fixed properties of resources, or to vary communication to the most convenient sensory channels is not a feature of brain organization invoked when things go wrong, in cases of pathology, disease, or environmental catastrophe, but a fundamental insight into how the brain normally works.

Basics of Brain Evo-Devo

Basics of Brain Anatomy and Neurogenesis

It is beyond the scope of this chapter to provide a complete introduction to neuroanatomy and neuro-embryology; for that the reader is referred to the introductory chapter on brain development in virtually any introductory neuroscience book (for those from a behavioral background, Purves *et al.* 2012; for those from a biology background, Kandel *et al.* 2013). Here we will describe very briefly the most central parts and concepts of the chapter; key terms may also be found in the glossary at the end of this chapter.

Brain development originally provided the basis for brain divisions and their names. The brain begins as a flat plate, the "neural plate," which rounds up along the posterior-to-anterior axis to form the "neural tube" with the plate's most lateral parts brought together and joined. The inside of the tube will become all of the fluid-filled ventricles of the brain. All neurons are generated next to this region, "the ventricular zone," and migrate away from the ventricles in a radial direction to their eventual mature location, usually on the scaffolding of non-neural "glial guides" (Boulder Committee 1970; Bystron *et al.* 2008). The immediate environment of the initial location in the neural plate provides critical information to specify each neuron's type and connections – for example, all motor neurons arise from the middle, "basal," part of the neural tube, while sensory components of the brain arise from the initially most lateral, "alar," part.

The neural tube gives rise to the entire brain, both the spinal cord and brain proper. The brain is composed of spinal cord, hindbrain (including pons and cerebellum), midbrain, and forebrain. Across the rostro-caudal axis, the central nervous system is segmented (Bergquist and Källén 1954). Segmental divisions of the spinal cord are easily recognizable as adult segments. In the brain proper, the segmentation persists (in the form of repetition of gene expression patterns and derived organization features) but individual segments become highly differentiated from each other, especially from extended neurogenesis in various regions. Hindbrain segments are called "rhombomeres" (Lumsden 1990), and forebrain segments "prosomeres" (Puelles and Ferran 2012). Across the central nervous system and within its segmental divisions are distinct gradients in the duration of neurogenesis (Rapaport *et al.* 1996; Rakic 2002; Caviness *et al.* 2009). The first neurons are produced relatively close to the same time everywhere, but neurogenesis ends first in the medial (basal) areas and last in the regions corresponding to the lateral neural plate (Langman and Haden 1970; Hollyday and Hamburger 1977). That is, lateral regions undergo terminal neurogenesis after more medial regions within the central nervous system. This contrast holds whether in the spinal cord, considering motor neurons versus sensory integrative neurons in the dorsal horn, or in the forebrain, contrasting basal forebrain nuclei with the cerebral cortex. A gradient in neurogenesis timing is also seen across the posterior-to-anterior axis of the brain, with the spinal cord maturing first and the forebrain maturing last (Workman *et al.* 2013).

We will be concerned mostly with the "cortex," which has unfortunately gathered multiple names over the years. The whole forebrain contains (at least) four different kinds of "cortex" including the olfactory bulb and olfactory cortices, the hippocampus, and the iso- or neocortex. "Neocortex," "isocortex," and sometimes just "cortex" refer to the six-layered structure dominating the surface view of the human brain (or any large brain, in fact). "Neocortex" is often used as the name for this structure, based on the erroneous assumption that it appeared *de novo* in mammals (Northcutt and Kaas 1995; Striedter 2005). "Isocortex" is the preferred name for this structure for comparative anatomists because "iso" references its consistent layered structure. Isocortex is the term we will use from this point onward.

The isocortex is formed similarly to other neural regions, but it is generated for an extended period, as is the cerebellum (for a more extended review see Finlay 2005; Workman *et al.* 2013). Early in development, proliferating cells are found in the ventricular zone. Some of these cells migrate radially superficial to the ventricular zone and form the "cortical plate" on the outer part of the brain (Angevine and Sidman 1961; Figure 10.1). In some brain regions such as the isocortex, some ventricular zone cells migrate superficial the ventricular zone and continue to proliferate in the subventricular zone. As development proceeds, the ventricular zone and subventricular zone wane and most of neurogenesis ends (Boulder Committee 1970; Bystron *et al.* 2008). Maturing neurons arriving at the cortical plate are deposited in an inside-out manner with respect to the order of their genesis with the last-generated neurons located on the outside (Rakic 1974, 2002, 2007).

The isocortex has three prominent organizational features: layers, columns, and "areas." Each cortical layer corresponds to a particular neuron type, order of generation, and type of synaptic connections (Rakic 1974, 2002). Inside to outside, lowest layers VI and V principally communicate outside the cortex; layer IV receives input from the thalamus, which relays primary sensory, motor, and visual information, and layers III–II communicate within the cortex (Gilbert and Kelly 1975). The cortex is made up of repeating columns comprised of these layers, a small number of columns in small brains and a large number in larger ones, each of which is integrating particular thalamic and cortical input, transforming it, and relaying it on (Mountcastle 1997). In large brains, cortical surface area increases the most, though its thickness increases a moderate amount as well (Rakic 1995; Hutsler *et al.* 2005). Cortical areas were recognized and numbered by Brodmann, one of the first to describe the isocortex, and correspond to a mosaic of topographically organized sensory and motor surfaces that all together make up the isocortex (Brodmann 1909). Each area, composed of multiple columns, has a particular pattern of input and output, and particular functional transformations of input to output. Primary sensory and motor regions are "special" within the array of areas, having unique recognition mechanisms and extremely strong trophic dependence in development on their specific thalamic input nucleus. For example, the lateral geniculate nucleus of the thalamus, relaying visual information from the retina, will project only to primary visual cortex, and the thalamus must make that connection for its neurons to survive (Mihailović *et al.* 1971; Wong-Riley 1972; Cowey and Stoerig 1989). The primary sensory and motor areas serve to anchor the other areas in the cortical sheet and anchor its anterior-to-posterior polarity.

Other areas ("association" and other) are less specific in their developmental requirements and adult function. As we gain more knowledge, the functions assigned to any one area are expanding rather than becoming more specified. For example, on the basis of neurological case studies on one patient each, "Broca's area" in the frontal lobe was originally assigned language syntax, and "Wernicke's area" in the temporal lobe was assigned semantics (Geschwind 1972). Current work, both electrophysiological studies in animals and human neuroimaging, has

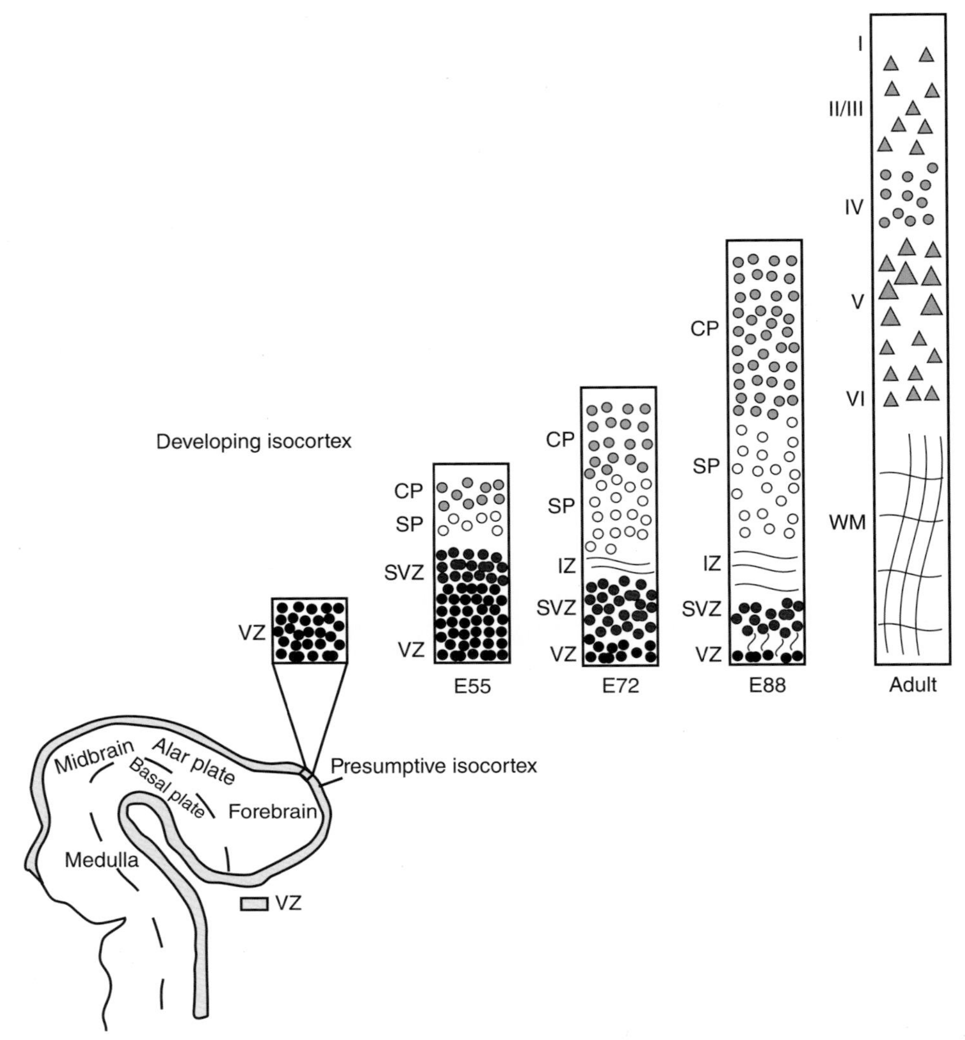

Figure 10.1. (A) Diagram of a developing embryo and the isocortex. Early in development, the ventricular zone (VZ) houses proliferating cells. VZ cells have processes that extend from the ventricular to the pial surface. Proliferating cells within the VZ undergo mitosis at the ventricular surface. (B) Diagram of isocortical development at different ages of development in a monkey (*Macaca fascicularis*). Early in development, the presumptive isocortex consists of a VZ. As development proceeds, proliferating cells exit the VZ and some cells migrate superficial to the VZ into the cortical plate (CP) and the subplate (SP). Other VZ cells migrate superficial to the VZ and form the subventricular zone (SVZ). As development progresses, the proliferating pool and the subplate wane. Cortical layers (I–VI) become evident by adulthood. In the cortical plate, late generated neurons migrate beyond earlier ones. That is, they migrate in an inside-out fashion. Legend: IZ, intermediate zone; WM, white matter. Parts of this diagram have been modified from Smart *et al.* 2002.

now shown these areas to be activated in multiple functions, including language functions and many others distantly related to language, if at all (Poldrack 2006; Anderson *et al.* 2013).

"Late Equals Large": The Relation of Neurogenesis to Brain Allometry

Understanding how brains vary systematically across species is fundamental to understanding what developmental mechanisms are candidates to produce adaptations (Gould 1975; Finlay and Darlington 1995; Striedter 2005; Carroll 2009; Charvet and Striedter 2011; Charvet *et al.* 2011; Finlay *et al.* 2011; Sylvester *et al.* 2011; Puelles and Ferran 2012; Workman *et al.* 2013). Looking at the gross divisions of the brain, and how they scale with brain size across mammals and vertebrates (i.e., sharks and rays, Figure 10.2, color plate) generally, a surprising consistency is its most obvious feature: evolution does not "choose" brain parts to enlarge at random. First, the size of brain parts is highly correlated with overall brain size, with about 96% of the variance accounted for by the single factor of brain size (Finlay and Darlington 1995; Yopak *et al.* 2010), and a second 3% related to the more independently varying olfactory bulb and limbic system (Reep *et al.* 2007). As Figure 10.2 (color plate) shows, the olfactory bulb and, to a lesser extent, the limbic system (e.g., olfactory cortex, subicular cortex) vary relatively independently of other brain regions in mammals as well as in sharks and rays (Yopak *et al.* 2010). Each major brain division, and, in fact, every subdivision and neuron type, has its own characteristic allometric scaling.

The telencephalic and cerebellar neuron number and volume particularly increase rapidly with brain size, such that the largest brains become disproportionately (but predictably) composed of these two structures (Finlay and Darlington 1995; Herculano-Houzel 2009). As an example, the natural-logged values of the telencephalon versus the brain scale with a slope of 1.08 but the natural-logged values of the medulla regressed against the brain scale with a slope of 0.7 in mammals. The Etruscan shrew has the smallest brain and the human has the largest brain of species shown in Figure 10.2 (color plate). The telencephalon and the medulla of Etruscan shrews occupy approximately 51% and 20% of their brain, respectively. The telencephalon and medulla of humans occupy approximately 85% and less than 1% of their brain, respectively. That is, brain regions scale allometrically with respect to other brain regions, and the allometric variation of brain regions is tied to variation in developmental duration across species.

During development, as mentioned in the prior section, there are gradients in duration of neurogenesis across the entire brain, and these have direct significance for the mechanisms of how some brain parts (e.g. cortex, cerebellum) disproportionately enlarge relative to others. Generally, segmental location, both from basal to alar (medial-to-lateral) and posterior to anterior, is associated with longer neurogenesis duration across mammals (Finlay and Darlington 1995; Finlay *et al.* 1998, 2001; Workman *et al.* 2013). That is, neurons in alar regions continue to be "born" after neurons in basal regions have concluded neurogenesis, and neurons in anterior regions continue to be born after neurons in the posterior regions

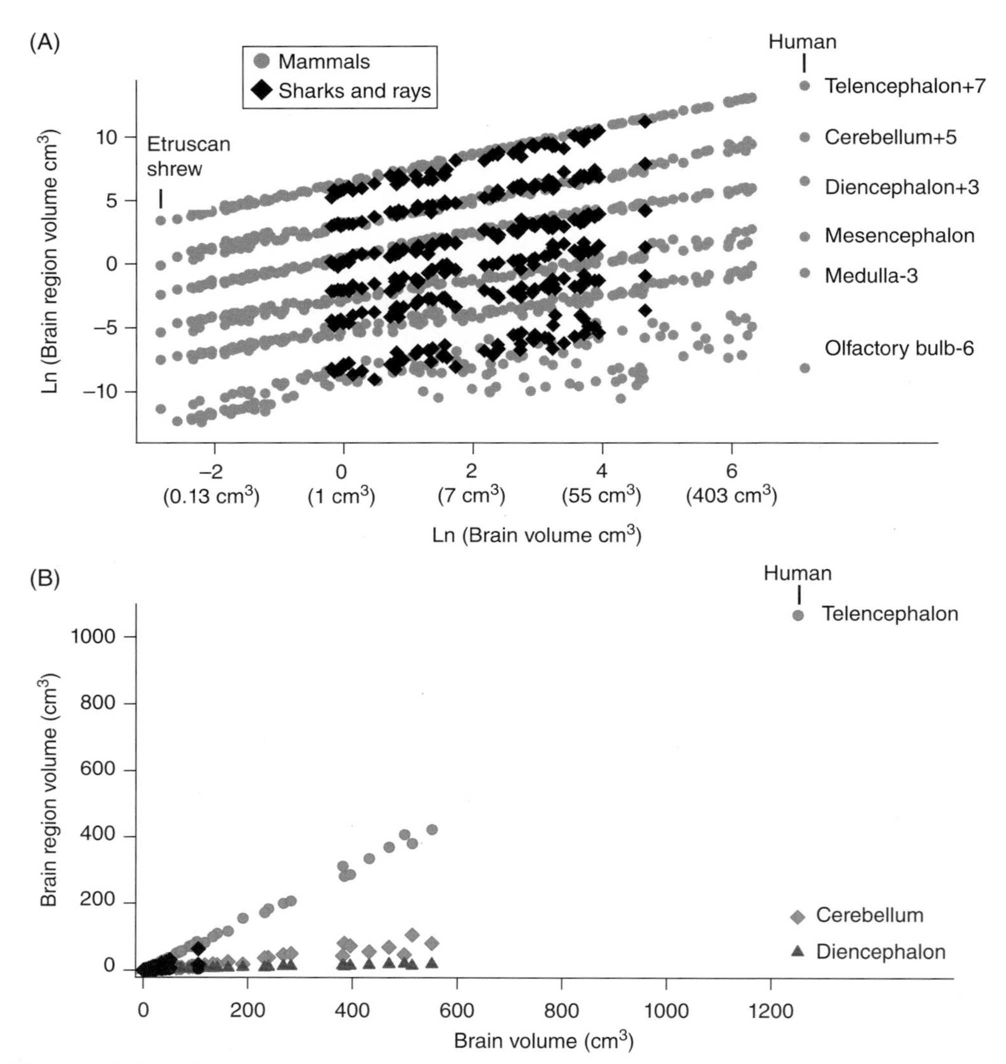

Figure 10.2. (A) The natural-logged values of brain region volumes are regressed against the natural-logged values of overall brain size in mammals, as well as in sharks and rays. Major brain regions strongly covary with overall brain size. One important exception is the olfactory bulb, which varies relatively independently of other brain regions. Brain regions also expand allometrically with respect to the other brain regions. In particular, the telencephalon and cerebellum become disproportionately enlarged relative to other brain regions, such as the medulla, as brains expand. For example, the Etruscan shrew has a brain volume around 50 mm^3. Its telencephalon and its medulla occupy approximately 51% and 20% of its brain, respectively. The human has a brain volume of approximately 1250 cm^3. Its telencephalon occupies approximately 85% of its brain but its medulla occupies less than 1% of its brain. (B) To appreciate the allometry of brain region volumes, the telencephalon, cerebellum, and diencephalon volumes are plotted against overall brain size in mammals and in sharks. The telencephalon becomes disproportionately larger than the cerebellum, which in turn becomes disproportionately larger than the diencephalon. A constant was added to the brain region volume in (A) to distinguish brain region volumes on the same graph. These constants are listed adjacent to each named structure. Data are from Stephan *et al.* 1981; Reep *et al.* 2007; and Yopak *et al.* 2009.

(Finlay *et al.* 1998; Koyama *et al.* 2011). The regions of the brain where neurogenesis continues late are the same regions that become disproportionately large in big brains (Finlay and Darlington 1995; Finlay *et al.* 1998; Charvet *et al.* 2011; Workman *et al.* 2013). Location of origin is by no means the only factor controlling neuron number, but it dominates at this general level of analysis. As an aside and cautionary note, please note that we are describing a mechanistic account of cell proliferation, not a developmental constraint on cell proliferation. As cell identity is typically determined after final cell division in neuro-embryology, it is as reasonable to suspect that high numbers of neurons with desirable properties (e.g., cortical neurons) have often been produced by selection over evolutionary time by "moving" those instructions to a pre-set area of extended proliferation or simply shifting a developmental border, rather than extending the duration of neurogenesis of the natal region of the desirable structure in the neural plate.

For those interested, the characteristic pattern of vertebrate brain evolution has the effect of allocating most new neurons in larger brains "off to the side" of basic sensorimotor command lines for approach and avoidance, to the cortex and cerebellum (Figure 10.2, color plate). The advantages of this arrangement are multiple. One major advantage is that sophisticated computations and predictions can be made in structures informed by the learning history of the organism and then re-entered into the stream of processing without compromising the integrity and speed of essential reflex and orienting pathways. The use of this arrangement for easy addition and subtraction of computing resources was discovered independently in the computer science literature, discussed elsewhere (Yopak *et al.* 2010; Finlay *et al.* 2011).

In addition to this whole-brain organization of gradients of neurogenesis duration, local gradients in neurogenesis within structures can be found nested within this whole-brain organization. The third section of this chapter will discuss the scaling and functional consequences of an anterior-posterior gradient of neurogenesis across the developing isocortex. We argue that an increase in cortex volume is not an indeterminate addition of neurons, but a particular, progressive amplification of a hierarchically organized information-processing strategy (Cahalane *et al.* 2012; Charvet *et al.* 2015).

If we examined the "allometry" of computer parts we would find a similar pattern of statistically common increases in the numbers of some elements, like RAM, and relative stability in others. Occasionally, additions will have a role in the improvement of a specific class of operations, like a graphics card, but for the most part, elements are added according to their generic function, whether or not the user intends to employ a computer's enhanced memory for bigger games or bigger data.

Isocortex Evolution and Development

General Issues in Allocating Structure to Function in the Isocortex

Across the isocortex of primates, neuron numbers "per column" or per unit of cortical surface vary considerably, and the sizes of cortical areas do so as well (Frahm *et al.* 1984; Kaskan *et al.* 2005; Rakic 2008; Collins *et al.* 2010; Charvet *et al.* 2015).

What predicts the variation? Neuron number per column has been shown to range fivefold across the primate isocortex, with most neurons per unit of cortical surface area found in the caudal cortex (i.e., primary visual cortex in primates; Collins *et al.* 2010; Cahalane *et al.* 2012; Charvet *et al.* 2015). From the beginning of research in this area, it has been hypothesized that neuron numbers in a cortical column, or the number of columns in the cortex, should be allocated according to the relative proportions of neurons in the sensory and motor periphery. The importance of the information to the animal is sometimes referred to as Jerison's "proper mass" (1973). For example, it seems reasonable that "visual" animals like ground squirrels or monkeys should allocate more cortex to vision by increase in the relative size of primary sensory areas, while nocturnal animals like mice or bats should dedicate more neurons to primary areas essential for smell, hearing, or touch. In fact, this generality holds fairly well for the smallest brains, such as the platypus, hedgehog, and shrew (reviewed in Krubitzer and Stozenberg, 2014). In larger brains, however, it is unclear whether the relative volume of a "named," genetically defined cortical area, for example, primary visual cortex, V1, is a good measure of how processing resources have been allocated, as functions play out over a much more extended cortical surface, functional allocation changing according to developmental history or immediate demand (Finlay and Anderson 2014).

An Enlarged Cortical Module for Improved Vision? Evaluation of a Specific Controversy

A very specific debate emerged about whether the large number of neurons per unit of cortical surface area in the primate (including human) V1 is a taxon-specific adaptation in that cortical area for better vision (Barton 1998; Krubitzer *et al.* 2011) or represents the normal, regionally disproportionate allometric expansion of the isocortex (Charvet *et al.* 2015). We choose this example because unlike other proposed modules in cognitive and social behavior, the features to be explained can be very precisely identified and quantified. The primary visual cortex of primates (anatomical "V1" or Brodmann Area 17) has many more neurons per unit of cortical surface area relative to all of its other cortical areas (Collins *et al.* 2010) and relative to any cortical area in other "non-visual" species, such as the laboratory mouse and rat (Charvet *et al.* 2015). It was suggested that the increased volume and neuron numbers in V1 have been under special selection, independent of overall isocortex size, because the V1 confers special visual capabilities in primates (Barton 1998). According to this view, there should be select changes in developmental parameters in V1 that account for the disproportionate V1 neuron numbers in primates, and some support has been found for this hypothesis (Dehay *et al.* 1993).

The essential problem in resolving the answer to this question was the lack of systematic comparisons of different cortical areas in a wide range of mammalian species that vary in brain size, developmental duration, and behavioral specializations. Previous studies had examined relatively very few species and few cortical areas (e.g., Rockel *et al.* 1980; O'Kusky and Colonnier 1982; Beaulieu and Colonnier 1989). That is, comparisons in neuron numbers would focus on select cortical areas (e.g., primary visual area, primary somatosensory cortex) in few model species (e.g., mice, cat, monkey). Since then, data on neuron numbers, cortical area, and volume of primary visual cortex have been described in a wider range of diurnal and nocturnal mammalian species (Skoglund *et al.* 1996; Collins *et al.* 2010; Naumann *et al.* 2012; Herculano-Houzel *et al.* 2013; Young *et al.* 2013; Seelke *et al.* 2014; Charvet *et al.* 2015). In particular, we found that the neuron number, area, and volume of visual cortices was predicted by neuron numbers, and taxon, but not nocturnal or diurnal niche, the latter an obvious measure of a species' dependence on color vision (Kaskan *et al.* 2005; Finlay *et al.* 2006; Finlay and Brodsky 2006; Anderson and Finlay 2014; Charvet *et al.* 2015).

The essential issues here are whether or not the size of a cortical area can be the object of targeted, unique selection independent of the rest of the cortex, and, whether such a change could improve a particular sensory or cognitive capacity (Barton and Harvey 2000; Finlay *et al.* 2001). We are belaboring this issue because it is central to the idea of cortex modularity. Cortical areas have traditionally been the prime candidates for "modules" likely to undergo special selection. We choose to investigate primary visual cortex because structure-function predictions are considerably less ambiguous than those that might be made for a "syntax module" or for "sociality." We have not been able to find evidence for such targeted selection in the case of primary visual cortex, though we will detail later a number of studies showing how the features of primary visual cortex can change as the result of changed retinal organization, or changed visual experience. We argue that the increase in neuron numbers in the primary visual cortex highlights an important, cross-cortical, scalable feature of cortex evolution not previously described.

From Gradients of Neurogenesis to Evolutionary Variations

A systematic analysis of the isocortex in several primate species shows that there is an anterior-posterior gradient in neuron number per unit of cortical surface area (Cahalane *et al.* 2012, Charvet *et al.* 2015; Figure 10.3, color plate). We found neuron numbers per unit of cortical surface area to be lowest in frontal poles and highest in the occipital pole (e.g., primary and secondary visual areas; Cahalane *et al.* 2012; Charvet *et al.* 2015). The systematic pattern of variation in neuron numbers across the primate isocortex corresponds to an anterior-posterior pattern of neurogenesis with neurons undergoing their final rounds of division earliest in the frontal pole and latest in the occipital pole of the cortex (i.e., V1; Rakic 1974, 2002). The large neuron number per unit of cortical surface area in the primary visual cortex of primates appears to be best described as a predictable outcome of the systematic increased duration of neurogenesis from anterior to posterior isocortex.

Why is this developmental pattern unlikely to be a special adaptation for vision in primates? First, it can be seen in multiple taxonomic groups. The gradient is not unique to primates, appearing also in rodents as well as in carnivores. Across a number of different primates, the gradient varies systematically with brain size, not niche. For example, the single nocturnal owl monkey, *Aotus*, does not differ in this feature from diurnal monkeys of the same brain size, although its retina and eye

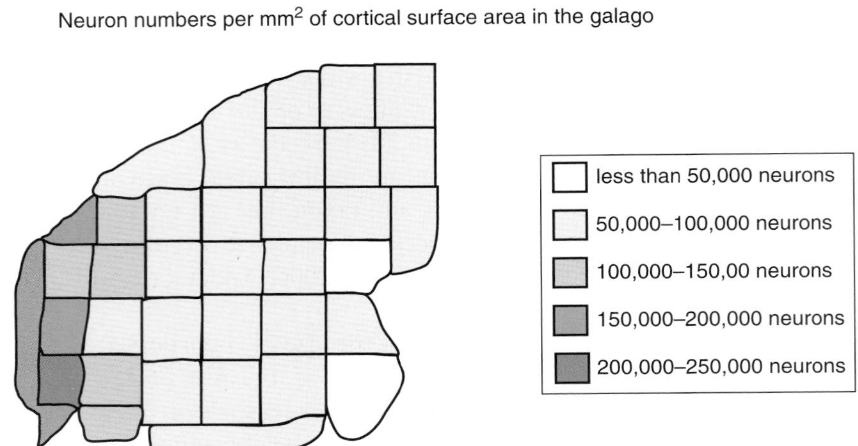

Figure 10.3. Neuron numbers per mm^2 of cortical surface area in a galago (*Otelemur crassicaudatus*) show that neuron numbers under a unit of cortical surface area increase across its rostro-caudal axis. Data are from Collins *et al.* 2010.

conformation are altogether different (Finlay *et al.* 2008). This pronounced gradient must certainly serve some function, and we argue its function is unlikely to be the simplistic one that "highly visual primates should have extra neurons in their visual cortex."

As with the whole-brain gradients described earlier, this hypothesis is a variation for the cortex of the "late equals large" principle we have discussed extensively elsewhere (Finlay *et al.* 2001; Finlay and Brodsky 2006). Those regions of the brain that generate neurons the longest and latest are the same regions likely to be allometrically favored, becoming relatively larger, when brains become larger (Figure 10.2, color plate). For the isocortex, it is essential to further investigate how the layering of the cortex, as briefly described earlier, interacts with the isocortical gradient, as it has consequences for cognition. Finally, we introduce the caveat here that the gradients of neurogenesis are *not* the only means of generating local or global differences in cell numbers, but one of a variety of possible influences yet to be quantitatively described.

Evidence for Developmental Gradients across the Entire Isocortex, and Their Consequences

Gradients in both the progression and the duration of isocortical neurogenesis have been described in several species of rodents, carnivores, and the rhesus monkey (Luskin and Shatz 1985; Jackson *et al.* 1989; Miyama *et al.* 1997; Rakic 2002; Smart *et al.* 2002). Most of these studies have relied on careful analyses of tritiated thymidine injected in embryos at different stages of embryonic development. These developmental schedules vary continuously across a rostro-caudal gradient.

The sequence of neurogenesis with regard to cortical layers is well established, with earliest generated cells populating the lower cortical layers, and later-generated cells the upper layers, as described earlier (Rakic 1974, 2002). This sequence interacts with the extension of neurogenesis within any one species' cortex, and across species, making layers IV through II the layers that get largest in isocortical regions with more time to proliferate (Charvet *et al.* 2011, 2015). Thus both posterior cortex and species with bigger brains overall, and particularly, both together, have proportionately more neurons in the upper cortical layers in their caudal isocortex (Figure 10.4, color plate).

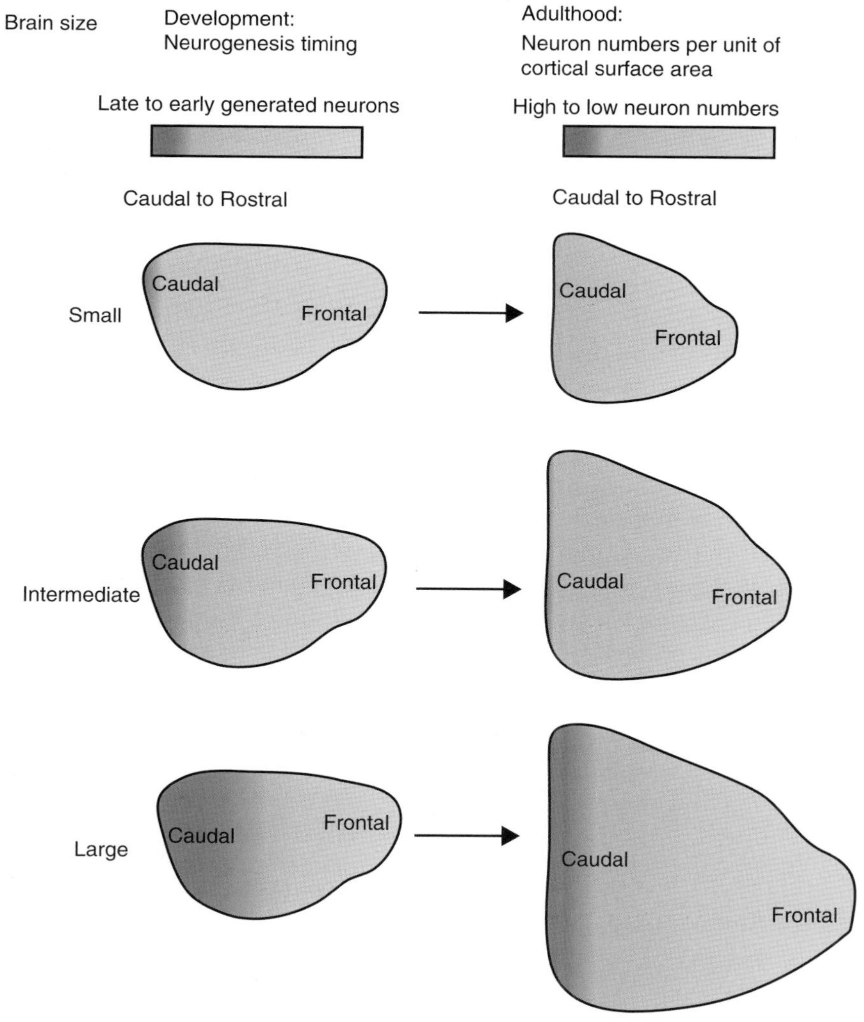

Figure 10.4. The diagram represents the gradient in neurogenesis timing in the isocortex during development and the gradient in neuron numbers under a unit of cortical surface area in adulthood. Gradients in neurogenesis timing vary across species, which would account for systematic variation in neuron numbers under a unit of cortical surface area in adulthood. Large-brained primate species would exhibit a greater disparity in neurogenetic schedules between the rostral and caudal poles leading to a greater disparity in neuron numbers per unit of cortical surface area across the isocortex in adulthood (Charvet *et al.* 2015).

Parenthetically, several additional features of cortical organization are important for understanding its changing organization as brains grow larger, as they interact with the gradient of neurogenesis, and we mention them here for the sake of completeness, and as areas for further inquiry. The first is the production of cortical areas. Similar to the segmental divisions of the central nervous system that are overlaid on the overall gradient of duration of neurogenesis across the central nervous system as described earlier, cortical areas (Kaas 1997) appear to be overlaid on the intrinsic gradient of neurogenesis. How much of what defines a "cortical area" in development is under strict genetic specification and how much is interactive is the subject of continued study (O'Leary *et al.* 1989, 2008, Rakic 2007; Kingsbury and Finlay 2001). As described in the introduction to this section, and the glossary at the end of the chapter, cortical areas are defined by a distinct pattern of thalamic or cortical connectivity, and usually represent a topographic map of some dimension, whether explicit, as in the sense of the visual field, or more abstract, as in the array-to-array maps of areas within the frontal pole. The number of cortical areas increases regularly with absolute isocortex size (Finlay *et al.* 2006; Kaskan *et al.* 2005).

Cortical areas, particularly primary sensory areas, may also accumulate other local genetically specified features, such as the axon-guidance and cell-cell recognition molecules that directly and tightly specify the connectivity of primary sensory regions (O'Leary 2008). The consequent activity-dependent features, like ocular dominance columns, dependent on that input will further canalize the development of cortical areas. Neuron numbers per unit of cortical surface area in primary sensory areas are higher than those found in adjacent regions (Cahalane *et al.* 2012), almost certainly due to a reduction of normal developmental neuronal death because of their relatively large dosage of thalamic input (Finlay and Slattery 1983). However, these cortical divisions are overlaid on a global gradient of neuron number produced by the neurogenetic gradient. The high neuron number-per-column of primary visual cortex thus likely represents the cross-cortex gradient, any local genetic modifications of neurogenesis that may have occurred, and the reduced developmental neuron death described earlier (Cahalane *et al.* 2012).

Overall then, a larger cortex has more cortical areas, those areas superimposed on a posterior-to-anterior gradient of cell number and density that becomes steeper and steeper in larger brains. Most of the variation in neuron numbers per unit of cortical surface area is accounted for by neuron numbers in layer IV, the thalamic input layer, and in layers II–III, the source of intra-cortical connectivity. Other features, both genetic and acquired through experience, should further differentiate cortical areas.

Feed-Forward and Feed-Back Pathways Align with the Isocortical Neuron Number Gradient

A very important feature of cortical organization that as far as is yet known is independent in origin from neurogenetic gradients, is a pervasive functional asymmetry of within-cortex connectivity that has been described in most detail for primary, secondary, and further visual cortices (Batardiere *et al.* 2002) but has also been noted in the frontal cortex (Badre 2008). Recall that layer II–III neurons primarily project intra-cortically, layer IV neurons primarily receive thalamic input, and layer V–VI neurons primarily project sub-cortically. There is a further distinction to be made, however, about intra-cortical connections. The "forward" projections of the primary visual cortex project in the manner of thalamic input to surrounding cortical regions, that is, to layer IV and the basal dendrites of layer III, while "backward" projections from V2 to V1 project in the generic manner usually described for intra-cortical projections to layers II–III and layer V. Given the posterior location of V1, "forward" and "backward" projections thus roughly align with the anterior-posterior order of neurogenesis, and produce a directionality in the flow of information, which is aligned with neuron numbers per unit of cortical surface area. This directionality or hierarchical ordering is not strict, but rather "degenerate" in the technical sense, with parallel components producing many ties in the ordering of cortical areas (Hilgetag *et al.* 2000). Most intra-cortical connections are local, with about 85% of variation in the presence or absence of a connection between cortical areas captured by the rule "connect to your nearest neighbor cortical area plus one further one" (Scannell *et al.* 1995). We emphasize this caveat to avoid the impression that information flows from back to front in the isocortex in a strict sequence.

The interaction of the anterior to posterior gradient in neurogenesis and eventual neuron numbers per unit of cortical area, and the asymmetry in connectivity along the same axis, makes the cortex a device which intrinsically and progressively "crunches," or more formally, "reduces the dimensionality" of its representations as sensory information flows from posterior to anterior (Cahalane *et al.* 2012; Badre 2008). This overall scheme progressively integrates multiple sensory modalities and motor representations in cortical areas that are progressively reduced in both surface area, and neuron numbers per unit of cortical surface area across the posterior-to-anterior axis. Moreover, the gradient in neuron number per unit of cortical surface area varies across primate species with the steepness of the gradient increasing with cortex size (Cahalane *et al.* 2012; Charvet *et al.* 2015). For instance, neuron numbers per unit of cortical surface area vary approximately fourfold in the galago but vary fivefold in the roughly 40 times bigger-brained baboon (Cahalane *et al.* 2012). We have very little information for the human isocortex (Shankle *et al.* 1998), but as neuron numbers and volume of the human cortex are predictable from other primate species (Hofman 1989; Herculano-Houzel 2009), there is no reason to predict deviation from other primate brains. Since the human brain is the largest of primates, however, the gradient in cell number should be the steepest, and the dimensionality reduction the greatest.

Progressive Changes in Neuron Morphology Are Further Evidence for Information Compression on the Posterior-Anterior Axis

The anterior-posterior gradient in neurogenesis during development is associated with gradients in anatomical and functional organization (Koyama *et al.* 2011; Cahalane *et al.* 2012). In the cortex, excitatory synapses are usually located on dendritic spines, small protuberances on the receptive dendritic arbor branching out from a pyramidal or stellate cell body. Larger dendritic arbors appear to require a somewhat larger cell body (nucleus and metabolic and transport machinery; Elston 2003a, 2003b). This positive correlation between size of neuron soma and dendritic arborizations, as well as spine numbers, across mammalian species is well documented (an initial observation, Bok 1959). In the isocortex, the largest cell bodies, the largest dendritic branches, and the greatest number of spines are found in frontal cortex, while the smallest neuronal cell bodies, smallest dendritic branches, and fewest spines are found at the occipital pole (Elston 2003a, 2003b; Elston *et al.* 2001; Bianchi *et al.* 2012; Cahalane *et al.* 2012). Neuronal cell body size of several New World monkeys in layers II–III steadily decreases from the frontal towards the occipital pole (Cahalane *et al.* 2012). Moreover, layer III pyramidal cells of rhesus monkeys in the anterior cingulate cortex are generally more spiny than neurons found more caudally in the posterior cingulate, the primary somatosensory and auditory areas, and are at least eight times more spiny than those in the occipital pole (i.e., V1; Elston *et al.* 2002, 2005; Elston and Rockland 2002). In marmosets, layer III pyramidal neurons exhibit a greater dendritic field area in rostral visual association areas than in the primary visual cortex, located more caudally (Elston *et al.* 1996). In humans, as in chimpanzees, dendrites of layer III pyramidal cells are larger in the frontal cortex than in more caudal regions (Bianchi *et al.* 2012). Overall, the larger number of spines, dendritic arbor, and cell body size reflects greater necessity and potential for integration in the anterior pole as information flows from large areas with many neurons per column toward this smaller target.

An association between position from posterior to anterior and increasingly more "abstract" processing has been noted (Badre 2008; Badre and D'Esposito 2009), comparing both within and across functional domains such as vision or motor commands. Information processing and integration appears to be hierarchically processed along a gradient, with more integration occurring in the frontal cortex. This functional gradient in abstraction has a general and scalable neurodevelopmental correlate arising from the interaction of the gradient of neurogenesis and the general flow of information in the isocortex. A further entailment should be noted: there is much current interest in the brain as a "prediction machine" (Clark 2013). In the reverse direction of "backward" flow of information, increasingly abstract predictions of the relationship between the current state of affairs and the immediate future can be used to directly influence the responses of neurons at lower levels.

The Consequences of Cortical Gradients for Longer Developing, Larger-Brained Species

The gradient in neuron numbers per unit of cortical surface area is steeper across the anterior-posterior axis of the isocortex in larger brains (Cahalane *et al.* 2012; Charvet *et al.* 2015), which is mirrored in species differences in the gradient of neurogenesis timing along this axis. As isocortices expand, there is a greater disparity in neuron numbers per unit of cortical surface area and a greater disparity in soma size, dendritic arbors, and spines across its poles. The variation in neuronal soma size in layers II–III across the rostro-caudal axis of the isocortex increases with neuron numbers in New World monkeys (Cahalane *et al.* 2012). Data from Cahalane *et al.* 2012 show that neuronal soma size varies approximately twofold in layer II–III of the tamarin cortex but that it varies threefold in layer II–III of the larger-brained capuchin. Also, the difference between dendritic spines of pyramidal neurons across the rostral and caudal poles increases as overall brains expand (Elston *et al.* 2001). For instance, the difference in spine number between the prefrontal cortex and area V2 is highest in humans, followed by macaques, and smaller in the two smallest primate brains samples (i.e., marmosets, owl monkeys; Elston 2003b).

Similar to the conserved alar-basal gradients of neurogenesis in the original neural plate that automatically generate "excess" neurons disproportionately to association regions such as the isocortex and cerebellum, the conserved cortical gradients described here transmute larger and larger versions of the cortex into an organization that intuitively corresponds to the varying demands of brains. This built-in ability to compress and abstract ever larger arrays of sensory information in space and time, to correlate with, and let them inform, action, corresponds to the correlated set of abilities (e.g. dietary innovation, tool use, complex social organization, ability to successfully colonize new habitats) associated with large brains (Lefebvre *et al.* 2004).

How Do Species-Specific Adaptations Find Their Place in This New Version of Cortical Organization?

We have demonstrated that there is systematic variation in neuron numbers and anatomy across the isocortex and that this pattern systematically varies across species. We suggested that a module-based account of this pattern is unlikely. Still, though, the assumption that evolution must select on the size and circuitry of cortical modules, corresponding to cortical areas (e.g., V1 for vision, or the orbitofrontal cortex for social behavior) remains pervasive in the field (Barton 1998; Duchaine *et al.* 2001; Schoenemann 2006; Dunbar and Shultz 2007). For instance, Barton (1998) correlated the size of the visual structures (e.g., V1) with diets in primate species with the assumption that an increased V1 allows species to better discriminate colors. Powell *et al.* (2012) argued that a selective increase in a region of the orbitofrontal cortex is a morphological correlate of the ability to manage larger social groups. These assumptions are increasingly at odds with experimental studies, for which we will give only a few examples.

New sensory specializations, whether induced experimentally, as in prostheses (Andersen *et al.* 2004), or those that vary between individuals within species, can quite often be employed directly within existing circuitry. For instance, an experimental study showed that the addition of a third opsin in adult red–green color-deficient adult marmosets was sufficient to produce trichromatic color vision supporting appropriate color discriminations (Mancuso *et al.* 2009). These findings show that the neural connections in adulthood can appropriately process a novel input and that evolutionary changes in the size of V1 or other visual areas are not necessary for an increased ability to discriminate colors. Rather, diversity and anatomical and behavioral adaptations can evolve by adding "plug-ins" to a general computational device. Returning to our computer metaphor of the introduction, it is as though if any "device" offers properly formatted input, the cortex stands ready to extract and employ any new information it provides, as a computer can employ streaming audio, visual, or internet signals without alterations to its core processors.

The assumption that particular, dedicated, and specially adapted cortical areas are subject to selection is also at odds with the observation of massive plasticity in the central nervous system. For instance, after sensory deprivation, some plastic changes lead to advantages in other sensory modalities, such as enhanced localization of sound sources and visual peripheral sensitivity in deaf animals (Merabet and Pascual-Leone 2009; Lomber *et al.* 2010). The brain changes in similar ways with exposure to much more mild forms of deprivation and/or opportunity. In "native" signers of American Sign Language, deaf or not, for whom a large visual angle must be monitored for communication compared to the more restricted face region of speakers and listeners, the response over this region is amplified in primary visual cortex (Bavelier and Neville 2002). Conversely, for exceptionally good readers, the representation of central gaze, the fovea, occupies a greater percentage of primary visual cortex (Verghese *et al.* 2014), and the functional organization of the cortex progressively changes with reading competence (Krafnick *et al.* 2014).

Finally, meta-analyses of neuroimaging studies in humans now demonstrate that the regions that are active during particular tasks bear little relationship to the names we have assigned to them according to what appears to be their most obvious sensory input or motor output (Poldrack 2006; Anderson *et al.* 2013). For example, "visual" cortex may routinely engage in decision tasks without a visual component (Sathian 2005; Merabet *et al.* 2008; Collignon *et al.* 2009). Overall, the notion that neuron number in a particular cortical area, whether visual, social, or motor, is an important locus of control for the elaboration of that function is becoming increasingly untenable. As we have stated in an earlier article on neuroplasticity (Anderson and Finlay 2014):

> Neuroplasticity, in turn, should inform us about what brain architectures have been selected over evolutionary time. If any current computer users were informed that their personal computers, which heretofore had been used only for word processing, could also store and transform images, few would be amazed. If the same people, however, were informed that their parents' video cameras, by simply adding new input and output devices, could function as word processors, they would probably be incredulous, and undertake a re-analysis of their presumptions about video camera technology. In the same way, understanding of how brains can change and understanding of neural architecture should bootstrap each other.

Mechanisms Available to Populate the Cortex for Adaptive Functions

If "mosaic" changes in neuron number in modules do not conform well to the evo-devo account of cortex adaptation described here, or the increasing literature on cortical plasticity, then how do new behavioral and cognitive abilities become resident in the cortex? Two kinds of evolutionary changes, of many possibilities, are particularly supported by the literature in development and plasticity at the moment. One is the alteration of the periphery, where there are many well-known examples. The second alteration relies on the coupling and uncoupling of motivational systems in the basal forebrain directing attention and reward. Finally, though we will not review this directly, the necessary coupling of large brain size with lengthy developmental schedules (Charvet and Finlay 2012; Workman *et al.* 2013) should elevate to prominence the study of changing life histories in primates and their role in organizing brain structure.

Changed Coupling of Preferences, Motivations, and Drives as Brain Organizers

Consider the development of face-recognition systems and "face areas" in the cortex, a field of study that now has extensive empirical description in children and adults. Human infants prefer to look at face-like configurations (Johnson 2011), and will work hard and learn quickly for social reinforcement from birth (Goldstein and Schwade 2008). The human eye has the special adaptation of a white sclera, allowing much better computation of direction of gaze in humans than in other primates (Striedter 2005; Provine *et al.* 2013), particularly useful in language and social learning in general. Even with the initial preference for faces, the specific adaptation of eye coloration, and enthusiasm for social learning, it still takes 7–10 years for the representation of faces in the cortex of human children to begin to approximate adult organization (Cohen *et al.* 2011). There is no innate "face module." That is, face processing in the isocortex is constructed. The significance of this for our argument about the organization of the isocortex is that the predisposing condition to produce a cortex with a substantial percentage of its volume devoted to faces and the nuances of emotional expression might only need a coupling of early eye and head movements toward expressive faces over long developmental time. The hierarchical extraction of this information by the cortex will necessarily follow.

Genetic alterations in the reward and reinforcement circuitry in the basal forebrain is a subject ripe for investigation in the case of primate evolution and is a promising venue of research. Change in goals sought by a developing organism will feed its developing brain a new and distinct pattern of input and behavioral consequences (Oudeyer and Kaplan 2007). In birds and in rodents, preference for particular territorial extents, group sizes, partners, and parenting procedures vary across species (Young and Wang 2004; Goodson *et al.* 2005) and can be manipulated experimentally by manipulating the coupling of the basal brain "extended amygdala" (Newman 1999) to dopaminergic behavioral reinforcement systems. The best known example of this is social monogamy in some vole species, compared to promiscuity in closely related species, dependent on minimal changes to oxytocin and vasopressin receptor systems (Young and Wang 2004). Coupling of potent reward systems with successful vocal manipulation of parental behavior, in the context of adequate isocortical capacity, might be necessary and sufficient to launch human language (Syal and Finlay 2009).

Information is Distributed

The final message for the anthropologist from the emerging field of evo-neuro-devo, therefore, is not to persist in looking for evidence in the gyral patterns of evolving hominins for hard-wired modules representing special capacities, but rather to appreciate and investigate how development and changing life histories in the context of amplified brain capacity might directly channel brain organization. Only recently has the profound effect of culture and life history on even the most basic perceptual and cognitive skills been appreciated (Henrich *et al.* 2010). This will require collaboration at every point between anthropologists studying human fossil and cultural remains, primatologists, cognitive and social scientists, and information scientists. We should let ourselves be influenced by the tremendous evolution of computing power happening before our eyes, from personal computing devices to the web, observing what becomes extinct, what adapts, and what survives.

Acknowledgements

This work was supported by National Science Foundation grant IBN-0138113 to BLF and a Eunice Kennedy Shriver National Institute of Child Health and Human Development fellowship F32HD067011 to CJC. The content is solely the responsibility of the authors and does not necessarily represent the official views of any supporting agency.

Glossary terms

Cortical column A column that is orthogonal to the cortical surface and extends from the cortical surface to the cortical white matter. A column is a layered structure and the individual layers (layers I–VI) of the isocortex have different patterns of connectivity.
Cortical layers The isocortex is typically divided into six layers (i.e., layers I–VI). These layers are distinguishable anatomically, by their birth dates, and by their patterns of connectivity. Layer I contains intra-cortical projections. Layer II–III neurons mainly project intra-cortically. Layer IV neurons mainly receive thalamic input. Layer V–VI neurons mainly project sub-cortically.
Cortical plate The cortical plate refers to a region within the developing isocortex and contains neurons that will segregate into layers VI–II.
Gyrus This is a bulged region of a convoluted brain region.
Isocortex The isocortex is distinguished from adjacent areas in that it consists of six layers. It is also called the neocortex although the term "neo" (i.e., new) implies that this structure evolved with the emergence of mammals. The observation that non-mammalian vertebrates exhibit a homolog of the isocortex entails that this structure did not evolve *de novo* in mammals.
Neurogenesis Neurogenesis occurs when a proliferating cell exits the cell cycle to become a neuron. A neuron is born when it exits the cell cycle.
Primary sensory areas These are cortical areas that receive thalamic input and process sensory information. Examples include the primary visual cortex (V1), primary somatosensory cortex (S1), and the primary auditory cortex (A1).
Pyramidal neuron A major neuron type that is located in parts of the brain such as the isocortex. The cell bodies of these neurons have a pyramidal shape.
Secondary visual cortex (V2) This is an association area and receives input from the primary visual cortex (V1).

Subventricular zone In development, this zone lies superficial to the ventricular zone and contains proliferating cells. A subventricular zone has been observed within the developing isocortex as well as in other regions (e.g., striatum).
Ventricular zone A proliferative region that gives rise to neurons and glia.

References

Akam, M. 1995. *Hox* genes and the evolution of diverse body plans. *Philosophical Transactions of the Royal Society of London, Series B: Biological Sciences* 349 (1329):313–319.

Andersen, R. A., Burdick J. W., Musallam S. *et al.* 2004. Cognitive neural prosthetics. *Trends in Cognitive Sciences* 8:486–493.

Anderson, M. L., and Finlay, B. L. 2014. Allocating structure to function: The strong links between neuroplasticity and natural selection. *Frontiers in Human Neuroscience* 7:918.

Anderson, M. L., Kinnison, J., and Pessoa L. 2013. Describing functional diversity of brain regions and brain networks. *NeuroImage* 73:50–58.

Angevine, J. B., Jr., and Sidman, R. L. 1961. Autoradiographic study of cell migration during histogenesis of cerebral cortex in the mouse. *Nature* 192:766–768.

Badre, D. 2008. Cognitive control, hierarchy, and the rostro-caudal organization of the frontal lobes. *Trends in Cognitive Sciences* 12(5):193–200.

Badre, D., and D'Esposito, M. 2009. Is the rostro-caudal axis of the frontal lobe hierarchical? *Nature Reviews Neuroscience* 10(9):659–669.

Barton, R. A. 1998. Visual specialization and brain evolution in primates. *Proceedings of the Royal Society of London, Series B: Biological Sciences* 265(1409):1933–1937.

Barton, R. A., and Harvey, P.H. 2000. Mosaic evolution of brain structure in mammals. *Nature* 405(6790):1055–1058.

Batardiere, A., Barone, P., Knoblauch, K. *et al.* 2002. Early specification of the hierarchical organization of visual cortical areas in the macaque monkey. *Cerebral Cortex* 12(5):453–465.

Bauernfeind, A. L., de Sousa, A. A., Avasthi, T. *et al.* 2013. A volumetric comparison of the insular cortex and its subregions in primates. *Journal of Human Evolution* 64(4):263–279.

Bavelier, D., and Neville, H. J. 2002. Cross-modal plasticity: Where and how? *Nature Reviews Neuroscience* 3(6):443–452.

Beaulieu, C., and Colonnier, M. 1985. A laminar analysis of the number of round-asymmetrical and flat-symmetrical synapses on spines, dendritic trunks, and cell bodies in area 17 of the cat. *Journal of Comparative Neurology* 231(2):180–189.

Bergquist, H., and Källén, B. 1954. Notes on the early histogenesis and morphogenesis of the central nervous system in vertebrates. *Journal of Comparative Neurology* 100(3):627–659.

Bianchi, S., Stimpson, C. D., Bauernfeind, A. L. *et al.* 2012. Dendritic morphology of pyramidal neurons in the chimpanzee neocortex: Regional specializations and comparison to humans. *Cerebral Cortex* 23(10):2429–2436.

Bok, S. T. 1959. *Histonomy of the cerebral cortex*. Amsterdam: Elsevier.

Boulder Committee. 1970. Embryonic vertebrate central nervous system: Revised terminology. *Anatomical Record* 166(2):257–261.

Brodmann, K. 1909. *Vergleichende lokalisationslehre der grosshirnrinde*. Leipzig: Barth.

Burke, A. C., Nelson, C. E., Morgan, B. A., and Tabin, C. 1995. *Hox* genes and the evolution of vertebrate axial morphology. *Development* 121(2):333–346.

Bystron, I., Blakemore, C., and Rakic, P. 2008. Development of the human cerebral cortex: Boulder Committee revisited. *Nature Reviews Neuroscience* 9:110–122.

Cahalane, D. J., Charvet, C. J., and Finlay, B. L. 2012. Systematic, balancing gradients in neuron density and number across the primate isocortex. *Frontiers in Neuroanatomy* 6:28.

Carroll, S. B. 2009. Endless forms: The evolution of gene regulation and morphological diversity. *Philosophy of Biology: An Anthology* 193.

Caviness, V. S., Nowakowski, R. S., and Bhide, P. G. 2009. Neocortical neurogenesis: Morphogenetic gradients and beyond. *Trends in Neuroscience* 32(8):443–450.

Charvet, C. J., Cahalane, D. J., and Finlay, B. L. 2015. Systematic, cross-cortex variation in neuron numbers in rodents and primates. *Cerebral Cortex* 25(1):147–160.

Charvet, C. J., Darlington, R. B., and Finlay, B L. 2013. Variation in human brains may facilitate evolutionary change toward a limited range of phenotypes. *Brain, Behavior and Evolution* 81(2):74–85.

Charvet, C. J., and Finlay, B. L. 2012. Embracing covariation in brain evolution: Large brains, extended development, and flexible primate social systems. *Progress in Brain Research* 195:71–87.

Charvet, C. J., and Striedter, G. F. 2011. Developmental modes and developmental mechanisms can channel brain evolution. *Frontiers in Neuroanatomy* 5:4.

Charvet, C. J., Striedter, G. F., and Finlay, B. L. 2011. Evo-devo and brain scaling: Candidate developmental mechanisms for variation and constancy in vertebrate brain evolution. *Brain, Behavior and Evolution* 78(3):248–257.

Clark, A. G. 2013. Whatever next? Predictive brains, situated agents, and the future of cognitive science. *Behavioral and Brain Sciences* 36(3):181–204.

Cohen, K. K. 2011. What can emerging cortical face networks tell us about mature brain organisation? *Developmental Cognitive Neuroscience* 1(3):246–255.

Collignon, O., Voss, P., Lassonde, M., and Lepore, F. 2009. Cross-modal plasticity for the spatial processing of sounds in visually deprived subjects. *Experimental Brain Research* 192(3):343–358.

Collins, C. E., Airey, D. C., Young, N. A., *et al.* 2010. Neuron densities vary across and within cortical areas in primates. *Proceedings of the National Academy of Sciences, USA* 107(36):15927–15932.

Cowey, A., and Stoerig, P. 1989. Projection patterns of surviving neurons in the dorsal lateral geniculate nucleus following discrete lesions of striate cortex: Implications for residual vision. *Experimental Brain Research* 75(3):631–638.

Dehay, C., Giroud, P., Berland, M. *et al.* 1993. Modulation of the cell cycle contributes to the parcellation of the primate visual cortex. *Nature* 366(6454):464–466.

Draghi, J., Wagner, G. P. 2008. Evolution of evolvability in a developmental model. *Evolution* 62(2):301–315.

Duchaine, B., Cosmides, L., and Tooby, J. 2001. Evolutionary psychology and the brain. *Current Opinion in Neurobiology* 11(2):225–230.

Dunbar, R. I. 1993. Coevolution of neocortical size, group size and language in humans. *Behavioral and Brain sciences* 16(4):681–693.

Dunbar, R. I., and Shultz, S. 2007. Evolution in the social brain. *Science* 317(5843):1344–1347.

Dyer, M. A., Martins, R., da Silva, F. M. *et al.* 2009. Developmental sources of conservation and variation in the evolution of the primate eye. *Proceedings of the National Academy of Sciences, USA* 106(22):8963–8968.

Elston, G. N. 2003a. Cortex, cognition and the cell: New insights into the pyramidal neuron and prefrontal function. *Cerebral Cortex* 13(11):1124–1138.

Elston, G. N. 2003b. Pyramidal cell heterogeneity in the visual cortex of the nocturnal New World owl monkey (*Aotus trivirgatus*). *Neuroscience* 117(1):213–219.

Elston, G. N., Benavides-Piccione, R., and DeFelipe, J. 2001. The pyramidal cell in cognition: A comparative study in human and monkey. *Journal of Neuroscience* 21:RC163.

Elston, G. N., Benavides-Piccione, R., and DeFelipe, J. 2005. A study of pyramidal cell structure in the cingulate cortex of the macaque monkey with comparative notes on inferotemporal and primary visual cortex. *Cerebral Cortex* 15:64–73.

Elston, G. N., Benavides-Piccione, R., DeFelipe, J., and Rockland, K. S. 2002. The pyramidal neuron in auditory, cingulate and prefrontal cortex of the macaque monkey: Regional variation in cell structure. *European Journal of Neuroscience* 3:222.

Elston, G. N., Rockland, K. S. 2002. The pyramidal cell in sensorimotor cortex of the macaque monkey: Phenotypic variation. *Cerebral Cortex* 10(12):1071.

Elston, G. N., Rosa, M. G., and Calford, M. B. 1996. Comparison of dendritic fields of layer III pyramidal neurons in striate and extrastriate visual areas of the marmoset: A Lucifer yellow intracellular injection. *Cerebral Cortex* 6(6):807–813.

Finlay, B. L. 2005. Brain and behavioral development (II): Cortical. *In:* B. Hopkins (ed.) *Cambridge encyclopedia of child development*. Cambridge: Cambridge University Press. pp.296–305.

Finlay, B. L. 2007. ***E pluribus unum***: Too many unique human capacities and too many theories. *In:* S. Gangestad and J. Simpson (eds) *The evolution of mind: Fundamental questions and controversies*. New York: Guilford Press. pp.301–304.

Finlay, B. L., and Brodsky, P. B. 2006. Cortical evolution as the expression of a program for disproportionate growth and the proliferation of areas. *In:* J. H. Kaas (ed.) *Evolution of nervous systems*. Oxford: Academic Press. pp.73–96.

Finlay, B. L., Cheung, D., and Darlington, R. B. 2006. Developmental constraints on or developmental structure in brain evolution? *In:* Y. Munakata and M. Johnson (eds) *Processes of change in brain and cognitive development: Attention and performance XXI*. New York: Oxford University Press. pp.131–162.

Finlay, B. L., and Darlington, R. B. 1995. Linked regularities in the development and evolution of mammalian brains. *Science* 268(5217):1578–1584.

Finlay, B. L., Darlington, R. B., and Nicastro, N. 2001. Developmental structure in brain evolution. *Behavioral and Brain Sciences* 24(2):263–278.

Finlay, B. L., Franco, E. C., Yamada, E. S. *et al.* 2008. Number and topography of cones, rods and optic nerve axons in New and Old World primates. *Visual Neuroscience* 25(3):289–299.

Finlay, B. L., Hersman, M. N., and Darlington, R. B. 1998. Patterns of vertebrate neurogenesis and the paths of vertebrate evolution. *Brain, Behavior and Evolution* 52(4–5):232–242.

Finlay, B. L., Hinz, F., and Darlington, R. B. 2011. Mapping behavioural evolution onto brain evolution: The strategic roles of conserved organization in individuals and species. *Philosophical Transactions of the Royal Society, Series B: Biological Sciences* 366(1574):2111–2123.

Finlay, B. L., and Slattery, M. 1983. Local differences in amount of early cell death in neocortex predict adult local specializations. *Science* 219(4590):1349–1351.

Frahm, H. D., Stephan, H., and Baron, G. 1984. Comparison of brain structure volumes in insectivora and primates. V. Area striata (AS). *Journal für Hirnforschung* 25(5):537–557.

Fodor, J. A. 1983. *The modularity of mind: An essay on faculty psychology*. Cambridge, MA: MIT Press.

Gerhart, J., and Kirschner, M. 1997. *Cells, embryos, and evolution: Toward a cellular and developmental understanding of phenotypic variation and evolutionary adaptability*. Oxford: Blackwell Science.

Gerhart, J., and Kirschner, M. 2007. The theory of facilitated variation. *Proceedings of the National Academy of Sciences, USA* 104(Suppl. 1):8582–8589.

Geschwind, N. 1972. Language and the brain. *Scientific American* 226(4):76–83.
Geschwind, N. 1974. *The development of the brain and the evolution of language*. Springer Netherlands.
Gilbert, C. D., and Kelly, J. P. 1975. The projections of cells in different layers of the cat's visual cortex. *Journal of Comparative Neurology* 163(1):81–105.
Goldstein, M. H., and Schwade, J. A. 2008. Social feedback to infants' babbling facilitates rapid phonological learning. *Psychological Science* 19(5):515–523.
Goodson, J. L., Evans, A. K., Lindberg, L., and Allen, C. D. 2005. Neuro-evolutionary patterning of sociality. *Proceedings of the Royal Society of London, Series B: Biological Sciences* 272(1560):227–235.
Gould, S. J. 1975. Allometry in primates, with emphasis on scaling and the evolution of the brain. *Contributions to Primatology* 5:244–292.
Gould, S. J. 1977. *Ontogeny and phylogeny*. Cambridge, MA: Harvard University Press.
Gould, S. J. 1980. The evolutionary biology of constraint. *Deadelus* 109(2):39–52.
Gould, S. J. 1987. The panda's thumb of technology. *Natural History* 1:14–23.
Gould, S. J. 1989. A developmental constraint in *Cerion*, with comments of the definition and interpretation of constraint in evolution. *Evolution* 43(3):516–539.
Gould, S. J., and Lewontin, R. C. 1979. The spandrels of San Marco and the panglossian paradigm: A critique of the adaptationist programme. *Proceedings of the Royal Society of London, Series B: Biological Sciences* 205(1161):581–598.
Heffer, A., and Pick, L. 2013. Conservation and variation in *Hox* genes: How insect models pioneered the evo-devo field. *Annual Reviews in Entomology* 58:161–179.
Henrich, J., Heine, S. J., and Norenzayan, A. 2010. The weirdest people in the world? *Behavioral and Brain Sciences* 33:61–85.
Hilgetag, C. C., O'Neill, M. A., and Young, M. P. 2000. Hierarchical organization of macaque and cat cortical sensory systems explored with a novel network processor. *Philosophical Transactions of the Royal Society of London, Series B: Biological Sciences* 355(1393):71–89.
Hofman, M. A. 1989. On the evolution and geometry of the brain in mammals. *Progress in Neurobiology* 32(2):137–158.
Hofman, M. A. 2014. Evolution of the human brain: When bigger is better. *Frontiers in Neuroanatomy* 8:15.
Holloway, R. L. 1968. The evolution of the primate brain: Some aspects of quantitative relations. *Brain Research* 7(2):121–172.
Hollyday, M., and Hamburger, V. 1977. An autoradiographic study of the formation of the lateral motor column in the chick embryo. *Brain Research* 132(2):197–208.
Hutsler, J. J., Lee, D.-G., Porter, K. K. 2005. Comparative analysis of cortical layering and supragranular layer enlargement in rodent carnivore and primate species. *Brain Research* 1052(1):71–81.
Jackson, C. A., Peduzzi, J. D., and Hickey, T. L. 1989. Visual cortex development in the ferret. I. Genesis and migration of visual cortical neurons. *Journal of Neuroscience* 9(4):1242–1253.
Jacobs, R. A. 1997. Nature, nurture and the development of functional specializations: A computational approach. *Psychonomic Bulletin and Review* 4(3):299–309.
Jerison, H. J. 1973. *Evolution of the brain and intelligence*. New York: Academic Press.
Herculano-Houzel, S. 2009. The human brain in numbers: A linearly scaled-up primate brain. *Frontiers in Human Neuroscience* 3:31.
Herculano-Houzel, S., Watson, C., and Paxinos, G. 2013. Distribution of neurons in functional areas of the mouse cerebral cortex reveals quantitatively different cortical zones. *Frontiers in Neuroanatomy* 7:35.

Iwaniuk, A. N., Dean, K. M., and Nelson, J. E. 2004. A mosaic pattern characterizes the evolution of the avian brain. *Proceedings of the Royal Society of London, Series B: Biological Sciences* 271(Suppl. 4):S148–S151.

Kaas, J. H. 1997. Theories of visual cortex organization in primates. In: K. S. Rockland, J. H. Kaas, and A. Peters (eds) *Cerebral cortex, vol. 12: Extrastriate cortex in primates*. New York: Plenum Press. pp.91–125.

Kandel, E., Schwartz, J., and Jessell, T. 2013. *Principles of neural science*, 5th edn. New York: McGraw-Hill Professional.

Kaplan, F., and Oudeyer, P.-Y. 2007. In search of the neural circuits of intrinsic motivation. *Frontiers in Neuroscience* 1:225–236.

Kaskan, P., Franco, C., Yamada, E. *et al.* 2005. Peripheral variability and central constancy in mammalian visual system evolution. *Proceedings of the Royal Society of London, Series B: Biological Sciences* 272:91–100.

Kingsbury, M. A., and Finlay, B. L. 2001. The cortex in multidimensional space: Where do cortical areas come from? *Developmental Science* 4(2):125–156.

Kirschner, M. W., and Gerhart, J. C. 2006. *The plausibility of life: Resolving Darwin's dilemma.* New Haven, CT: Yale University Press.

Koyama, M., Kinkhabwala, A., Satou, C. *et al.* 2011. Mapping a sensory-motor network onto a structural and functional ground plan in the hindbrain. *Proceedings of the National Academy of Sciences, USA* 108(3):1170–1175.

Krafnick, A. J., Flowers, L. D., Luetje, M. M. *et al.* 2014. An investigation into the origin of anatomical differences in dyslexia. *Journal of Neuroscience* 34:901–908.

Krubitzer, L., Campi, K. L., and Cooke, D. F. 2011. All rodents are not the same: A modern synthesis of cortical organization. *Brain, Behavior and Evolution* 78(1):51–93.

Krubitzer, L., and Stolzenberg, D. S. 2014. The evolutionary masquerade: Genetic and epigenetic contributions to the neocortex. *Current Opinion in Neurobiology* 24:157–165.

Langman, J., and Haden, C. C. 1970. Formation and migration of neuroblasts in the spinal cord of the chick embryo. *Journal of Comparative Neurology* 138(4):419–431.

Lee, S., and Stevens, C. F. 2007. General design principle for scalable neural circuits in a vertebrate retina. *Proceedings of the National Academy of Sciences, USA* 104(31):12931–12935.

Lefebvre, L. 2013. Brains, innovations, tools and cultural transmission in birds, non-human primates, and fossil hominins. *Frontiers in Human Neuroscience* 7:245.

Lefebvre, L., Reader, S. M., and Sol, D. 2004. Brains, innovations and evolution in birds and primates. *Brain, Behavior and Evolution* 63(4):233–246.

Lefebvre, L., Reader, S. M., and Sol, D. 2013. Innovating innovation rate and its relationship with brains, ecology and general intelligence. *Brain, Behavior Evolution* 81(3):143–145.

Lehman, J., and Stanley, K. O. 2013. Evolvability is inevitable: Increasing evolvability without the pressure to adapt. *PLoS One* 8(4):e62186.

Lemons, D., and McGinnis, W. 2006. Genomic evolution of *Hox* gene clusters. *Science* 313(5795):1918–1922.

Lewis, E. B. 1978. A gene complex controlling segmentation in *Drosophila*. *Nature* 276(5688): 565–570.

Linde-Medina, M., and Diogo, R. 2014. Do correlation patterns reflect the role of development in morphological evolution? *Evolutionary Biology* 41(3):494–502.

Lomber, S. G., Meredith, A. M., and Kral, A. 2010. Cross-modal plasticity in specific auditory cortices underlies visual compensations in the deaf. *Nature Neuroscience* 13(11):1421–1427.

Lumsden, A. 1990. The cellular basis of segmentation in the developing hindbrain. *Trends in Neuroscience* 13(8):329–335.

Luskin, M. B., and Shatz, C. J. 1985. Neurogenesis of the cat's primary visual cortex. *Journal of Comparative Neurology* 242(4):611–631.

Mancuso, K., Hauswirth, W. W., Li, Q. *et al.* 2009. Gene therapy for red-green colour blindness in adult primates. *Nature* 461(7265):784–787.

Masel, J., and Trotter, M. V. 2010. Robustness and evolvability. *Trends in Genetics* 26(9):406–414.

Merabet, L. B., Hamilton, R., Schlaug, G. *et al.* 2008. Rapid and reversible recruitment of early visual cortex for touch. *PLoS One* 3(8):e3046.

Merabet, L. B., and Pascual-Leone, A. 2009. Neural reorganization following sensory loss: The opportunity of change. *Nature Reviews Neuroscience* 11(1):44–52.

Mihailović, L. T., Čupić, D., and Dekleva, N. 1971. Changes in the numbers of neurons and glial cells in the lateral geniculate nucleus of the monkey during retrograde cell degeneration. *Journal of Comparative Neurology* 142(2):223–229.

Miyama, S., Takahashi, T., Nowakowski, R. S., and Caviness, V. S., Jr. 1997. A gradient in the duration of the G1 phase in the murine neocortical proliferative epithelium. *Cerebral Cortex* 7(7):678–689.

Moriarty, D. E., and Miikkulainen, R. 1997. Forming neural networks through efficient and adaptive coevolution. *Evolutionary Computation* 5(4):373–399.

Mountcastle, V. B. 1997. The columnar organization of the cortex. *Brain* 120:701–722.

Nagarajan, N., and Stevens, C. F. 2008. How does the speed of thought compare for brains and digital computers? *Current Biology* 18(17):R756–R758.

Naumann, R. K., Anjum, F., Roth-Alpermann, C., and Brecht, M. 2012. Cytoarchitecture, areas, and neuron numbers of the Etruscan Shrew cortex. *Journal of Comparative Neurology* 520(11):2512–2530.

Newman, S. W. 1999. The medial extended amygdala in male reproductive behavior. A node in the mammalian social behavior network. *Annals of the New York Academy of Sciences* 877(1):242–257.

Northcutt, R. G., and Kaas, J. H. 1995. The emergence and evolution of mammalian neocortex. *Trends in Neuroscience* 18(9):373–379.

O'Kusky, J., and Colonnier, M. 1982. Postnatal changes in the number of neurons and synapses in the visual cortex (area 17) of the macaque monkey: A stereological analysis in normal and monocularly deprived animals. *Journal of Comparative Neurology* 210(3):291–306.

O'Leary, D. D. M. 1989. Do cortical areas emerge from a protocortex? *Trends in Neuroscience* 12(10):400–406.

O'Leary, D. D. M., and Sahara, S. 2008. Genetic regulation of arealization of the neocortex. *Current Opinion in Neurobiology* 18(1):90–100.

Oudeyer, P.-Y., and Kaplan, F. 2007. What is intrinsic motivation? A typology of computational approaches. *Frontiers in Neurorobotics* 1:6.

Penn, D. C., Holyoak, K. J., and Povinelli, D. J. 2008. Darwin's mistake: Explaining the discontinuity between human and nonhuman minds. *Behavioral and Brain Sciences* 31(2):109–178.

Poldrack, R. A. 2006. Can cognitive processes be inferred from neuroimaging data? *Trends in Cognitive Sciences* 10(2):59–63.

Powell, J., Lewis, P. A., Roberts, N. *et al.* 2012. Orbital prefrontal cortex volume predicts social network size: An imaging study of individual differences in humans. *Proceedings of the Royal Society of London, Series B: Biological Sciences* 279(1736):2157–2162.

Provine, R. R., Cabrera, M. O., and Nave-Blodgett, J. 2013. Red, yellow, and super-white sclera. *Human Nature* 24(2):126–136.

Puelles, L., and Ferran, J. L. 2012. Concept of neural genoarchitecture and its genomic fundament. *Frontiers in Neuroanatomy* 6:47.

Purves, D., Cabeza, R., Huettel, S. A. *et al.* 2012. *Principles of cognitive neuroscience*. Boston: Sinauer Associates.

Raff, E. C., and Raff, R. A. 2000. Dissociability, modularity, evolvability. *Evolution and Development* 2(5):235–237.

Raff, R. A. 2000. Evo-devo: The evolution of a new discipline. *Nature Reviews Genetics* 1(1):74–79.

Rakic, P. 1974. Neurons in rhesus monkey visual cortex: Systematic relation between time of origin and eventual disposition. *Science* 183(4123):425–427.

Rakic, P. 1995. A small step for the cell, a giant leap for mankind: A hypothesis of neocortical expansion during evolution. *Trends in Neuroscience* 18(9):383–388.

Rakic, P. 2002. Pre- and post-developmental neurogenesis in primates. *Clinical Neuroscience Research* 2(1):29–39.

Rakic, P. 2007. The radial edifice of cortical architecture: From neuronal silhouettes to genetic engineering. *Brain Research Reviews* 55:204–219.

Rakic, P. 2008. Confusing cortical columns. *Proceedings of the National Academy of Sciences, USA* 105(34):12099–12100.

Rapaport, D. H., Rakic, P., and Lavail, M. M. 1996. Spatiotemporal gradients of cell genesis in the primate retina. *Perspectives in Developmental Neurobiology* 3(3):147.

Reep, R. L., Finlay, B. L., and Darlington, R. B. 2007. The limbic system in mammalian brain evolution. *Brain Behavior and Evolution* 70(1):57–70.

Rockel, A. J., Hiorns, R. W., and Powell, T. P. S. 1980. Basic uniformity in the structure of the cerebral cortex. *Brain* 103:221–243.

Rubenstein, J. L. R., Martinez, S., Shimamura, K., and Puelles, L. 1994. The embryonic vertebrate forebrain: The prosomeric model. *Science* 266 (5185):578–580.

Rutledge, J. J., Eisen, E. J., and Legates, J. E. 1974. Correlated response in skeletal traits and replicate variation in selected lines of mice. *Theoretical and Applied Genetics* 45(1):26–31.

Sathian, K. 2005. Visual cortical activity during tactile perception in the sighted and the visually deprived. *Developmental Psychobiology* 46(3):279–286.

Scannell, J. W., Blakemore, C., and Young, M. P. 1995. Analysis of connectivity in the cat cerebral cortex. *Journal of Neuroscience* 15(2):1463–1483.

Schoenemann, P. T. 2006. Evolution of the size and functional areas of the human brain. *Annual Review of Anthropology* 35:379–406.

Seelke, A. M., Dooley, J. C., and Krubitzer, L. A. 2014. The cellular composition of the marsupial neocortex. *Journal of Comparative Neurology* 522(10):2286–2298.

Shankle, W. R., Landing, B. H., Rafii, M. S. *et al.* 1998. Evidence for a postnatal doubling of neuron number in the developing human cerebral cortex between 15 months and 6 years. *Journal of Theoretical Biology* 191(2):115–140.

Skoglund, T. S., Pascher, R., and Berthold, C. H. 1996. Heterogeneity in the columnar number of neurons in different neocortical areas in the rat. *Neuroscience Letters* 208(2):97–100.

Smart, I. H. M., Dehay, C., Giroud, P. *et al.* 2002. Unique morphological features of the proliferative zones and postmitotic compartments of the neural epithelium giving rise to striate and extrastriate cortex in the monkey. *Cerebral Cortex* 12(1):37–53.

Snider, J., Pillai, A., and Stevens, C. F. 2010. A universal property of axonal and dendritic arbors. *Neuron* 66(1):45–56.

Sperber, D. 2001. In defense of massive modularity. *In:* E. Dupoux and J. Mehler (eds) *Language, brain and cognitive development*. Cambridge, MA: MIT Press. pp.46–57.

Stanley, K. O., and Miikkulainen, R. 2003. A taxonomy for artificial embryogeny. *Artificial Life* 9(2):93–130.

Stephan, H., Frahm, H., and Baron, G. 1981. New and revised data on volumes of brain structures on insectivores and primates. *Folia Primatologica* 35(1):1–29.

Stevens, C. F. 2002. Predicting functional properties of visual cortex from an evolutionary scaling law. *Neuron* 36(1):139–142.

Stout, D., Toth, N., Schick, K., and Chaminade, T. 2008. Neural correlates of Early Stone Age toolmaking: Technology, language and cognition in human evolution. *Philosophical Transactions of the Royal Society, Series B: Biological Sciences* 363(1499):1939–1949.

Striedter, G. F. 2005. *Principles of brain evolution*. Sunderland: Sinauer Associates.

Syal, S., and Finlay, B. L. 2011. Thinking outside the cortex: Social motivation in the evolution and development of language. *Developmental Science* 14(2):417–430.

Sylvester, J. B., Pottin, K., and Streelman, T. J. 2011. Integrated brain diversification along the early neuraxes. *Brain, Behavior Evolution* 78(3):237–247.

Verghese, A., Kolbe, S. C., Anderson, A. J. *et al.* 2014. Functional size of human visual area V1: A neural correlate of top-down attention. *NeuroImage Pt* 1:47–52.

Wagner, A. 2005. Robustness, evolvability, and neutrality. *FEBS Letters* 579(8):1772–1778.

West-Eberhard, M. J. 2003. *Developmental plasticity and evolution.* New York: Oxford University Press.

Wickett, J. C., Vernon, P. A., and Lee, D. H. 1994. In vivo brain size, head perimeter, and intelligence in a sample of healthy adult females. *Personality and Individual Differences* 16(6):831–838.

Wong-Riley, M. T. T. 1972. Changes in the dorsal lateral geniculate nucleus of the squirrel monkey after unilateral ablation of the visual cortex. *Journal of Comparative Neurology* 146:519–547.

Workman, A. D., Charvet, C. J., Clancy, B. *et al.* 2013. Modeling transformations of neurodevelopmental sequences across mammalian species. *Journal of Neuroscience* 33(17):7368–7383.

Yopak, K. E., Lisney, T. J., Darlington, R. B. *et al.* 2010. A conserved pattern of brain scaling from sharks to primates. *Proceedings of the National Academy of Sciences, USA* 107(29): 12946–12951.

Young, L. J., and Wang, Z. 2004. The neurobiology of pair bonding. *Nature Neuroscience* 7(10):1048–1054.

Young, N. A., Collins, C. E., and Kaas, J. H. 2013. Cell and neuron densities in the primary motor cortex of primates. *Frontiers in Neural Circuits* 7:30.

Chapter 11

Growing Up Fast, Maturing Slowly: The Evolution of a Uniquely Modern Human Pattern of Brain Development

Philipp Gunz
Department of Human Evolution, Max-Planck-Institute for Evolutionary Anthropology, Leipzig, Germany

Introduction

The developmental mechanisms that set up brain and body plans are highly conserved across vertebrates (Charvet *et al.* 2011). All mammals therefore share a common brain architecture and Finlay and Darlington (1995) have shown that, with the exception of the olfactory bulb, brain components scale with brain size across a wide range of mammals. In many ways a human brain is therefore simply a scaled version of a typical primate brain (Jerison 1973). As there are no new anatomical structures or novel cortical areas that distinguish human brains from the brains of our primate relatives, the cognitive differences between members of our own species and our closest cousins are usually interpreted as a consequence of differences in brain size (Gibson 2002). Likewise, the discussions about the cognitive evolution of the human genus usually emphasize the evolutionary increase of adult endocranial volume (Figure 11.1, color plate). Brain size increased within the genus *Homo*, particularly within the species *Homo erectus* and its descendants (Ruff *et al.* 1997; Rightmire 2004; Schoenemann 2006). Much of this increase, however, is linked to increases in body size (Martin 1981; Wood and Collard 1999). Brain volume increases independent from body size increases are only documented during the last 0.5 million years (Ruff *et al.* 1997; Hublin *et al.* 2015). Importantly, this increase of brain volume evolved separately in the African lineage ancestral to modern humans, and in the European lineage leading to the Neanderthals (Bruner *et al.* 2003). There is no doubt that brain size plays a major role in the anatomical organization and the connectivity of the brain (Striedter 2005). However, if, or to what extent, variation in absolute or relative brain size among mammalian groups (Roth and Dicke 2005) can be linked to

Developmental Approaches to Human Evolution, First Edition. Edited by Julia C. Boughner and Campbell Rolian.

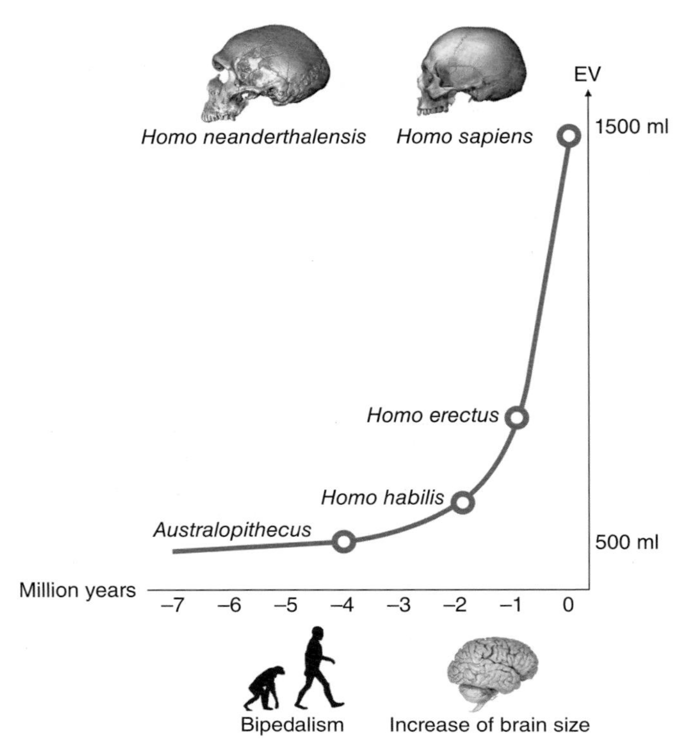

Figure 11.1. Brain-size evolution based on endocranial volumes. Evolutionary increases of brain size occurred within the genus *Homo*, particularly within the species *Homo erectus* and its descendants. Much of this dramatic increase, however, can be related to evolutionary increases in body size.

intelligence *directly* is a controversial and unresolved issue, and there is mounting evidence that brain size alone is not the only relevant factor contributing to cognitive abilities. The absolute size of the nervous system affects the number of neurons, their spacing and density, and their connectivity (Sporns 2011a); evolutionary changes in brain size therefore inevitably change the network topology of the brain. On top of these size-related changes, the internal architecture of the brain is also shaped by the tempo and mode of brain development and external stimuli. The focus of the present chapter is therefore to (1) explore the effects of brain volume increase in hominins on brain organization, and (2) identify those evolutionary changes to the hominin brain that cannot be explained by size increase alone.

The primary way to describe the evolutionary processes leading to the human brain is to compare modern human brains to those of our closest living relatives, the African apes (Figure 11.2), as well as to our fossil relatives and ancestors. Approaching this issue from an evo-devo perspective, this chapter explores the evolutionary changes of prenatal and postnatal brain development that are responsible for increased adult brain size in hominins, and how they affect life history and internal brain organization. Species differences in growth rate and timing are usually discussed in the framework of heterochrony, a term originally coined by Haeckel (1866) to describe an

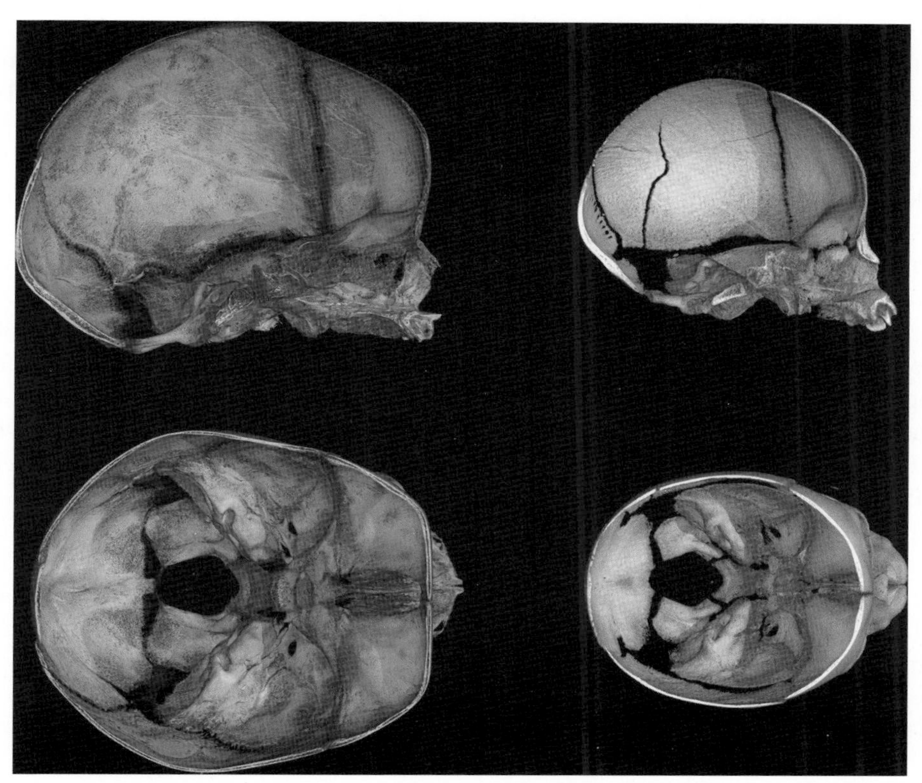

Figure 11.2. Newborn crania of a modern human (left) and a chimpanzee (right). High resolution computed tomographic scans illustrate that modern humans and chimpanzees have different brain volumes and cranial shapes at the time of birth. Note that in both species the cranial sutures are still wide open to accommodate the expansion of the developing brain.

exception to his theory of recapitulation. The concept was subsequently adapted and redefined independently of recapitulation by de Beer (1930), Gould (1977), and Alberch *et al.* (1979). Today's usage of this term usually refers to evolutionary changes in growth rates, and shifts in the onset and cessation of growth (Klingenberg 1998; Mitteroecker *et al.* 2005). Here we will consider the evolutionary trajectories of brain size, brain organization, and brain shape, and the interaction of these three variables. In particular, we will discuss the novel insights made possible by quantifying developmental changes of endocranial size together with changes of endocranial shape using statistical shape analysis (geometric morphometrics; Bookstein 1991; Slice 2007; Mitteroecker and Gunz 2009). In contrast to Gould's (1977) definition of heterochrony that relies on just a single variable to describe shape, geometric morphometric methods quantify shape using multiple variables. The classic terminology of heterochrony can therefore only describe a subset of all possible scenarios (see Mitteroecker *et al.* 2005 for an in-depth discussion of this topic).

As brains do not fossilize, we will discuss brain evolution based on "endocasts," that is, imprints of the brain and its surrounding structures in the fossilized internal braincase. These impressions can either be taken from the actual fossil specimens using latex and plaster (Holloway 1970; Holloway *et al.* 2004), or digitally from computed tomographic (CT) scans (Conroy *et al.* 1998, 1999, 2000; Falk *et al.* 2000; Neubauer *et al.* 2004, 2012; Schoenemann *et al.* 2007; Zollikofer and Ponce de León 2013). Two big advantages of the CT-based extraction of endocasts are that one can (1) digitally remove non-brain-derived fossilized sediments (i.e. morphological "noise") from the endocranial cavity (Prossinger *et al.* 2003; Weber and Bookstein 2011), and (2) correct for fossil distortions and displacements of anatomical structures without risk of harming the original specimen (Zollikofer *et al.* 1998; Zollikofer 2002; Ponce de León *et al.* 2008; Gunz *et al.* 2009; Weber and Bookstein 2011). Endocasts provide information about endocranial volume (as a proxy for brain size), about external surface details related to brain organization and blood supply, and about the overall shape of the brain. As brains develop during prenatal and postnatal ontogeny, the bones of the skull accommodate the expanding brain, largely via depositional growth at the cranial sutures (Moss and Young 1960; Enlow 1968; Morriss-Kay and Wilkie 2005; Richtsmeier *et al.* 2006; Frazier *et al.* 2008; Heuzé *et al.* 2010, 2011). The pressure created by the growing brain is redirected by the endocranial connective tissues, such as the *falx cerebri*, and the *tentorium cerebelli* (Moss and Young 1960; Bruner 2014). The internal shape of the braincase therefore reflects the complex interplay of tempo and mode of early brain development and the growth of cranial bones. Endocranial shape changes throughout development are thus potentially informative about growth rate and timing of early brain development in extant and fossil groups.

An Evolutionary Shift in Life History

Compared with our closest living relatives, and compared with earliest hominins, modern human adults have brains that are more than three times larger (Figure 11.1, color plate) in absolute terms. Understanding the broader context of modern human brain evolution requires us to look back 6 million years, close to the split between hominin and chimpanzee lineages. This journey into our evolutionary past, however, does not start with brains. It starts with feet and hips. Around 6 million years ago hominins evolved a peculiar way of locomotion (Senut *et al.* 2001; Brunet *et al.* 2002): walking upright on their hind limbs. The fossil evidence is scarce, so the exact timing and many details of the origin of bipedalism remain unclear (Richmond and Strait 2000; Richmond *et al.* 2001; Richmond and Jungers 2008; Kivell and Schmitt 2009; Lovejoy *et al.* 2009). However, there is no doubt that hominin bipedalism was established fully in the genus *Australopithecus* at 3.6 million years ago in east Africa (Raichlen *et al.* 2010). It is even possible to provide an exact date here, because paleoanthropologists uncovered fossilized footprints of upright walking hominins at the site of Laetoli in Tanzania (Leakey and Hay 1979; Leakey 1981; White and Suwa 1987). These footprints in a layer of wet volcanic ash were most likely created by *Australopithecus afarensis*, a group of hominins with brain and body sizes comparable to those found among chimpanzees today (McHenry 1992, 1994; Green *et al.* 2007; Kimbel and Delezene 2009). The evolution of bipedalism therefore preceded the evolutionary increase of brain size by as much as 4 million years.

This sequence of events is important, because the evolutionary adaptations to bipedal locomotion entail fundamental changes to the entire postcranial skeleton (Lovejoy 1975; Stern and Susman 1983; Tague and Lovejoy 1986; Hunt 1994; Ruff 1995; Whitcome *et al.* 2007). Most importantly, the relative width of the pelvis decreased, thereby narrowing the birth canal (Rosenberg and Trevathan 1996, 2001, 2002). The dimensions and shape of the female pelvis constrain the size and shape the fetus can reach *in utero*, as at the time of birth it has to pass through the birth canal. Birth in non-human primates is generally a solitary event without complications for either mother or baby: the fetus passes through the birth canal without significant changes in orientation, and the process of parturition is quick (Hirata *et al.* 2011). In humans, however, the large size of the fetal head and shoulder combined with a narrow birth canal require multiple movements and rotations of the fetus (Rosenberg and Trevathan 1996). The "obstetrical dilemma" hypothesis (Krogman 1951; the term was coined by Washburn 1960) posits that in all hominins, including our own species, the neonatal sizes of head and brain are constrained by the biomechanics of upright walking: in bipedal hominins with their narrow birth canals, the selection for large brains with their high energetic costs required changing the typical primate pattern of life history. Life history shifts in the hominin lineage changed the "precocial" primate pattern to a pattern of "secondary altriciality" (*sensu* Portmann 1944), characterized by offspring that depends entirely on parental care for a long time period. Alternative explanations for the relatively altricial state of human neonates have emphasized metabolic constraints and the evolution of cooperative breeding. Pregnancy places a heavy metabolic burden on mothers, as they must support the energetic requirements of the ever-increasing metabolic rates of the growing fetuses (Dunsworth *et al.* 2012). These authors have therefore argued that limits to maternal metabolism impose constraints on human gestation length and fetal growth that are more important than obstetric constraints. In particular, Dunsworth *et al.* (2012) questioned the claim central to the obstetric dilemma hypothesis that the width of the female pelvis is constrained by locomotor efficiency. Instead, these authors stressed that "energetic constraints of both mother and fetus are the primary determinants of gestation length and fetal growth in humans." Whereas the fact that multiple pregnancies (especially twins and triplets) are not uncommon in humans seems to be at odds with this emphasis on an energetic dilemma, there is little doubt that metabolic demands play an important role during gestation and lactation. It has been argued that cooperative breeding in early hominins may have allowed for the evolution of large brains, through energy subsidies for mothers and weaned children (Isler and van Schaik 2009, 2012; Navarrete *et al.* 2011). Rather than invoking an obstetrical dilemma, Isler and van Schaik (2012) therefore argued that the relatively altricial state of human neonates could also be explained by the evolution of increased efficiency of postnatal allomaternal care in hominins.

To overcome either obstetric or energetic constraints (or both), humans are characterized by an evolutionary change in life history: shifting a significant portion of brain development to the time after birth (Figure 11.3, color plate).

A larger percentage of brain growth and maturation happens postnatally in humans than in apes. Humans are born with only approximately 28% of their adult

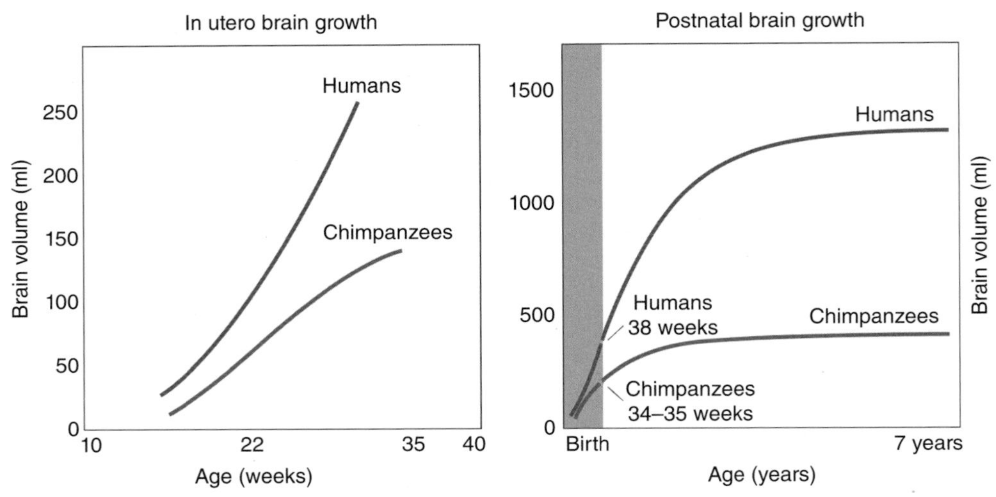

Figure 11.3. Prenatal (left side) and postnatal brain growth in modern humans and chimpanzees. Prenatal data are based on ultrasound (US) images of fetal brain development. The right panel shows endocranial volumes measured from computed tomographic scans; fetal US data (gray background) are the same as in left panel. Data based on Neubauer *et al.* (2012); Neubauer and Hublin (2012); Sakai *et al.* (2012).

volume completed, compared with more than 40% in chimpanzees (DeSilva and Lesnik 2006), yet the brains of modern human neonates are already almost as large as in adult chimpanzees. As the typical gestation length of humans (38 weeks) is only slightly longer than in chimpanzees (34–35 weeks; Sakai *et al.* 2012), differences in brain development between humans and chimpanzees are already evident prenatally. Based on a longitudinal 3D ultrasound study of chimpanzee fetuses, Sakai *et al.* (2012) showed that at 16 weeks of gestation the brain volume in humans is already twice that of a chimpanzee fetus (left panel in Figure 11.3, color plate). Moreover, the rate of brain growth decreases in chimpanzees after 22 weeks of gestation, whereas brain growth in humans accelerates until the 32nd gestational week (Sakai *et al.* 2012). Human brains continue to grow at almost fetal rates for almost a year postpartum (right panel in Figure 11.3, color plate). Subsequently, the rate of brain growth slows down; however, the adult endocranial volume is not reached before the age of 6 (Lenroot *et al.* 2007). The brains of chimpanzees also continue to grow after birth, but they reach adult brain volumes earlier than modern humans (Neubauer and Hublin 2012), although there is debate about the extent and significance of that difference in duration of growth (Leigh 2004, 2012). With regards to absolute brain growth, human brains grow faster both prenatally and postnatally than those of chimpanzees. At the same time, human brains mature more slowly than those of chimpanzees. It is reasonable to posit that this temporal variation may factor into the development of uniquely human cognitive abilities. Compared with chimpanzees, modern human infants learn longer, at the expense of being dependent on allocare for a longer period of time. In humans, a large portion of brain growth and maturation occurs after birth when the infant is exposed to the stimuli of the extra-uterine environment; this period is longer in humans than in chimpanzees. As we will discuss below, differences between humans and chimpanzees in brain maturation can also be found studying the internal organization of the brain; these species differences are more dramatic than those inferred from brain volume data.

It is not known yet when this protracted pattern of human brain development evolved. Based on a comparative analysis of the fossil of a 3-year-old *Australopithecus afarensis* from Dikika, Alemseged *et al.* (2006) argued that the pattern of brain growth in *A. afarensis* was not exactly like the one in great apes, but rather shifted towards the modern human pattern. This would imply that some changes in life history preceded the evolutionary increase of endocranial size. An assessment of the *Homo erectus* subadult from Mojokerto (Coqueugniot *et al.* 2004; Hublin and Coqueugniot 2006), however, revealed that the *Homo erectus* brain growth trajectory was intermediate between chimpanzees and modern humans. It thus seems that the patterns of brain development are variable among hominins, and do not follow a simple linear evolutionary trajectory from an ape pattern towards a modern human pattern. Such variability of brain growth trajectories among closely related species has been demonstrated by Leigh (2007) for papionin monkeys. Reducing an evolving system to a dichotomy of "human-like" versus "chimpanzee-like" is therefore too simplistic and obscures the complexity of actual evolutionary trajectories within the hominin lineage.

Changes in Brain Organization in Living Apes and Fossil Hominins

In addition to brain size, modern human brains also differ from those of great apes in their organization. The overall organization of the brain can be inferred from the relative sizes of different brain areas, as well as group differences in brain morphology, such as brain convolutions: hominid brains fold during fetal development creating a characteristic pattern of gyri, which are surrounded by grooves either called sulci (if they are shallow) or fissures (if they are large furrows). Sulcal patterns give insights about brain areas and external morphology – and thus potentially brain function – within and across primate taxa. There is debate whether the differences in brain morphology and organization between apes and humans are merely by-products of the evolutionary size increase, or whether they were selected for in the hominin lineage. Human brains largely conform to primate scaling rules (Sacher 1982; Deacon 1997; Laughlin and Sejnowski 2003), but Rilling and colleagues maintain that several aspects of the human brain deviate from allometric predictions (Rilling and Insel 1999; Rilling 2006, 2014): they argue that based on a non-human anthropoid scaling pattern, modern humans have larger temporal lobes than predicted for their brain volume, larger relative size of the neocortex, and greater gyrification in the prefrontal cortex. Interestingly, these brain reorganizations are not necessarily reflected at the cellular level (Herculano-Houzel 2012): counting neurons in brains of four adult human males, Azevedo *et al.* (2009) found an average neuron number of approximately 86 billion cells for the entire brain. This neuron count matches the scaling predictions for a primate of human proportions (these authors used 75 kg as the human male average), based on a regression model derived from six primate species (Herculano-Houzel *et al.* 2007). Herculano-Houzel (2012) therefore argued that, at least with regards to neuron number, modern human brains could not be considered outliers among primates. In line with Shea (1983), Herculano-Houzel and her colleagues instead suggested that great apes might deviate from the scaling pattern of primates in having evolved larger than expected body sizes for their brain size (Azevedo *et al.* 2009).

Central to the discussion about the relationship between brain reorganization and brain size are the convolutional details seen on endocasts of the ape-sized (both brain and body size) early hominin fossils. The external morphology of the brain can leave impressions in the bone; however, not all features can be identified reliably on endocasts. This lack of clarity, even with CT scan images, makes it notoriously hard to trace the evolution of sulcal patterns based on fossil endocasts. The basic sulcal patterns are very similar between apes and humans (Connolly 1950), despite that there are substantial differences in cognitive abilities between the two groups. Even subtle differences between modern humans and apes may therefore be significant and deserve attention and scrutiny. Several authors have argued that *Australopithecus* shows signs of brain reorganization that precede the evolutionary increase of brain size in hominins (Dart 1925; Holloway *et al.* 2004; Carlson *et al.* 2011; Falk 2014). Historically, this debate about brain reorganization has focused on two issues: (1) whether it is possible to identify the position of an impression called the "lunate sulcus" on hominin fossils (Smith 1903, 1904), and (2) whether human frontal lobes are unusually large for a primate of their body size (Brodmann 1909). In brains of apes, the lunate sulcus marks the anterior limit of the occipital lobe and corresponds to the anterior lateral boundary of the primary visual cortex. Grafton Elliot Smith (1903, 1904) argued that a homologous structure could be identified in some modern human brains in a more posterior position than in apes. As a consequence, the identification of an "ape-like" anteriorly placed lunate sulcus or a "human-like" posteriorly placed lunate sulcus in fossil hominins has been used to infer evolutionary changes of the underlying brain structure, based on the idea that (i) the position of the lunate sulcus is informative about the size of the primary visual cortex, and (ii) that a relative evolutionary expansion of the brain's parietal and temporal cortices had shifted the lunate sulcus posteriorly. Announcing the first *Australopithecus* fossil ever discovered, Dart (1925) described a posterior position of a lunate sulcus on the natural endocast of the Taung child, and used this feature to link *Australopithecus* (translated "the Southern ape") to the human lineage. This sparked a lengthy and still unresolved controversy about whether the lunate sulcus can be identified reliably in Taung and other australopith fossils, and its importance for cognitive evolution (Falk 1980a, 1980b, 1983, 1985, 1989 *vs.* Holloway 1981, 1984, 1985, 1991). Notably, a re-evaluation using magnetic resonance imaging of the lunate sulcus in a large human sample showed that this sulcus is typically absent in modern humans (Allen *et al.* 2006): in line with other researchers, these authors questioned the phylogenetic significance of the lunate sulcus, and this debate continues today.

For more than a century, the frontal lobe has received much attention as a likely candidate for some of the unique cognitive abilities that evolved in the human lineage.

However, classic reports based on the pioneering comparative anatomy work of Brodmann (1909) and Blinkov and Glezer (1968) that human frontal lobes are expanded compared to other primates have not been confirmed by more recent neuroimaging studies of larger comparative samples (Semendeferi and Damasio 2000; Semendeferi *et al.* 2002). That being said, there is evidence for the evolutionary reorganization of the frontal lobes in modern humans relative to extant apes and fossil hominins, particularly in the prefrontal cortex (Semendeferi *et al.* 2011; Teffer and Semendeferi 2012). Regarding sulcal anatomy of the brain's frontal lobe, the human-ape differences (Connolly 1950) in the "middle frontal sulcus" have received attention, because this feature reproduces well on australopith endocasts (Falk 2014). Like anatomically modern human brains and endocasts, several australopith specimens from South Africa differ from the "ape pattern" in that australopiths have a separate "middle frontal sulcus" that might result from expanded prefrontal cortices (Falk 2014).

Taken together, there is consensus among paleoneurologists that early hominins depart from the sulcal patterns found in apes, and therefore show signs of brain reorganization despite their ape-like endocranial volumes. Thus, even though debate and controversy about the identity of features on fossil endocasts persist, evolutionary changes in brain organization appear to precede increase in brain size in the hominin lineage (Hublin *et al.* 2015). This in turn means that hominins evolved patterns of brain development other than those directly related to brain size increase, and that human brain organization is not entirely a by-product of enlarged brain size.

Endocranial Shape Changes during Development

The globular neurocranial shape of adult *Homo sapiens* differs from the elongated braincases of all archaic human groups (Lieberman *et al.* 2002, 2004). These neurocranial shape differences reflect brain shape differences between fossil and modern humans. Neanderthals have elongated braincases like all other fossil *Homo* species, yet the endocranial volume ranges of modern humans and Neanderthals overlap. The evolutionary increase of brain size in the human lineage does therefore not suffice as an explanation of the characteristic globular brain shape of *Homo sapiens*. In a series of publications, we have recently shown that the globular shape of *Homo sapiens* is the result of a postnatal phase of brain development that distinguishes modern humans from chimpanzees and Neanderthals: using geometric morphometrics on virtual endocasts extracted from CT scans, Neubauer *et al.* (2010) quantified brain ontogenetic shape changes between birth and adulthood in recent modern humans and chimpanzees. They showed that while *Homo sapiens* and *Pan* differ in endocranial shape throughout postnatal ontogeny, these two groups share a pattern of endocranial shape changes following the eruption of the deciduous dentition. Recently, Scott *et al.* (2014) compared ontogenetic endocranial shape changes in a cross-sectional sample of hominoids (modern humans, chimpanzees, gorillas, orangutans, and gibbons): their findings confirmed that the patterns of endocranial development from the eruption of the deciduous dentition to adulthood are similar among *all* extant hominoids, and argued that "the resilient pattern of endocranial

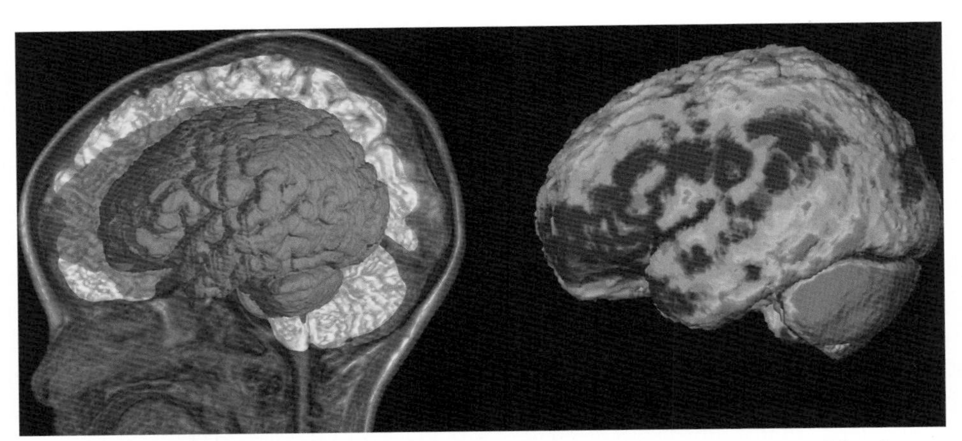

Figure 11.4. Globularization phase in modern humans. During early postnatal development, human brains do not only change their size, but also their shape. Magnetic resonance image scans of neonatal brain (blue) compared to a three year old. On the right hand side, the amount of shape change color coded from blue (small shape differences) to red (large shape differences). Developmental shape changes are most dramatic in the cerebellum, and the parietal lobe. Figure created using MRI data from ALBERT brain atlas (Gousias *et al.* 2012, 2013).

development within the *Hominoidea* superfamily is most likely due to stabilizing selection rather than convergent evolution." Some aspects of brain development are therefore shared by all extant hominoids (Neubauer *et al.* 2010; Scott *et al.* 2014). Importantly, however, in the first year of life, modern humans depart from this pattern. During a postnatal "globularization phase" (Neubauer *et al.* 2009, 2010, 2012; Gunz *et al.* 2010) modern human brains (Figure 11.4, color plate) change from an elongated shape to the distinctive globular shape of *Homo sapiens.* These neurocranial shape changes from birth to the eruption of the deciduous dentition include a bulging of the parietal bones and occipital bones, and flexion of the cranial base. Notably, they are reminiscent of the shape differences between braincases of adult modern humans and Neanderthals (Bruner *et al.* 2003). In Gunz *et al.* (2010, 2012) we therefore tested if Neanderthal infants have a globularization phase during their brain development like that in human infants. Central to these studies were comparisons of virtual reconstructions of the brains of two Neanderthal neonates, one from Le Moustier, France (Maureille 2002b, 2002a), and the other from Mezmaiskaya, Russia (Golovanova *et al.* 1999), to other Neanderthal subadults. Figure 11.5 (color plate) shows that, at birth, Neanderthal and modern human neonates can be distinguished based on facial and dental characteristics (Tillier 1995, 1996; Ponce de León *et al.* 2008; Zollikofer and Ponce de León 2010) even while the shapes of their relatively elongated braincases (and therefore endocasts) are almost identical. Further, newborn Neanderthals and modern humans have brains that are about the same size (Ponce de León *et al.* 2008; Gunz *et al.* 2010, 2012). After birth, however, the brain developmental trajectories of modern humans and Neanderthals diverge. Specifically, the globularization phase of early brain development is absent from chimpanzees and Neanderthals. The different endocranial shape trajectories of modern humans and Neanderthals corroborate the notion, originally put forward by

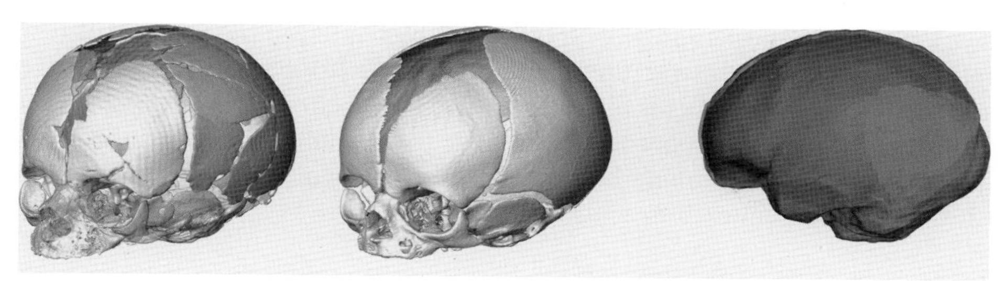

Figure 11.5. Virtual reconstruction of a Neanderthal neonate (left; based on Mezmaiskaya – Gunz *et al.* 2012) compared to the computed tomographic scan of a modern human baby (middle). Whereas Neanderthal facial characteristics like a large projecting face, large nose, and tall orbits are already visible, the braincases and endocasts (right panel) of Neanderthals (red) and modern humans (blue) are almost identical at the time of birth. The pronounced endocranial shape differences between *Homo sapiens* and *Homo neanderthalensis* develop in a postnatal globularization phase directly after birth.

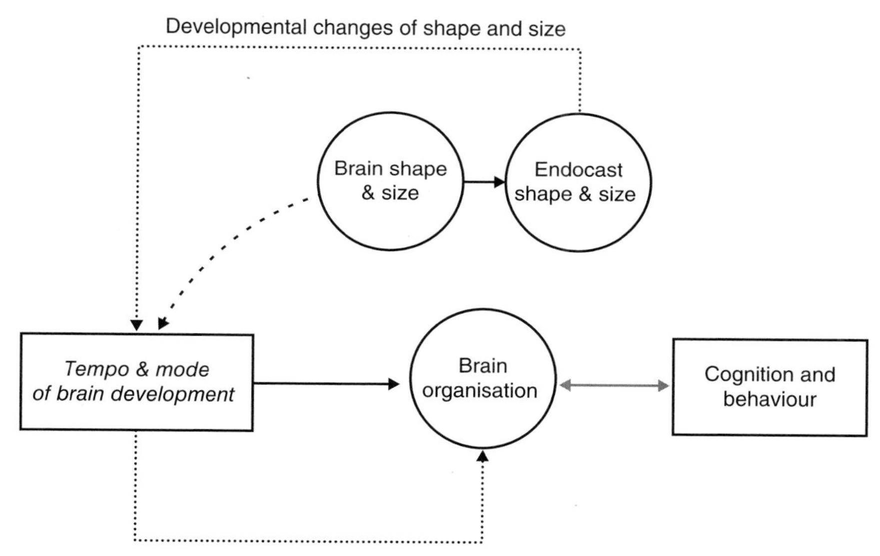

Figure 11.6. Tempo and mode of brain development affect the pattern of neural wiring, and thereby behavior and cognition. Tempo and mode of brain development also affect brain shape and size. Brain-shape changes during development reflect the developmental speed and timing of the underlying brain circuitry. Redrawn after Gunz *et al.* (2012).

Emiliano Bruner and colleagues (Bruner *et al.* 2003; Bruner 2004, 2010; Bruner and Manzi 2007), that modern humans and Neanderthals reached similar endocranial capacities through different developmental pathways. The timing aspect in early postnatal ontogeny is informative about what causes the braincase to bulge, and also helps to elucidate the phase's potential importance for cognitive development: as the globularization phase coincides with the timing of rapid postnatal brain expansion, it is likely that species differences in the tempo and mode of brain development contribute to the endocranial shape differences between modern humans and Neanderthals (Figure 11.6). Differences between modern humans and Neanderthals in developmental patterns and growth rates have been inferred from studies of dental eruption and dental microstructure (Smith *et al.* 2007, 2010; Smith 2013, but see e.g. Macchiarelli *et al.* 2006). As we will review in the next section, such group differences in the tempo of early brain ontogeny are likely to have implications for the internal organization of the developing brain, and thus for cognitive abilities (Langbroek 2012; Boeckx and Benítez-Burraco 2014).

Evolution and Development of the Brain's Networks

Recent work on the evolution and development of cognition has emphasized the importance of the internal organization of the brain (Rapoport 1999; Schenker *et al.* 2005; Sporns 2011a, 2011b; Bianchi *et al.* 2012; Bullmore and Sporns 2012; Buckner and Krienen 2013; van den Heuvel and Sporns 2013; Kim *et al.* 2014) and how its parts are connected during early development. Greenough *et al.* (1987) emphasized the importance of brain plasticity of mammalian brains, which require certain stimuli for normal maturation. Such "experience expectant" processes can evolve when certain environmental inputs, such as seeing contrast borders between objects, are ubiquitous and reliably occur for all members of a species. They suggested that this principle is particularly important for shaping the sensory and motor systems, and typically works via intrinsically governed excess generation of synaptic connections, which are subsequently pruned based on experiential input. Mammalian brains have also evolved the capacity to store information that is "experience dependent," via generation of synaptic connections as a response to stimuli (Greenough *et al.* 1987).

Histological studies of brain tissue and magnetic resonance imaging (MRI) in humans have revealed that the postnatal internal changes of the brain's wiring pattern are dramatic and extend long past the time of reaching adult brain volume (Figure 11.7, color plate). Whereas the number of neurons does not change significantly during postnatal brain development (Huttenlocher 1979), the brain's wiring network is initially sparse. In the first years of life the networks of the brain are established in dramatic bursts of synaptogenesis, reinforced by myelination, and subsequently optimized by pruning (Low and Cheng 2006). These dynamic biological changes continue throughout childhood and adolescence (Giedd *et al.* 1996a, 1996b, 1996c, 1999; Paus *et al.* 1999; Casey *et al.* 2000; Thompson *et al.* 2000; Lenroot *et al.* 2007; Giedd and Rapoport 2010; Hagmann *et al.* 2010). Genes affect the brain's wiring pattern by controlling precisely timed sequences of synaptogenesis and the subsequent pruning of connections. However, the network topology itself is neither intrinsically predetermined nor static. Sporns (2011b) stressed that brain connectivity changes dynamically throughout life, and that processes of self-organization underlie "the highly variable yet robust nature of brain dynamics" (Sporns 2011b, p.253).

Patterns of brain development and their genetic underpinnings are extremely conserved across mammals (Reichert 2009; Finlay and Darlington 1995). It is therefore remarkable that Liu *et al.* (2012) discovered a human-specific expression change in genes associated with synaptic functions. Studying cortical synaptic development of humans, chimpanzees, and macaques on a gene expression level – specifically the

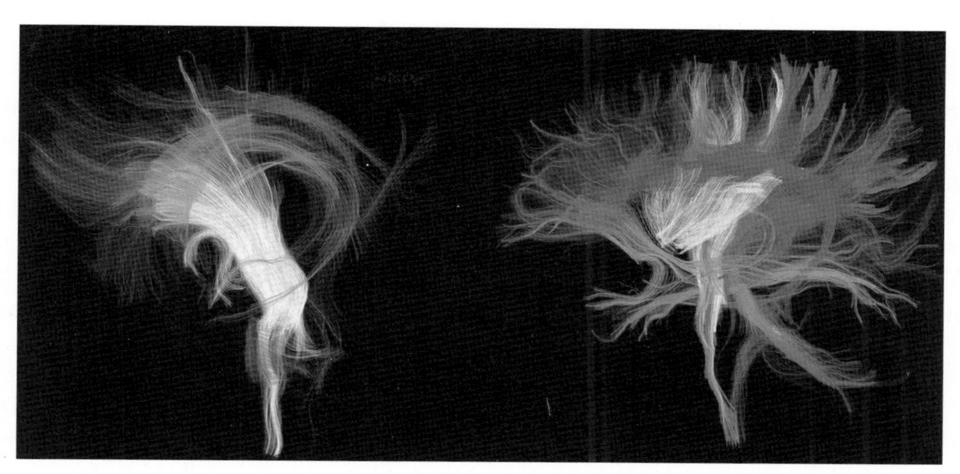

Figure 11.7. The brain's internal architecture – symbolized here by white matter connections – is the result of precisely timed sequences of synaptogenesis and the subsequent pruning of connections. However, the exact network topology is neither intrinsically predetermined nor static. Clinical data show that timing and rate of growth affect the wiring pattern of the brain. Fiber bundles (drawn in different colors) of white matter connections of a fetus in gestational week 17 (left panel), and a 3 year old (right panel). Figure created using MRI data are from the BrainSpan Atlas of the Developing Human Brain (Miller *et al.* 2014; http://brainspan.org).

timing of synaptic development in the prefrontal cortex (PFC) – they found that expression in the PFC peaked at younger than 1 year of age in chimpanzees and macaques but much later, at 5 years of age, in humans. This is consistent with the findings discussed in the previous section, that the overall maturation of the human brain is protracted compared to other primates. The delay of PFC maturation in humans is even more dramatic than in other brain regions, suggesting that brain plasticity is higher in human infants for a longer time than in those of our primate relatives. The PFC is a key region for higher executive functions and plays an integrative role for planning, reasoning, and problem solving; it is also important for social behavior. A matching pattern emerges from comparative studies of myelination: based on histological data, Miller *et al.* (2012) showed that chimpanzees and humans differ in the developmental pattern of neocortical myelination: at the time of birth human brains are less myelinated than those of chimpanzees. Myelin growth in humans was found well beyond adolescence, whereas in chimpanzees myelination stops earlier, that is, around the time of sexual maturity (Miller *et al.* 2012). Analyzing the development of white matter using MRI scans, Sakai *et al.* (2011, 2013) found that chimpanzees and humans differ from macaques (presumably representing the ancestral primate condition) in the maturation of the brain's prefrontal portion. Specifically, white-matter volume increased from infancy to adulthood in both hominids, but not the monkey. However, during infancy, prefrontal white-matter volume increased at a higher rate in humans than in chimpanzees. Sakai *et al.* (2011) suggested that this protracted human brain development enhances the impact of postnatal experiences on neural connectivity and that "the rapid development of the human prefrontal white matter during infancy may help the development of complex social interactions."

Future Directions

The developmental processes of the brain that underlie the modern human brain's globularization phase are still unknown, although future research using longitudinal data – which would be feasible to do using current imaging tools – would help characterize this process during childhood brain development. For instance, MRI scanning the same individuals as they grow would make it possible to assess, even in the first years of life, the relationship between overall brain shape and brain structural connectivity, as well as with the speed and timing of brain development.

Comparisons of Neanderthal, Denisovan, and modern human genomes (Burbano *et al.* 2010; Green *et al.* 2010; Castellano *et al.* 2014; Pääbo 2014; Prüfer *et al.* 2014) indicated positive selection on several brain regions in *Homo sapiens* that occurred after the split with Neanderthals. Several among these genes are likely to be critical for brain development, as they affect mental and cognitive development in living people. Studying the nuclear DNA of a Denisovan fossil specimen, Meyer *et al.* (2012) identified several changes that are likely to affect the brain and the nervous system that are unique to modern humans in genes that are extremely conserved among primates: *SLITRK1* and *KATNA1* are involved in axonal and dendritic growth; *ARHGAP32* and *HTR2B* affect synaptic transmission; *ADSL* and *CNTNAP2* are autism candidate genes; *CNTNAP2* has also been linked to a susceptibility to language disorders (Meyer *et al.* 2012). These authors therefore speculated that "crucial aspects of synaptic transmission may have changed in modern humans" (2012, p.225).

The number of single nucleotide differences that are unique to modern humans and differ from Neanderthals and Denisovans is surprisingly small: these genetic changes would affect only 87 proteins. Of these, a larger than expected number affect neuron formation and are expressed during fetal development of the cerebral cortex (Prüfer *et al.* 2014), and might therefore point to some aspect of cortex development that is unique to modern humans (Pääbo 2014). It has, for example, been hypothesized that three of these genes (*CASC5*, *KIF18A*, and *SPAG5*) might affect which type of neuronal precursor cell is formed (Pääbo 2014; Prüfer *et al.* 2014).

Whereas any inferences about Neanderthal cognition can but be speculative at this time, the morphological and developmental contrasts between modern humans and Neanderthals provide the context necessary to help understand the role of the genes that are derived in modern humans and thus driving uniquely human patterns of brain growth and, subsequently, modern human cognition.

One question that remains is when in the evolutionary history of our species the globularization phase evolved. Unfortunately, to date no fossil remains of early modern human newborns or infants have been discovered. However, even in the absence of these subadult fossils, it will be possible to narrow down the origin of the globularization phase and explore its evolutionary trajectory by studying adult *Homo sapiens* fossils from different time periods. As recent *Homo sapiens* and Neanderthals have almost identical endocranial shapes at the time of birth, it is reasonable to expect the same to be true for fossil *Homo sapiens* neonates. The appearance of adult fossils with globular braincases in the fossil record should therefore mark the origin of this developmental phase. If the globularization phase evolved gradually within *Homo sapiens*, we would expect adult endocranial shapes to also gradually become more globular over evolutionary time. Alternatively, if globularization evolved relatively fast then we would expect a sudden, sharp, and consistent change from elongated endocranial shapes to globular shapes that fall within recent modern variation.

Summary

Taken together, the evidence reviewed in this chapter shows that human brains largely correspond to primate scaling rules. However, not all differences in brain organization between humans and apes can be explained by allometric scaling. Human brain evolution therefore involved aspects that are not related to the large brain size exclusively. Compared to other primates, humans give birth to relatively immature babies, which are dependent on allocare for a long period of time. Brain maturation takes longer in humans than in chimpanzees, and is particularly delayed in regions associated to higher executive functions, such as the prefrontal cortex. With regards to the pattern of brain development, modern humans stand out not only among primates, but also among large-brained hominins. During a globularization phase in the first year of life, modern humans deviate from the generalized hominoid pattern of brain development of postnatal ontogeny. Subsequently, extant and fossil hominoids including *Homo sapiens* develop their brains along a shared developmental trajectory of brain shape changes. A better understanding of early postnatal brain development in living people will therefore be critical for answering several long-standing questions about the evolution of human cognition.

Acknowledgements

I am grateful to Julia Boughner and Campbell Rolian for inviting me to contribute to this book, and for organizing a fantastic symposium on Evolutionary Developmental Anthropology at the annual meeting of the American Association of Physical Anthropologists 2014 in Calgary; moreover, they have both offered constructive criticism and helped editing this chapter. Many of the concepts and ideas discussed here were developed in collaborative projects with Simon Neubauer and Jean-Jacques Hublin; I want to thank them for their support. Heiko Temming and Martin Dockner CT-scanned the specimens shown in Figure 11.2; Figure 11.4 (color plate) is based on data from the ALBERT Brain atlas, Copyright Imperial College of Science, Technology and Medicine and Ioannis S. Gousias 2012 and 2013, all rights reserved. I am grateful to Lubov Golovanova and Vladimir Doronichev for providing access to the fossil shown in Figure 11.5 (color plate). Figure 11.7 (color plate) is based on data from BrainSpan: Atlas of the Developing Human Brain (http://brainspan.org). This research was funded by the Max Planck Society.

References

Alberch, P., Gould, S. J., Oster, G. F., and D. B. Wake. 1979. Size and shape in ontogeny and phylogeny. *Paleobiology* 5(3):296–317.

Alemseged, Z., Spoor, F., Kimbel, W. H. *et al.* 2006. A juvenile early hominin skeleton from Dikika, Ethiopia. *Nature* 443(7109):296–301.

Allen, J. S., Bruss, J., and Damasio, H. 2006. Looking for the lunate sulcus: A magnetic resonance imaging study in modern humans. *Anatomical Record Part A: Discoveries in Molecular, Cellular, and Evolutionary Biology* 288(8):867–876.

Azevedo, F. A. C., Carvalho, L. R. B., Grinberg, L. T. *et al.* 2009. Equal numbers of neuronal and nonneuronal cells make the human brain an isometrically scaled-up primate brain. *Journal of Comparative Neurology* 513(5):532–541.

de Beer, G. R. 1930. *Embryology and evolution.* Oxford: Clarendon Press.

Bianchi, S., Stimpson, C. D., Bauernfeind, A. L. *et al.* 2012. Dendritic morphology of pyramidal neurons in the chimpanzee neocortex: Regional specializations and comparison to humans. *Cerebral Cortex* doi:10.1093/cercor/bhs239.

Blinkov, S. M., and Glezer, I. I. 1968. *The human brain in figures and tables: A quantitative handbook.* New York: Basic Books.

Boeckx, C., and Benítez-Burraco, A. 2014. The shape of the human language-ready brain. *Frontiers in Psychology* 5:282.

Bookstein, F. L. 1991. *Morphometric tools for landmark data: Geometry and biology.* Cambridge: Cambridge University Press.

Brodmann, K. 1909. *Vergleichende Lokalisationslehre der Grosshirnrinde in Ihren Prinzipien Dargestellt auf Grund des Zellenbaues.* Leipzig: Barth.

Bruner, E. 2004. Geometric morphometrics and paleoneurology: Brain shape evolution in the genus *Homo*. *Journal of Human Evolution* 47(5):279–303.

Bruner, E. 2010. Morphological differences in the parietal lobes within the human genus. *Current Anthropology* 51(Suppl. 1):77–88.

Bruner, E. 2014. Functional craniology, human evolution, and anatomical constraints in the Neanderthal braincase. *In:* T. Akazawa, N. Ogihara, H. C. Tanabe, and H. Terashima (eds) *Dynamics of Learning in Neanderthals and Modern Humans Volume 2.* Tokyo: Springer Japan.

Bruner, E., and Manzi, G. 2007. Landmark-based shape analysis of the archaic *Homo* calvarium from Ceprano (Italy). *American Journal of Physical Anthropology* 132(3):355–366.

Bruner, E., Manzi, G., and Arsuaga, J. L. 2003. Encephalization and allometric trajectories in the genus *Homo*: Evidence from the Neandertal and modern lineages. *Proceedings of the National Academy of Sciences, USA* 100(26):15335–15340.

Brunet, M., Guy, F., Pilbeam, D. *et al.* 2002. A new hominid from the Upper Miocene of Chad, Central Africa. *Nature* 418(6894):145–151.

Buckner, R. L., and Krienen, F. M. 2013. The evolution of distributed association networks in the human brain. *Trends in Cognitive Sciences* 17(12):648–665.

Bullmore, E., and Sporns, O. 2012. The economy of brain network organization. *Nature Reviews Neuroscience* 13(5):336–349.

Burbano, H. A., Hodges, E., Green, R. E. *et al.* 2010. Targeted investigation of the Neandertal genome by array-based sequence capture. *Science* 328(5979):723–725.

Carlson, K. J., Stout, D., Jashashvili, T. *et al.* 2011. The endocast of MH1, *Australopithecus sediba*. *Science* 333(6048):1402–1407.

Casey, B. J., Giedd, J. N., and Thomas K. M. 2000. Structural and functional brain development and its relation to cognitive development. *Biological Psychology* 54(1–3):241–257.

Castellano, S., Parra, G., Sánchez-Quinto, F. A. *et al.* 2014. Patterns of coding variation in the complete exomes of three Neandertals. *Proceedings of the National Academy of Sciences, USA* 111(18):6666–6671.

Charvet, C. J., Striedter, G. F., and Finlay, B. L. 2011. Evo-devo and brain scaling: Candidate developmental mechanisms for variation and constancy in vertebrate brain evolution. *Brain, Behavior and Evolution* 78(3):248–257.

Connolly, C. J. 1950. *External morphology of the primate brain*. Springfield, IL: C. C. Thomas.

Conroy, G. C., Weber, G. W., Seidler, H., and Tobias P. V. 1999. Endocranial capacity of early hominids. *Science* 283(5398):9.

Conroy, G. C., Weber, G. W., Seidler, H. *et al.* 1998. Endocranial capacity in an early hominid cranium from Sterkfontein, South Africa. *Science* 280(5370):1730.

Conroy, G. C., Weber, G. W., Seidler, H. *et al.* 2000. Endocranial capacity of the Bodo cranium determined from three-dimensional computed tomography. *American Journal of Physical Anthropology* 113(1):113–118.

Coqueugniot, H., Hublin, J.-J., Veillon, F. *et al.* 2004. Early brain growth in *Homo erectus* and implications for cognitive ability. *Nature* 431(7006):299–302.

Dart, R. A. 1925. *Australopithecus africanus*: The man-ape of South Africa. *Nature* 115:195–199.

Deacon, T. W. 1997. What makes the human brain different? *Annual Review of Anthropology* 337–357.

DeSilva, J. and Lesnik, J. 2006. Chimpanzee neonatal brain size: Implications for brain growth in *Homo erectus*. *Journal of Human Evolution* 51(2):207–212.

Dunsworth, H. M., Warrener, A. G., Deacon, T. *et al.* 2012. Metabolic hypothesis for human altriciality. *Proceedings of the National Academy of Sciences, USA* doi:10.1073/pnas.1205282109.

Enlow, D. H. 1968. *The human face*. New York: Harper & Row.

Falk, D. 1980a. Hominid brain evolution: The approach from paleoneurology. *Yearbook of Physical Anthropology* 23:93–107.

Falk, D. 1980b. A reanalysis of the South African australopithecine natural endocasts. *American Journal of Physical Anthropology* 53(4):525–539.

Falk, D. 1983. The Taung endocast: A reply to Holloway. *American Journal of Physical Anthropology* 60(4):479–489.

Falk, D. 1985. Apples, oranges, and the lunate sulcus. *American Journal of Physical Anthropology* 67(4):313–315.

Falk, D. 1989. Ape-like endocast of "ape-man" Taung. *American Journal of Physical Anthropology* 80(3):335–339.

Falk, D. 2014. Interpreting sulci on hominin endocasts: Old hypotheses and new findings. *Frontiers in Human Neuroscience* 8:134.

Falk, D., Redmond, J. C., Guyer, J. *et al.* 2000. Early hominid brain evolution: A new look at old endocasts. *Journal of Human Evolution* 38(5):695–717.

Finlay, B. L., and Darlington, R. B. 1995. Linked regularities in the development and evolution of mammalian brains. *Science* 268(5217):1578–1584.

Frazier, B. C., Mooney, M. P., Losken, H. W. *et al.* 2008. Comparison of craniofacial phenotype in craniosynostotic rabbits treated with anti-Tgf-beta2 at suturectomy site. *Cleft Palate-Craniofacial Journal* 45(6):571–582.

Gibson, K. R. 2002. Evolution of human intelligence: The roles of brain size and mental construction. *Brain, Behavior and Evolution* 59(1–2):10–20.

Giedd, J. N., Blumenthal, J., Jeffries, N. O. *et al.* 1999. Brain development during childhood and adolescence: A longitudinal MRI study. *Nature Neuroscience* 2(10):861–863.

Giedd, J. N., and Rapoport, J. L. 2010. Structural MRI of pediatric brain development: What have we learned and where are we going? *Neuron* 67(5):728–734.

Giedd, J. N., Rumsey, J. M., Castellanos, F. X. *et al.* 1996a. A quantitative MRI study of the corpus callosum in children and adolescents. *Developmental Brain Research* 91(2):274–280.

Giedd, J. N., Snell, J. W., Lange, N. *et al.* 1996b. Quantitative magnetic resonance imaging of human brain development: Ages 4–18. *Cerebral Cortex* 6(4):551–560.

Giedd, J. N., Vaituzis, A. C., Hamburger, S. D. *et al.* 1996c. Quantitative MRI of the temporal lobe, amygdala, and hippocampus in normal human development: Ages 4–18 years. *Journal of Comparative Neurology* 366(2):223–230.

Golovanova, L. V., Hoffecker, J. F., Kharitonov, V. M., and Romanova, G. P. 1999. Mezmaiskaya Cave: A Neanderthal occupation in the Northern Caucasus. *Current Anthropology* 40(1):77–86.

Gould, S. J. 1977. *Ontogeny and phylogeny*. Cambridge: Belknap Press.

Gousias, I. S., Edwards, A. D., Rutherford, M. A. *et al.* 2012. Magnetic resonance imaging of the newborn brain: Manual segmentation of labelled atlases in term-born and preterm infants. *Neuroimage* 62(3):1499–1509.

Gousias, I. S., Hammers, A., Counsell, S. J. *et al.* 2013. Magnetic resonance imaging of the newborn brain: Automatic segmentation of brain images into 50 anatomical regions. *PLoS One* 8(4):e59990.

Green, D. J., Gordon, A. D., and Richmond, B. G. 2007. Limb-size proportions in *Australopithecus afarensis* and *Australopithecus africanus*. *Journal of Human Evolution* 52(2):187–200.

Green, R. E., Krause, J., Briggs, A. W. *et al.* 2010. A draft sequence of the Neandertal genome. *Science* 328(5979):710–722.

Greenough, W. T., Black, J. E., and Wallace, C. S. 1987. Experience and brain development. *Child Development* 58(3):539–559.

Gunz, P., Mitteroecker, P., Neubauer, S. *et al.* 2009. Principles for the virtual reconstruction of hominin crania. *Journal of Human Evolution* 57(1):48–62.

Gunz, P., Neubauer, S., Golovanova, L. *et al.* 2012. A uniquely modern human pattern of endocranial development. Insights from a new cranial reconstruction of the Neandertal newborn from Mezmaiskaya. *Journal of Human Evolution* 62(2):300–313.

Gunz, P., Neubauer, S., Maureille, B., and Hublin, J.-J. 2010. Brain development after birth differs between Neanderthals and modern humans. *Current Biology* 20(21):R921–922.

Haeckel, E. 1866. *Generelle Morphologie der Organismen.* Berlin: Georg Reimer.

Hagmann, P., Sporns, O., Madan, N. *et al.* 2010. White matter maturation reshapes structural connectivity in the late developing human brain. *Proceedings of the National Academy of Sciences, USA* 107(44):19067–19072.

Herculano-Houzel, S. 2012. Neuronal scaling rules for primate brains: The primate advantage. *Progress in Brain Research* 195:325–340.

Herculano-Houzel, S., Collins, C. E., Wong, P., and Kaas, J. H. 2007. Cellular scaling rules for primate brains. *Proceedings of the National Academy of Sciences, USA* 104(9):3562.

Heuzé, Y., Boyadjiev, S. A., Marsh, J. L. *et al.* 2010. New insights into the relationship between suture closure and craniofacial dysmorphology in sagittal nonsyndromic craniosynostosis. *Journal of Anatomy* 217(2):85–96.

Heuzé, Y., Martínez-Abadías, N., Stella, J. M. *et al.* 2011. Unilateral and bilateral expression of a quantitative trait: Asymmetry and symmetry in coronal craniosynostosis. *Journal of Experimental Zoology. Part B, Molecular and Developmental Evolution* doi:10.1002/jez.b.21449.

Hirata, S., Fuwa, K., Sugama, K. *et al.* 2011. Mechanism of birth in chimpanzees: Humans are not unique among primates. *Biology Letters* 7(5):686–688.

Holloway, R. L. 1970. New endocranial values for the australopithecines. *Nature* 227(5254):199–200.

Holloway, R. L. 1981. Revisiting the South African Taung australopithecine endocast: The position of the lunate sulcus as determined by the stereoplotting technique. *American Journal of Physical Anthropology* 56(1):43–58.

Holloway, R. L. 1984. The Taung endocast and the lunate sulcus: A rejection of the hypothesis of its anterior position. *American Journal of Physical Anthropology* 64(3):285–287.

Holloway, R. L. 1985. The past, present, and future significance of the lunate sulcus in early hominid evolution. *In Hominid evolution: Past, present, and future.* New York: Liss.

Holloway, R. L. 1991. On Falk's 1989 accusations regarding Holloway's study of the Taung endocast: A reply. *American Journal of Physical Anthropology* 84(1):87–91.

Holloway, R. L., Broadfield, D. C., and Yuan, M. S. 2004. *The human fossil record: Brain endocasts, the paleoneurological evidence.* Hoboken, NJ: Wiley-Liss.

Hublin, J.-J. and Coqueugniot, H. 2006. Absolute or proportional brain size: That is the question. A reply to Leigh's (2006) comments. *Journal of Human Evolution* 50(1):109–113.

Hublin, J.-J., Neubauer, S., and Gunz, P. 2015. Brain ontogeny and life history in Pleistocene hominins. *Philosophical Transactions of the Royal Society of London, Series B: Biological Sciences doi:10.1098/rstb.2014.0062.*

Hunt, K. D. 1994. The evolution of human bipedality: Ecology and functional morphology. *Journal of Human Evolution* 26(3):183–202.

Huttenlocher, P. R. 1979. Synaptic density in human frontal cortex – developmental changes and effects of aging. *Brain Research* 163(2):195–205.

Isler, K., and van Schaik, C. P. 2009. The expensive brain: A framework for explaining evolutionary changes in brain size. *Journal of Human Evolution* 57(4): 392–400.

Isler, K., and van Schaik, C. P. 2012. Allomaternal care, life history and brain size evolution in mammals. *Journal of Human Evolution* 63(1):52–63.

Jerison, H. J. 1973. *Evolution of the brain and intelligence.* New York: Academic Press.

Kim, D.-J., Davis, E. P., Sandman, C. A. *et al.* 2014. Longer gestation is associated with more efficient brain networks in preadolescent children. *Neuroimage* doi:10.1016/j.neuroimage.2014.06.048.

Kimbel, W. H., and Delezene, L. K. 2009. "Lucy" redux: A review of research on *Australopithecus afarensis. American Journal of Physical Anthropology* 140(S49):2–48.

Kivell, T. L., and Schmitt, D. 2009. Independent evolution of knuckle-walking in African apes shows that humans did not evolve from a knuckle-walking ancestor. *Proceedings of the National Academy of Sciences, USA* 106(34):14241–14246.

Klingenberg, C. P. 1998. Heterochrony and allometry: The analysis of evolutionary change in ontogeny. *Biological Reviews* 73:79–123.

Krogman, W. M. 1951. The scars of human evolution. *Scientific American* 185(6):54–57.

Langbroek, M. 2012. Trees and ladders: A critique of the theory of human cognitive and behavioural evolution in Palaeolithic archaeology. *Quaternary International* 270:4–14.

Laughlin, S. B., and Sejnowski, T. J. 2003. Communication in neuronal networks. *Science* 301(5641):1870–1874.

Leakey, M. D. 1981. Tracks and tools. *Philosophical Transactions of the Royal Society of London, B: Biological Sciences* 292(1057):95–102.

Leakey, M. D., and Hay, R. L. 1979. Pliocene footprints in the Laetolil Beds at Laetoli, northern Tanzania. *Nature* 278(5702):317–323.

Leigh, S. R. 2004. Brain growth, life history, and cognition in primate and human evolution. *American Journal of Primatology* 62(3):139–164.

Leigh, S. R. 2007. Homoplasy and the evolution of ontogeny in papionin primates. *Journal of Human Evolution* 52(5):536–558.

Leigh, S. R. 2012. Brain size growth and life history in human evolution. *Evolutionary Biology* doi:10.1007/s11692-012-9168-5.

Lenroot, R. K., Gogtay, N., Greenstein, D. K. *et al.* 2007. Sexual dimorphism of brain developmental trajectories during childhood and adolescence. *Neuroimage* 36(4):1065–1073.

Lieberman, D. E., Krovitz, G. E., and McBratney-Owen, B. 2004. Testing hypotheses about tinkering in the fossil record: The case of the human skull. *Journal of Experimental Zoology. Part B, Molecular and Developmental Evolution* 302(3):284–301.

Lieberman, D. E., McBratney, B. M., and Krovitz, G. 2002. The evolution and development of cranial form in *Homo sapiens*. *Proceedings of the National Academy of Sciences, USA* 99(3):1134–1139.

Liu, X., Somel, M., Tang, L. *et al.* 2012. Extension of cortical synaptic development distinguishes humans from chimpanzees and macaques. *Genome Research* 22(4):611–622.

Lovejoy, C. O. 1975. Biomechanical perspectives on the lower limb of early hominids. *In:* R. Tuttle (ed.) *Primate Functional Morphology and Evolution*. The Hague: Mouton. pp.291–326.

Lovejoy, C. O., Suwa, G., Simpson, S. W. *et al.* 2009. The great divides: *Ardipithecus ramidus* reveals the postcrania of our last common ancestors with African apes. *Science* 326(5949):100–106.

Low, L. K., and Cheng, H.-J. 2006. Axon pruning: An essential step underlying the developmental plasticity of neuronal connections. *Philosophical Transactions of the Royal Society, Series B: Biological Sciences* 361(1473):1531–1544.

Macchiarelli, R., Bondioli, L., Debenath, A. *et al.* 2006. How Neanderthal molar teeth grew. *Nature* 444:748–751.

Martin, R. D. 1981. Relative brain size and basal metabolic rate in terrestrial vertebrates. *Nature* 293(5827):57–60.

Maureille, B. 2002a. La redécouverte du nouveau-né néandertalien Le Moustier 2. *Paléo* 14:221–238.

Maureille, B. 2002b. A lost Neanderthal neonate found. *Nature* 419(6902):33–34.

McHenry, H. M. 1992. How big were early hominids? *Evolutionary Anthropology: Issues, News, and Reviews* 1(1):15–20.

McHenry, H. M. 1994. Behavioral ecological implications of early hominid body size. *Journal of Human Evolution* 27(1–3):77–87.

Meyer, M., Kircher, M., Gansauge, M.-T. *et al.* 2012. A high-coverage genome sequence from an archaic Denisovan individual. *Science* 338(6104):222–226.

Miller, D. J., Duka, T., Stimpson, C. D. *et al.* 2012. Prolonged myelination in human neocortical evolution. *Proceedings of the National Academy of Sciences, USA* 109(41):16480–16485.

Miller, J. A., Ding, S.-L., Sunkin, S. M. *et al.* 2014. Transcriptional landscape of the prenatal human brain. *Nature* 508(7495):199–206.

Mitteroecker, P., and Gunz, P. 2009. Advances in geometric morphometrics. *Evolutionary Biology* 36(2):235–247.

Mitteroecker, P., Gunz, P., and Bookstein, F. L. 2005. Heterochrony and geometric morphometrics: A comparison of cranial growth in *Pan paniscus* versus *Pan troglodytes*. *Evolution and Development* 7(3):244–258

Morriss-Kay, G. M., and Wilkie, A. O. M. 2005. Growth of the normal skull vault and its alteration in craniosynostosis: Insights from human genetics and experimental studies. *Journal of Anatomy* 207(5):637–653.

Moss, M. L., and Young, R. W. 1960. A functional approach to craniology. *American Journal of Physical Anthropology* 18:281–292.

Navarrete, A., Van Schaik, C. P., and Isler, K. 2011. Energetics and the evolution of human brain size. *Nature* 480:90–93.

Neubauer, S., Gunz, P., and Hublin, J.-J. 2009. The pattern of endocranial ontogenetic shape changes in humans. *Journal of Anatomy* 215(3):240–255.

Neubauer, S., Gunz, P., and Hublin, J.-J. 2010. Endocranial shape changes during growth in chimpanzees and humans: A morphometric analysis of unique and shared aspects. *Journal of Human Evolution* 59(5):555–566.

Neubauer, S., Gunz, P., Mitteroecker, P., and Weber, G. W. 2004. Three-dimensional digital imaging of the partial *Australopithecus africanus* endocranium MLD 37/38. *Canadian Association of Radiologists Journal* 55(4):271–278.

Neubauer, S., Gunz, P., Weber, G. W., and Hublin, J.-J. 2012. Endocranial volume of *Australopithecus africanus*: New CT-based estimates and the effects of missing data and small sample size. *Journal of Human Evolution* doi:10.1016/j.jhevol.2012.01.005.

Neubauer, S., and Hublin, J.-J. 2012. The evolution of human brain development. *Evolutionary Biology* doi:10.1007/s11692-011-9156-1.

Pääbo, S. 2014. The human condition – a molecular approach. *Cell* 157(1):216–226.

Paus, T., Zijdenbos, A., Worsley, K. *et al.* 1999. Structural maturation of neural pathways in children and adolescents: *In vivo* study. *Science* 283(5409):1908–1911.

Ponce de León, M. S., Golovanova, L., Doronichev, V. *et al.* 2008. Neanderthal brain size at birth provides insights into the evolution of human life history. *Proceedings of the National Academy of Sciences, USA* 105(37):13764–13768.

Portmann, A. 1944. *Biologische Fragmente Zu Einer Lehre Vom Menschen.* Basel: Schwabe.

Prossinger, H., Seidler, H., Wicke, L. *et al.* 2003. Electronic removal of encrustations inside the Steinheim cranium reveals paranasal sinus features and deformations, and provides a revised endocranial volume estimate. *Anatomical Record, Part B: New Anatomist* 273(1):132–142.

Prüfer, K., Racimo, F., Patterson, N. *et al.* 2014. The complete genome sequence of a Neanderthal from the Altai Mountains. *Nature* 505(7481):43–49.

Raichlen, D. A., Gordon, A. D., Harcourt-Smith, W. E. H. *et al.* 2010. Laetoli footprints preserve earliest direct evidence of human-like bipedal biomechanics. *PLoS One* 5(3):e9769.

Rapoport, S. I. 1999. How did the human brain evolve? A proposal based on new evidence from *in vivo* brain imaging during attention and ideation. *Brain Research Bulletin* 50(3):149–165.

Reichert, H. 2009. Evolutionary conservation of mechanisms for neural regionalization, proliferation and interconnection in brain development. *Biology Letters* 5(1):112–116

Richmond, B. G., Begun, D. R., and Strait, D. S. 2001. Origin of human bipedalism: The knuckle-walking hypothesis revisited. *American Journal of Physical Anthropology Suppl.* 33:70–105.

Richmond, B. G., and Jungers, W. L. 2008. *Orrorin tugenensis* femoral morphology and the evolution of hominin bipedalism. *Science* 319(5870):1662–1665.

Richmond, B. G., and Strait, D. S. 2000. Evidence that humans evolved from a knuckle-walking ancestor. *Nature* 404(6776):382–385.

Richtsmeier, J. T., Aldridge, K., DeLeon, V. B. *et al.* 2006. Phenotypic integration of neurocranium and brain. *Journal of Experimental Zoology. Part B, Molecular and Developmental Evolution* 306(4):360–378.

Rightmire, G. P. 2004. Brain size and encephalization in early to mid-Pleistocene *Homo*. *American Journal of Physical Anthropology* 124(2):109–123.

Rilling, J. K. 2006. Human and nonhuman primate brains: Are they allometrically scaled versions of the same design? *Evolutionary Anthropology* 15(2):65.

Rilling, J. K. 2014. Comparative primate neuroimaging: Insights into human brain evolution. *Trends in Cognitive Sciences* 18(1):46–55.

Rilling, J. K., and Insel, T. R. 1999. The primate neocortex in comparative perspective using magnetic resonance imaging. *Journal of Human Evolution* 37(2):191–223.

Rosenberg, K. and Trevathan, W. 1996. Bipedalism and human birth: The obstetrical dilemma revisited. *Evolutionary Anthropology: Issues, News, and Reviews* 4:161–168.

Rosenberg, K., and Trevathan, W. 2001. The evolution of human birth. *Scientific American* 285(5):72–77.

Rosenberg, K., and Trevathan, W. 2002. Birth, obstetrics and human evolution. *BJOG* 109(11):1199–1206.

Roth, G., and Dicke, U. 2005. Evolution of the brain and intelligence. *Trends in Cognitive Sciences* 9(5):250–257.

Ruff, C. B. 1995. Biomechanics of the hip and birth in early *Homo*. *American Journal of Physical Anthropology* 98(4):527–574.

Ruff, C. B., Trinkaus, E., and Holliday, T. W. 1997. Body mass and encephalization in Pleistocene *Homo*. *Nature* 387(6629):173–176.

Sacher, G. A. 1982. The role of brain maturation in the evolution of the primates. In: E. Armstrong, D. Falk (eds) *Primate Brain Evolution*. Springer.

Sakai, T., Hirata, S., Fuwa, K. *et al.* 2012. Fetal brain development in chimpanzees versus humans. *Current Biology* 22(18):R791–R792.

Sakai, T., Matsui, M., Mikami, A. *et al.* 2013. Developmental patterns of chimpanzee cerebral tissues provide important clues for understanding the remarkable enlargement of the human brain. *Proceedings of the Royal Society of London, Series B: Biological Sciences* 280(1753):20122398.

Sakai, T., Mikami, A., Tomonaga, M. *et al.* 2011. Differential prefrontal white matter development in chimpanzees and humans. *Current Biology* 21(16):1397–1402.

Schenker, N. M., Desgouttes, A.-M., and Semendeferi, K. 2005. Neural connectivity and cortical substrates of cognition in hominoids. *Journal of Human Evolution* 49(5):547–569.

Schoenemann, P. T. 2006. Evolution of the size and functional areas of the human brain. *Annual Review of Anthropology* 35:379–406.

Schoenemann, P. T., Gee, J., Avants, B. *et al.* 2007. Validation of plaster endocast morphology through 3D CT image analysis. *American Journal of Physical Anthropology* 132(2):183–192.

Scott, N., Neubauer, S. Hublin, J.-J., and Gunz, P. 2014. A shared pattern of postnatal endocranial development in extant hominoids. *Evolutionary Biology* doi:10.1007/s11692-014-9290-7.

Semendeferi, K., and Damasio, H. 2000. The brain and its main anatomical subdivisions in living hominoids using magnetic resonance imaging. *Journal of Human Evolution* 38(2):317–332.

Semendeferi, K., Lu, A., Schenker, N., and Damasio, H. 2002. Humans and great apes share a large frontal cortex. *Nature Neuroscience* 5(3):272–276.

Semendeferi, K., Teffer, K., Buxhoeveden, D. P. *et al.* 2011. Spatial organization of neurons in the frontal pole sets humans apart from great apes. *Cerebral Cortex* 21(7):1485–1497.

Senut, B., Pickford, M., Gommery, D. *et al.* 2001. First hominid from the Miocene (Lukeino Formation, Kenya). *Comptes Rendus de l'Académie des Sciences* 332:137–144.

Shea, B. T. 1983. Phyletic size change and brain/body allometry: A consideration based on the African pongids and other primates. *International Journal of Primatology* 4(1):33–62.

Slice, D. E. 2007. Geometric morphometrics. *Annual Review of Anthropology* 36:261–281.

Smith, G. E. 1903. The so-called "Affenspalte" in the human (Egyptian) brain. *Anatomischer Anzeiger* 24:74–83.

Smith, G. E. 1904. The morphology of the occipital region of the cerebral hemisphere in man and the apes. *Anatomischer Anzeiger* 24:436–451.

Smith, T. M. 2013. Teeth and human life-history evolution. *Annual Review of Anthropology* 42(1):191–208.

Smith, T. M., Tafforeau, P., Reid, D. J. *et al.* 2010. Dental evidence for ontogenetic differences between modern humans and Neanderthals. *Proceedings of the National Academy of Sciences, USA* 107(49):20923–2098.

Smith, T. M., Toussaint, M., Reid, D. J. *et al.* 2007. Rapid dental development in a Middle Paleolithic Belgian Neanderthal. *Proceedings of the National Academy of Sciences, USA* 104(51):20220–20225.

Sporns, O. 2011a. *Networks of the brain.* Cambridge, MA: MIT Press.

Sporns, O. 2011b. The human connectome: A complex network. *Annals of the New York Academy of Sciences* 1224:109–125.

Stern, J. T., and Susman, R. L. 1983. The locomotor anatomy of *Australopithecus afarensis*. *American Journal of Physical Anthropology* 60(3):279–317.

Striedter, G. F. 2005. *Principles of brain evolution.* Sunderland, MA: Sinauer Associates.

Tague, R. G., and Lovejoy, C. O. 1986. The obstetric pelvis of A.L. 288–1 (Lucy). *Journal of Human Evolution* 15(4):237–255.

Teffer, K., and Semendeferi, K. 2012. Human prefrontal cortex: Evolution, development, and pathology. *Progress in Brain Research* 195:191–218.

Thompson, P. M., Giedd, J. N., Woods, R. P. *et al.* 2000. Growth patterns in the developing brain detected by using continuum mechanical tensor maps. *Nature* 404(6774):190–193.

Tillier, A. M. 1995. Neanderthal ontogeny: A new source for critical analysis. *Anthropologie* 33(1–2):63–68.

Tillier, A. M. 1996. The Pech de l'Azé and Roc de Marsal children (middle paleolithic, France): Skeletal evidence for variation in Neanderthal ontogeny. *Human Evolution* 11(2):113–119.

van den Heuvel, M. P., and Sporns, O. 2013. Network hubs in the human brain. *Trends in Cognitive Sciences* 17(12):683–696.

Washburn, S. L. 1960. Tools and human evolution. *Scientific American* 203:62–75.

Weber, G. W., and Bookstein, F. L. 2011. *Virtual anthropology: A guide to a new interdisciplinary field.* London: Springer.

Whitcome, K. K., Shapiro, L. J., and Lieberman, D. E. 2007. Fetal load and the evolution of lumbar lordosis in bipedal hominins. *Nature* 450(7172):1075–1078.

White, T. D., and Suwa, G. 1987. Hominid footprints at Laetoli: Facts and interpretations. *American Journal of Physical Anthropology* 72(4):485–514.

Wood, B., and Collard, M. 1999. The human genus. *Science* 284(5411):65–71.

Zollikofer, C. P. E. 2002. A computational approach to paleoanthropology. *Evolutionary Anthropology* 11(Suppl. 1):64–67.

Zollikofer, C. P. E., and Ponce de León, M. S. 2010. The evolution of hominin ontogenies. *Seminars in Cell & Developmental Biology* 21:441–452.

Zollikofer, C. P. E., and Ponce De León, M. S. 2013. Pandora's growing box: Inferring the evolution and development of hominin brains from endocasts. *Evolutionary Anthropology* 22(1):20–33.

Zollikofer, C. P. E., Ponce De León, M. S., and Martin, R. D. 1998. Computer-assisted paleoanthropology. *Evolutionary Anthropology* 6(2):41–54.

Chapter 12

FOXP2 and the Genetic and Developmental Basis of Human Language

Carles Lalueza-Fox
Institut de Biologia Evolutiva, CSIC-Universitat Pompeu Fabra, Barcelona, Spain

The Discovery

The scientific story of the *FOXP2* gene starts back over two decades ago, with the description of an English family with members over three generations affected by a speech and language disorder, called "developmental verbal dyspraxia" (Vargha-Khadem *et al.* 1995). The family's identity is kept anonymous, but is known scientifically as KE. The mode of inheritance of the condition followed a typical pattern of a single, dominant-autosomal mutation, starting in the grandparents' generation and affecting 15 people, about half the family members (Figure 12.1). It is still the only reported example of a simple inherited Mendelian marker affecting language. The disorder was not specific to grammar or speech, and affected members have problems in understanding complex syntactical structures, in processing words according to grammatical rules (marking number, gender, and tense) and in controlling complex coordinated facial and oral movements, which affect their speech (Vargha-Khadem *et al.* 1995). However, some aspects of the precise cognitive impairments of the affected KE family members, such as a claimed reduction in verbal and non-verbal IQ scores as compared to the non-affected members (Vargha-Khadem *et al.* 1995), remain controversial among biolinguists and neurobiologists because of the precise definition of what is a non-verbal test, and also general considerations about the notion of intelligence (Piattelli-Palmarini and Uriagereka 2011).

A few years later, Fisher *et al.* (1998) were able to circumscribe the location of the mutation affecting KE family members to a region in the long arm of chromosome 7. Later on, Lai *et al.* (2001) located the mutation in a particular gene of that region,

Developmental Approaches to Human Evolution, First Edition. Edited by
Julia C. Boughner and Campbell Rolian.

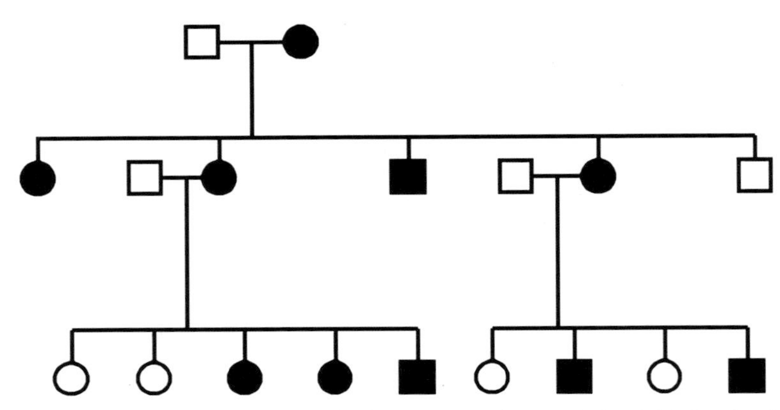

Figure 12.1. Partial representation of three generations of the KE family. Individuals affected by the language impairment associated to *FOXP2* malfunction are represented in black. Unaffected individuals represented as white symbols (circles: females; squares: males). The condition follows an inheritance mode typical of a dominant autosomal mutation (modified from Fisher *et al.* 1998).

FOXP2, thanks to the identification of an individual not related to the KE family who had a similar condition accompanied by a chromosomal translocation affecting the *FOXP2* gene. What was more interesting than the identification of the gene itself, however, was the nature of it. *FOXP2* have 17 exons (DNA sequences encoded by a gene) and contain a forkhead-binding domain typical of a family of genes, known as the *FOX* genes, that are transcription factors. This allows the FOX proteins to regulate the expression of a wide range of target genes by binding to the promoter regions of these genes to trigger their transcription from DNA to RNA. Thus, the *FOX* genes play crucial roles in development and are implicated as causative agents in different immunological and metabolical disorders. Intriguingly, the number of *FOX* genes also correlates with the anatomical complexity of the organism (e.g., 4 in *Saccharomyces*, 20 in *Drosophila*, 15 in *Caenorhabditis elegans*, 29 in *Ciona*, and 39 in humans). In the case of the KE family, all affected members display an amino acid change in the critical and highly conservative forkhead region (Lai *et al.* 2001).

With mouse embryo experiments, it was soon discovered that *FOXP2* is preferentially expressed in the developing brain (but also in other tissues such as the lungs, the guts, and the heart) (Enard *et al.* 2009). Disrupting the *FOXP2* gene in mice causes severe motor impairment and problems with the pups' vocalizations. Thus, it seemed that *FOXP2* had a role in the generation of a neural substrate in different brain regions probably related to aspects of motor control that in humans is also involved in the acquisition of language. Specifically, the analysis of transgenic mice bearing the humanized version of the *FOXP2* gene suggests that the evolutionary changes that took place in the hominin lineage affect brain regions that are connected via cortico-basal ganglia circuits (Reimers-Kipping *et al.* 2011).

The Role of a Transcription Factor

The *FOXP2* gene is a transcription factor (Lai *et al.* 2001). Like other FOX proteins, the FOXP2 protein carries a "forkhead box," a small string of 80–100 amino acids that binds to the promoter region of other genes and triggers their transcription from DNA to RNA. Thus, the FOX proteins (and their coding *FOX* genes) are important elements in regulating the expression of many other genes and thus are crucial in aspects such as the development of growing embryos, playing a critical role in tissue and organ formation. FOX proteins also regulate gene activities in the eyes, lungs, and brain, and cardiovascular, digestion, and immune systems, as well as the cell-division cycle in adults. Consequently, mutations in this gene family (especially those mutations placed in the critical forkhead box) have been associated experimentally to multiple developmental disorders, including tumor development, immune dysregulation, and metabolic problems (Friedman and Kaestner 2006).

For understanding the evolutionary importance and the precise role of *FOXP2* in the evolution of modern human language, we need to know precisely which genes *FOXP2* targets in humans and also which developmental processes are involved. This complicated task was not accomplished until 2009, when a working list of the network of genes regulated by *FOXP2* was published (Konopka *et al.* 2009). This work, mainly based on *in vitro* microarrays, showed that 61 genes were upregulated and 55 downregulated by the human *FOXP2* version, as compared to the chimpanzee *FOXP2* gene. *In vivo* assays with human and chimpanzee brain samples from three brain regions (caudate nucleus, frontal pole, and hippocampus) showed a significant overlap with a subset of the previous list of genes (Table 12.1). Ten genes from the original *in vitro* set (so-called "hub" genes) are highly connected in the network. Interestingly, two of these genes, *DLX5* and *SYT4*, have important roles in brain development and function. However, other "hub" genes (Table 12.2) seem only marginally related to language or even to cerebral functions. Additional genes in the network, such as *RUNX1T1*, *COL9A1*, *CNTNAP2*, *FGF14*, *PPP2R2B*, and *GJA12*, are linked to mental retardation, language impairments, or cranial abnormalities. This heterogeneity of functions is nevertheless expected due to the multifunctional impact of any key transcription factor.

Another study with a different methodological approach (*in vivo* arrays and *in situ* hybridization on brains) pointed out a list of 264 target genes for *FOXP2*, including, for instance, *DISC1*, a candidate gene for schizophrenia, bipolar disorder, and recurrent major depression (Vernes *et al.* 2011, Walker *et al.* 2012). The evidence of these studies suggests that the key function of *FOXP2* is to regulate in the embrionic brain the neurite outgrowth and axon guidance pathways (Vernes *et al.* 2011).

FOXP2 Molecular Evolution

Due to its critical role in brain development, *FOXP2* is an extremely conserved gene: in fact, it is among the 5% most conserved genes in the genome (Enard *et al.* 2002). In 2002, Enard *et al.* (2002) sequenced *FOXP2* in different primates (chimpanzee, gorilla,

Table 12.1. Differentially expressed genes up- and downregulated by *FOXP2*, determined by overlapping of both *in vitro* and *in vivo* microarray assays (Konopka *et al.* 2009). Neuronal, cranial, or language impairment phenotypes associated to genes from this list are included.

Upregulated	Associated phenotype
ADAMTS9	
BCAN	
COL9A1	Stickler syndrome (craniofacial abnormalities)
EXPH5	
FRZB	
IGFBP4	
ISLR2	
MGST1	
NPTX2	
PDGFRA	
PRICKLE1	myoclonus epilepsy with ataxia
RUNX1T1	
SLC30A3	
Downregulated	
ACCN2	temporal lobe epilepsy
B3GNT1	
C6orf48	
C8orf13	
CACNB2	
DCN	
ELMO1	
ENPP2	
FAM43A	bipolar disorder?
FAM43B	
FGF14	spinocerebellar ataxia type 27 (motor-related speech defects)
FLJ11286	
GJA12	ataxia, nystagmus, motor impairments, and mental retardation
GLRX	
HIST2H2BE	
IFIT2	
IGFBP3	
MAOB	different neurological and psychiatric disorders
PPP2R2B	spinocerebellar ataxia type 12 (motor-related speech defects)

orangutan, and macaque), including humans from all continents (Figure 12.2). They found that the nucleotide sequence is 93.5% identical in mouse and humans, while its protein sequence differs by just one amino acid change between mouse and most primates (the protein has 715 amino acids in total). In contrast, humans and chimpanzees

Table 12.2. "Hub" genes regulated by *FOXP2*, as determined by *in vitro* assays (Konopka *et al.* 2009).

Gene	Location	Function	Possible disease association
FAM43A	3q29	unknown	bipolar disorder?
ADM	11p15.4	hypotensive/vasodilator agent	vascular diseases
ZCCHC12	Xq24	transcriptional co-activator	X-linked mental retardation
ROR2	9q22	formation of the chondrocytes	brachydactyly type B1, Robinow syndrome (RRS)
ZNF556	19p13.3	transcriptional regulation?	–
DLX5	7q22	transcriptional factor bone development	hand/foot-split malformation, cranial dismorphogenesis
SYT4	18q12.13	Ca(2+)-dependent exocytosis, synaptic growth class-III neuronal intermediate filament protein	–
PRPH	12q12–q13		sporadic ALS
EBF3	10q26.3	transcriptional activator	cell carcinoma?
ADAMTS9	3p14.3–p14.2	cleaves the large aggregating proteoglycans	gastric, colorectal, pancreatic cancer

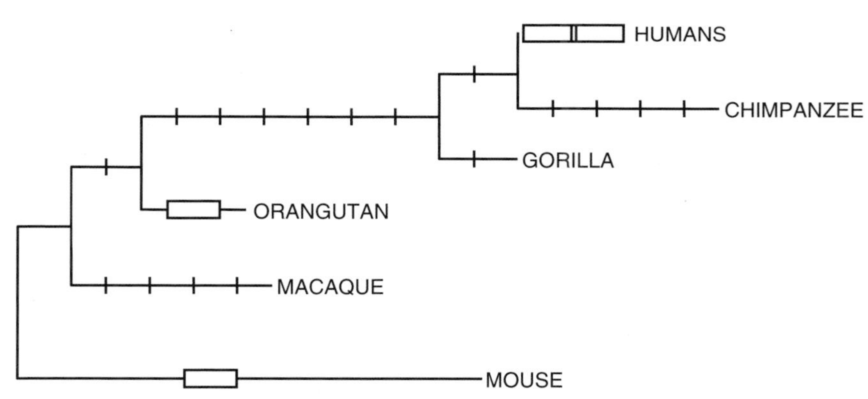

Figure 12.2. Phylogenetic tree of the *FOXP2* gene in mouse and in different primates, showing silent (synonymous) mutations as vertical lines and functional (amino acid changes) mutations as boxes. The extreme conservation of the gene along the primates branch and the remarkable acceleration of changes in the human lineage can be observed. Numerous synonymous changes along the mouse branch are not displayed for clarity (modified from Enard *et al.* 2002).

differ by two amino acid changes (at positions 303 and 325), despite being evolutionarily much more closely related (sharing a common ancestor about 4.6–6.2 million years ago (Enard *et al.* 2002)) than both species to mice (primate and mouse lineages having diverged around 70 million years ago). In addition, no amino acid changes were found among all modern humans investigated. This lack of change suggests these two substitutions took place after the divergence of humans from their common ancestor with chimpanzees and the derived allele has become fixed in the human lineage.

Transgenic mice carrying the humanized version of the *FOXP2* gene (Enard *et al.* 2009) show alterations in neuronal morphology (increased dendrite lengths and increased synaptic plasticity) and in ultrasonic vocalization, as well as decreased dopamine concentrations in the brain. The inference of this functional study is that the changes in *FOXP2* played a role in the emergence of human language by adapting cortico-basal ganglia circuits (i.e., neurons in the cortex, the striatum, and the thalamus) through enlarging dendritic trees and increasing synaptic plasticity in medium spiny neurons (Reimers-Kipping *et al.* 2011).

Neanderthals and the Selective Sweep on *FOXP2*

Neanderthals were an archaic hominin group (some paleontologists may call them a different human species) that inhabited Western Europe and Central Asia from about 250,000 years ago until their extinction with the arrival of modern humans, about 30,000 years ago (Lalueza-Fox and Gilbert 2011). They had a distinct skeletal morphology that sometimes has been interpreted as resulting from drift and cold-adaption. The extent of their cognitive abilities and symbolic behavior, including the capacity for language, has long been debated among researchers and it is still a matter of controversy (Lalueza-Fox and Gilbert 2011). Recently, paleogenetic techniques have provided direct information on the genomics of Neanderthals and also Denisovans (Meyer *et al.* 2012), an intriguing hominin group discovered at Denisova Cave in the Altai Mountains), providing a new source of evidence on this issue.

A new mutation occurs in a particular genetic background. If this mutation is beneficial and thus positively selected, the surrounding region is also selected. Thus the mutation as well as its genetic background can become fixed. Together, this lowers the variation in that region compared to other genomic regions under neutral evolution. This process is known as selective sweep. After the selective process has taken place, additional, random mutations occur, producing a characteristic pattern of rare variants (e.g., not shared among individuals). Again, this other process is different from a neutral variation pattern. With enough time, the signature of the sweep will become unrecognizable. However, it is theoretically possible to estimate when it took place just by looking at the pattern of variation around the candidate mutation.

Initially, the date of the selective sweep for the two human amino acid substitutions in the *FOXP2* gene was estimated to be around 260,000 years ago (Enard *et al.* 2002). At that time the archaeological and paleontological record indicated that humans and Neanderthals diverged about 400,000 years ago. Therefore, *FOXP2* was assumed to be unique to modern humans.

Thus it was a surprise when, in 2007, the same *FOXP2* variants were retrieved (Krause *et al.* 2007) from two exceptionally well-preserved Neanderthals from the 49,000-year-old El Sidrón site in Asturias (Northern Spain) (Lalueza-Fox *et al.* 2011). Neanderthal samples usually have some degree of inadvertent modern human contamination via contact with the modern human workers. To control these exogenous DNA sequences, two novel significant steps were taken: first, the samples were retrieved with an anticontamination protocol implemented at the excavation site itself, designed to prevent any human handling of the samples; the samples were retrieved with sterile instruments, handled with gloves, and immediately frozen for subsequent genetic analyses (Fortea *et al.* 2008) (Figure 12.3, color plate); and second, the diagnostic *FOXP2* nucleotide positions were co-amplified with ancestral (e.g., shared with chimpanzee but not with modern humans) X- and Y-chromosome variants, thus showing that contamination was negligible in those samples. Later on, after sequencing the Neanderthal genome via specimens from the Vindija cave in Croatia (Green *et al.* 2010) and the Denisova genome (Reich *et al.* 2010; Meyer *et al.* 2012), the presence of the modern human version of *FOXP2* was confirmed in both Neanderthals and Denisovan hominins. All three lineages shared a common ancestor around 800,000 years ago (Meyer *et al.* 2012). Discarding the unlikely scenario that *FOXP2* was acquired by hybridization with these different hominins (we have to remember that modern Africans do not show signs of Neanderthal or Denisovan interbreeding and yet they carry the same version of the gene), the most plausible

Figure 12.3. The excavation protocol designed to obtain uncontaminated Neanderthal samples for DNA analysis at the El Sidrón site (Asturias, Northern Spain). The initial retrieval of the key amino acid substitutions at the *FOXP2* gene was done with two bone samples belonging to two different Neanderthal males from this site.

explanation is that the "modern human" version was already present in the common ancestor of these three hominin species. Therefore, the modern human version of the *FOXP2* gene is not so modern, as once thought.

Originally, the presence of the modern human *FOXP2* version in Neanderthals was seen as an inconsistency of the selective sweep time estimate. This would not be impossible because all these estimates have large error margins. However, further evidence suggested that, in fact, two selective events in or around the *FOXP2* gene were superimposed: one, older event associated to the two amino acid changes; and a second, younger event likely associated to a functional element around exon 7 of the gene (Ptak *et al.* 2009). The second, more recent event would be associated to the original 260,000-year-ago sweep that likely took place after the separation of the modern human and the Neanderthal lineages. We do not have, however, enough functional evidence to understand the specific role of each of the two selective events in the emergence of modern human language.

Additional DNA sequencing in and around the *FOXP2* gene in the El Sidrón SD 1253 Neanderthal sample showed a nucleotide difference downstream of exon 7, in the binding motif of a transcriptional factor POU3F2 that was fixed between this Neanderthal and the majority of modern human samples studied (Maricic *et al.* 2013). Thus Neanderthals and Denisovans (Meyer *et al.* 2012) carry the ancestral position of this nucleotide, while most (98%) of modern humans carry a derived allele (the exception being some sub-Saharan African individuals). Functional studies in mouse with a reporter construct (e.g., an engineered DNA sequence that attaches to a regulatory region of interest) showed that the ancestral allele at POU3F2 activates transcription of the *FOXP2* gene more efficiency than the derived allele (Maricic *et al.* 2013). Therefore, this experimentally modeled change likely also affects the expression of the *FOXP2* gene in modern humans, and is thus to date the most obvious candidate associated to the second, more recent selective event detected at *FOXP2*.

However, it is unclear how carrying the ancestral variant at this regulatory region could have translated to the modern language phenotype. The frequency of individuals with the ancestral allele is low, but high enough for discarding any population-level language impairment. Obviously, the issue of potential differences in the regulation of *FOXP2* associated to possible functional differences needs further research.

The Role of *FOXP2* in the Emergence of Language

Shortly after its discovery, different groups of scientists called into question the importance and the role of *FOXP2* in the evolution and development of language, most notably researchers in biolinguistics and cognitive and evolutionary linguistics. Berwick and Chomsky, authors of a chapter in the book *The Biolinguistic Enterprise* (2011), make a computer metaphor and state that "*FOXP2* is more akin to the blueprint that aids in the construction of a properly functioning input-output system for a computer, like its printer, rather than the construction of the computer's central processor itself." But in my view, the role of *FOXP2* as a transcription factor acting upon a large gene network fits better with the analogy of a computer's central processor than with that of a peripheral device.

Robotics researcher and evolutionary linguist Luc Steels says in a paper (Steels 2011), "However, despite original enthusiasm, this gene is no longer considered to be 'the' or 'a' language gene partly because the KE-family impairments are certainly not restricted to language but to many other cognitive capacities including facial motor control, and partly because *FOXP2* is a regulatory gene that impacts a wide range of phenotypic features all over the body." However, I think that the role of *FOXP2* in regulating a gene network does not invalidate it as a candidate for having played a crucial role in the development of language.

These claims suggest that much of the current criticism is focused on the unspecific and broad role of the *FOXP2* gene. It is surprising that, even though dysfunction in this gene clearly impacts language, it doesn't even deserve the attribution of "*a* language gene" (it is certainly "a language-impairment gene"). Does this mean that a gene with a much more restricted effect on some particular aspect of language, such as, say, tense aspects or grammar aspects, would be a better candidate to play a crucial role in the initial emergence of language from an evolutionary point of view? Again, in my opinion, probably not.

Many researchers would agree that language as a human phenotype has involved the modification of tens, if not hundreds, of genes, in what is contemplated as a gradual process of transformation. This is likely the case for complex traits such as morphological structural modifications (e.g., the musculoskeletal system in relation to bipedalism). But why should this also be the case for language? It is difficult to imagine a long and gradual evolutionary process that involved the progressive modification of hundreds of genes by progressively increasing the selective advantage of human language capacity. Does the existence of many intermediate steps in language abilities –like those seen, for instance, in a complex structure like the eye – really make sense? Of course the problem is that all humans now share this final phenotype and it is impossible to observe these potential intermediate steps, if they ever existed. However, I think it is more parsimonious to assume that the emergence of language was triggered by changes in only a few, critical genes. In this scenario, a transcription factor such as *FOXP2* is what you would expect from an evolutionary point of view, more than the usual gradualistic assumption of "a complex trait needs the sequential modification of many independent genes."

Some animal adaptations, such as echolocation or visual structures, have arisen many times along different evolutionary lineages under similar selective pressures. Convergent evolution runs along particular gene networks that allow recurrent genetic modifications and thus trigger the repeated opportunity for natural selection on the traits involved. These traits, known as orthologous phenotypes or phenologs, are defined as phenotypes related by the orthology of the associated genes in different species (McGary *et al.* 2010). That is, some gene networks have a high evolvability, while others are very conservative. In this context, we should ask ourselves why the language phenotype, as we know it, has not evolved in any other primate lineages. It is likely that the genes involved are under very strong selective constraints, probably because their function is crucial for the right development and survival of the organism and thus almost no genetic changes are allowed. Then, it is again more plausible to think that human language is related to evolutionary modifications in just a few crucial genes that affected an otherwise very conservative gene network.

Future Directions

The recent sequencing of Neanderthal and Denisovan genomes at higher quality (and also the possible retrieval of more archaic hominins such as *Homo heidelbergensis*) will help elucidate some aspects of the evolution of language that could not be soundly analyzed with the low coverage genome drafts because of gaps and problems associated to sequencing errors and *postmortem* DNA damage. For instance, it would be possible to explore the existence of functional differences in regulatory elements affecting the expression of additional genes that could be found to be associated with language. The publication of the Denisova genome at 30x coverage (Meyer *et al.* 2012) has already shown the existence of functional changes in a gene, *CNTNAP2* ("Contactin-associated protein-like 2"), not previously detected in the low coverage draft. *CNTNAP2* codes for a protein that functions in the vertebrate nervous system as cell adhesion molecules and receptors and shows enriched expression in language-related circuits of the brain. Interestingly, this gene is downregulated by *FOXP2* and has been associated in several studies to a specific language impairment, to Gilles de la Tourette syndrome, and to schizophrenia, epilepsy, and autism (Newbury *et al.* 2010). A recent study suggests that a common variant of *CNTNAP2* (rs7794745) is relevant for sentence processing (Kos *et al.* 2012). The genetic *FOXP2–CNTNAP2* pathway is one example of gene interactions that will need further functional research.

Finally, the main problem is, as usual, connecting genotypes to phenotypes; in the case of *FOXP2* this is exacerbated by its large multifunctional role and also by the complexity of the phenotype involved (Preuss 2012). The future functional study using mouse models of the large network of genes regulated by *FOXP2* will help establish the functional pathways involved in the development of speech and language. By comparing these findings to the sequences of high-coverage ancient hominin genomes, we would be able to reconstruct the evolution of the molecular processes underlying this unique human phenotype.

Acknowledgments

I am grateful to Arcadi Navarro (Institute of Evolutionary Biology, CSIC-Universitat Pompeu Fabra), who provided useful comments on this manuscript. The author is supported by a grant (BFU2012-34157) from the Ministerio de Economía y Competitividad of Spain.

References

Berwick, R. C., and Chomsky, N. 2011. The biolinguistic program: The current state of its development. *In:* A. M. Di Sciullo and C. Boeckx (eds) *The biolinguistic enterprise*. New York: Oxford University Press. pp.19–41.

Enard, W., Gehre, S., Hammerschmidt, K. *et al.* 2009. A humanized version of *Foxp2* affects cortico-basal ganglia circuits in mice. *Cell* 137(5):961–971.

Enard, W., Przeworski, M., Fisher, S. E. *et al.* 2002. Molecular evolution of *FOXP2*, a gene involved in speech and language. *Nature* 418:869–872.

Fisher, S. E., Vargha-Khadem, F., Watkins, K. E. *et al.* 1998. Localisation of a gene implicated in a severe speech and language disorder. *Nature Genetics* 18(2):168–170.

Fortea, J., de la Rasilla, M., García-Tabernero, A. *et al.* 2008. Excavation protocol of bone remains for Neandertal DNA analysis in El Sidrón cave (Asturias, Spain). *Journal of Human Evolution* 55(2):353–357.

Friedman, J. R., and Kaestner, K. H. 2006. The *Foxa* family of transcription factors in development and metabolism. *Cellular and Molecular Life Sciences* (19–20):2317–2328.

Green, R. E., Krause, J., Briggs, A. W. *et al.* 2010. A draft sequence of the Neandertal genome. *Science* 328:710–722.

Konopka, G., Bomar, J. M., Winden, K. *et al.* 2009. Human-specific transcriptional regulation of CNS development genes by *FOXP2*. *Nature* 462:213–217.

Kos, M., van der Brink, D., Snijders, T. M. *et al.* 2012. *CNTNAP2* and language processing in healthy individuals as measured with ERPs. *PLoS ONE* 7(10):e46995.

Krause, J., Lalueza-Fox, C., Orlando, L. *et al.* 2007. The derived *FOXP2* variant of modern humans was shared with Neanderthals. *Current Biology* 17(21):1908–1912.

Lai, C. S. L., Fisher, S. E., Hurst, J. A. *et al.* 2001. A forkhead-domain gene is mutated in a severe speech and language disorder. *Nature* 413:519–522.

Lalueza-Fox, C., and Gilbert, M. T. P. 2011. Paleogenomics of archaic hominins. *Current Biology* 21(24):R1002–R1009.

Lalueza-Fox, C., Rosas, A., Estalrrich, A. *et al.* 2011. Genetic evidence for patrilocal mating behavior among Neandertal groups. *Proceedings of the National Academy of Sciences, USA* 108:250–253.

Maricic, T., Gunther, V., Georgiev, O. *et al.* 2013. A recent evolutionary change affects a regulatory element in the human *FOXP2* gene. *Molecular Biology and Evolution* 30(4):844–852.

McGary, K. L., Park, T. J., Woods, J. O. *et al.* 2010. Systematic discovery of nonobvious human disease models through orthologous phenotypes. *Proceedings of the National Academy of Sciences, USA* 107(14):6544–6549.

Meyer, M., Kircher, M., Gansauge, M.-T. *et al.* 2012. A high-coverage genome sequence from an archaic Denisovan individual. *Science* 338(6104):222–226.

Newbury, D. F., Fisher, S. E., and Monaco, A. P. 2010. Recent advances in the genetics of language impairment. *Genome Medicine* 2:6.

Piattelli-Palmarini, M., and Uriagereka, J. 2011. A geneticist's dream, a linguist's nightmare: The case of *FOXP2*. *In:* A. M. Di Sciullo and C. Boeckx (eds) *The biolinguistic enterprise*. New York: Oxford University Press. pp.100–125.

Preuss, T. M. 2012. Human brain evolution: From gene discovery to phenotype discovery. *Proceedings of the National Academy of Sciences, USA* 109:10709–10716.

Ptak, S. E., Enard, W., Wiebe, V. *et al.* 2009. Linkage disequilibrium extends across putative selected sites in *FOXP2*. *Molecular Biology and Evolution* 26:2181–2184.

Reich, D., Green, R. E., Kircher, M. *et al.* 2010. Genetic history of an archaic hominin group from Denisova Cave in Siberia. *Nature* 468:1053–1060.

Reimers-Kipping, S., Hevers, W., Pääbo, S., and Enard, W. 2011. Humanized *Foxp2* specifically affects cortico-basal ganglia circuits. *Neuroscience* 175:75–84.

Steels, L. 2011. Modeling the cultural evolution of language. *Physics of Life Reviews* 8:339–356.

Vargha-Khadem, F., Watkins, K., Alcock, K. *et al.* 1995. Praxic and nonverbal cognitive deficits in a large family with a genetically transmitted speech and language disorder. *Proceedings of the National Academy of Sciences, USA* 92:930–933.
Vernes, S. C., Oliver, P. L., Spiteri, E. *et al.* 2011. *Foxp2* regulates gene networks implicated in neurite outgrowth in the developing brain. *PLoS Genetics* 7:e1002145.
Walker, R. M., Hill, A. E., Newman, A. C. *et al.* 2012. The DISC1 promoter: Characterization and regulation by *FOXP2*. *Human Molecular Genetics* 21:2862–2872.

Chapter 13
Assembly Instructions Included

Kenneth Weiss[1] and Anne Buchanan[2]
[1] *Departments of Anthropology and Biology, Pennsylvania State University, University Park, PA, USA*
[2] *Department of Anthropology, Pennsylvania State University, University Park, PA, USA*

Introduction: A New Perspective on Development and Evolution

Charles Darwin was greatly interested in embryological development, but primarily as it showed sequences or stages in form among contemporary species that might reflect their history of shared ancestry, and the inverse problem of using developmental similarity to argue for descent with modification from common ancestry. His reasoning, so elegantly developed and presented in *On the Origin of Species*, was that if evolution were true, it would be a *connected* materialistic historical process, and if so, the traits of organisms that are here today will have evolved over time. The same must also apply to development (including life history, homeostasis, and response to environments) of those traits: the trait at one stage of development is connected to the stages that preceded it, as are the responsible genetic mechanisms. This nested historical connectedness between related species and among tissues during the development of individuals means that comparative genetics and embryology should help to elucidate the evolution of these processes. But the evolution of development was an area that Darwin left largely untouched, as did most of biology for almost a century after Darwin's ideas were published.

Many wonderfully informative experiments were done in the early 20th century, by manipulating embryonic tissues in various ways (Hall 1999). But the study of development lacked the kind of formal theory analogous to population genetics. Even without knowledge of what a "gene" – an inherited causal element – was, one could say that newly arising variation would increase or decrease in subsequent frequency in a population, depending on whether it helped the organism bearing it. This could be expressed in mathematical terms (Provine 1971), whereas developmental

Developmental Approaches to Human Evolution, First Edition. Edited by
Julia C. Boughner and Campbell Rolian.

studies at the time were more purely experimental, and included such procedures as cutting up and rearranging bits of early embryos to understand the future fate of specific tissues, but without such a single, coherent theory. As a result, leading biologists and evolutionary thinkers largely divorced development from the key component of evolution, inheritance (Gilbert *et al.* 1996). If you only inherit your genome, then how can that gene list explain the four-dimensional dance by which the fertilized egg, a single cell, turns into a complex organism?

Until not that long ago, and despite its manifest importance in determining what organisms actually are, development was almost wholly missing even from the major histories of biology (Dunn 1965; Provine 1971; Mayr and Provine 1980; Mayr 1982; Gould 2003). And yet the evolution of development is just as central as inheritance to our understanding of life, and the identification several decades ago of genes involved in embryological processes, and centrally, of the differential, context-specific usage of subsets of the repertoire of genes, finally brought the study of development into modern biology. Now called evo-devo, the field has flourished along with genetic methods, and understanding of the molecular basis of development is well advanced.

It is important to make clear that in this chapter we refer strictly to biological processes when we use the word "development" – embryology, or the study of growth from conception and how that is guided by genes, homeostasis, or the maintenance of internally stable conditions, and response to environments, how environmental conditions influence gene expression. We are not concerned here with cultural development over time, child development in the sense of learning and acculturation, nor economic development among nations. The word has different meaning and epistemology – and different sorts of controversy – in those contexts.

Thanks to modern molecular experimental and genomic technologies like the polymerase chain reaction (PCR) or DNA sequencing methodologies, as well as powerful methods for analyzing genetic and genomic data, evo-devo, has become a hot research area and enormous progress has already been made in the understanding of the role of developmental processes in evolution, as well as of the role evolution plays in shaping developmental processes. The evolution of mineralization is one example that we have worked on for many years (Kawasaki and Weiss 2003, 2006; Kawasaki *et al.* 2004, 2007, 2009).

A family of what we named secretory calcium-binding phosphoprotein (SCPP) genes includes, in any vertebrate species, many different genes. The genes are generally located adjacent to each other on a single chromosome. An SCPP gene codes for proteins that are secreted from the cell capture calcium ions that are floating by to facilitate the formation of various crystal structures. Each gene works in a different way, to form the particular structures in bone, scales, and teeth. They are expressed in the tissue and time when a newly formed structure in the embryo is ready to be hardened. Bones individually harden at appropriate stages, and dermal bones (like the cranial vault) are different from those that form from a cartilage mold (like long bones). SCPP genes are also involved in the composition of saliva, tears, and milk.

The evolution of this family of genes, and their specific use in each species, involves gene duplication and loss of individual members, and changes in the nearby DNA sequence that is used to control their context-specific expression. The first SCPP genes arose early in cartilaginous ancestors of jawed and boned vertebrates. So, new SCPP gene sets that were first involved in mineralizing exoskeletal scales later were used for internal bones, and similarly to enable the development of teeth, with their differently structured mineralized dentine and enamel.

But what about birds? They don't have teeth, but they do form calcified eggshells. They have exploited SCPP genes for the latter, but lost the expression of any of these genes in the developing jaws. And mammals also evolved to produce milk, a calcium-rich nutrient in which the calcium must not form large hardened clumps. The evolution of lactation involved SCPP gene recruitment in milk production. Similarly with tears and saliva.

So much is now understood from many examples like this that it's possible to propose a synthesis of a set of principles that unify a kind of theoretical understanding of life on the developmental, evolutionary, and ecological time scales. A categorization of these principles is given in Table 13.1, and we discuss them below. They reflect many kinds of causal symmetry, but also some important differences.

From Darwin's time to the present, evolutionary thinking has mainly concerned achieved, generally adult, forms and behavior, and could be studied primarily by observational methods. Those methods could attempt to identify genotypes whose bearers did better, or worse, in competitive terms. While natural selection in the wild was generally too slow to document directly, countless volumes of selection experiments in the laboratory and agricultural breeding have sped up the same kind of process and clearly shown us, as they did to Darwin himself, how it works.

Among the first dramatic examples showing that genetic expression in particular structures in embryos of different species reflect shared evolutionary ancestry involved the linearly arrayed *Hox* genes that are expressed, one by one, from tail to head, as an embryo develops (Gilbert 2010). Interrupting the genes experimentally prevented the proper segmentation of the embryo. Then the same genes were shown to be just as central to the segmentation of limb elements. In both instances, the same (homologous) sets of genes exist in insects and vertebrates, with very similar expression patterns that, in combination with other genes, lead fly and human embryos to end up with similar types of modularity of quite different resulting structures. But in a sense, a fly's leg and your leg are related by the evolution of the development of segmented axial structures.

Table 13.1. Some general principles of life (for explanation, see text).

1. *Inheritance with Memory*: *Life is one continuous history*
2. *Modularity*: *Life as LEGOs*
3. *Sequestration*: *Life is isolation – almost*
4. *Coding*: *The action is in the interaction*
5. *Contingency*: *Working only from today*
6. *Chance*: *Change without direction*
7. *Adaptability in the Face of Changing Circumstances*
8. *Cooperation*: *Getting along as well as getting ahead*

Source: Weiss and Buchanan, 2009b.

Evo-devo has perforce generally been centered on practical model systems on which experiments could be done; mice, zebrafish, fruit flies, and the worm *C. elegans*, among many others. The laboratory mouse is, for anthropologists, probably the most relevant but by no means the most important to the broader understanding of the genomic nature of life. What has been done are basically "static" experiments about genetic mechanisms in the model system at hand, but by the indirect method of comparison among model systems we have rapidly inferred an understanding of the evolutionary course of the mechanisms being discovered. If evolutionary theory is correct (as it seems very clearly to be), a model experimental system can be extrapolated (even if just imperfectly) to other species *only* because of the evolutionary connection among species. Again, insight into the evolution of a trait like mineralization, building bone and teeth with calcium, is one example, but there are many.

Specifically, it is our evolutionary connections that make model systems such as mice relevant to anthropology, and an evo-devo perspective is more widely being adopted by biological anthropology as a way of thinking about traditional problems in this field. Questions such as how humans evolved, how we differ from non-human primates, what makes humans unique, and more general questions such as how evolution constrains development and how relevant genetic variation arises and is maintained can now be addressed with this broader perspective. If large heads with a certain shape and cognitive abilities enabled our ancestors to do important fitness-related things, and that required changes in genotype frequencies, we can now begin to understand what the relevant genetic mechanisms may have been. Of course, we have to keep in mind that we are not, in fact, mice – and that we do not as yet have a good formal theory of the degree to which that matters to any given research problem. We need to bear in mind that any model system must be used with circumspection.

Comparative and descriptive morphology have been a focus of physical anthropology since its inception, and many early practitioners ("physical" anthropologists) were trained in anatomy. Early in its history, anthropologists were using phrenology and other craniometric studies to identify group traits, or indicators of traits such as intelligence (Gould 1981). The argument was that the brain is compartmentalized locally in function and therefore that the amount of that function in an individual will be reflected in the relative size to which the relevant part of the brain grows (and that this would affect the shape of the initially pliable cranial plates). Today, anthropologists use genes to make inferences about brain development and function, such as genes involved in developmental processes like signaling between tissues, or between neurons that connect synaptically during development. To some extent these mechanistic studies are interpreted in the same, often culpable, group-related terms as in the past.

A Broader Evo-Devo

Much, though by no means all, of anthropological evo-devo is the application of new tools to old questions. Like Linnaeus and Darwin, anthropologists have been, and to a large extent remain, primarily interested in net results; comparative morphology and what it tells us about the evolution of the hominin lineage. Linnaeus made it his life's

Figure 13.1. Comparative embryology from von Baer, as sketched out schematically by Darwin in his working notebook.

work to catalogue God's creations and Darwin's was to explain evolution from a common ancestor by a strictly historical process that he centered around common ancestry and natural selection. They were both interested in the results, but not primarily in mechanism, *how* a trait or organism arose, though Darwin mused about inheritance in a general way (his theory of gemmules). This is not a fault of theirs, but a limitation of the technologies and knowledge available to them. Indeed, Darwin was quite interested in *comparative* embryology, and his four volumes on barnacles reflect this. Figure 13.1 shows his famous notebook sketch of the branching nature of observed variation, based on work by the prominent contemporary embryologist Karl von Baer (1792–1876). But he was using this to show the historical course of development of embryos in a given species, to compare the final state among species, the idea being that the stages of a more "complete" development in one species reflected the "stopping points" in other species as the different species had evolved to date. Even he realized he wasn't *explaining* development.

Notice that Darwin's diverging tree-like structure of relationships is of exactly the same *type* as that of variation among species that he would later build into his theory of evolution of species. The importance of logical, historical ancestor-descendant, and developmental branching structures to the nature of life has proven to be as fundamental to our understanding of the evolution of development as it was for Darwin and his theory of evolution in general.

Indeed, much of what biological anthropologists do today still centers on these traditional questions and approaches, and molecular techniques have much to offer. If we can identify genetic mechanisms of development of, say, some craniofacial trait in present- day studies, can we use fossil or comparative traits (or in some cases even DNA from ancient specimens) to account for the *reason* that they have the different traits that they have? And can this approach be extended to language, opposable thumbs, upright posture, brain size, and so on, in ways that are important as well as merely edifying?

Of course, there are here two meanings of "reason." One has to do with the mechanism of development. We can, if we're circumspect enough, extrapolate a mechanistic explanation for trait differences between close species, such as that slight head-shape changes are due to different timing of expression of homologous genes in the species. The other meaning has to do with evolutionary history and the adaptive or whatever events that may have brought the trait about, or kept it, in different lineages. There, we are mostly guessing.

However, evo-devo is more than a set of ever-evolving molecular techniques to infer some specific mechanism. It also provides a new and broader *logical* perspective on development and evolution, and in particular how Darwinian processes working on the net results have molded the mechanism. Darwin's mechanism for genetic change was the generic idea of competition favoring one or the other variant, however it came about. But we can now begin looking at the "how" question. For example, his idea was comparative variation, and if common ancestry were a correct assumption, that implied branching. This suggests the value of looking at the kinds of general principles we listed in Table 13.1; we have learned in the last century or so that genes are carriers of information, that development is the hierarchical use of genes, that life is modular at all levels, and so on; knowledge which provides a broader perspective from which to synthesize these principles (see Table 13.1).

These and perhaps other similar kinds of principles are a foundation for evo-devo, whether explicitly or implicitly, and provide a framework for the specific questions anthropologists ask as well. They, or some equivalent enumeration, also form the basis of experimental work in development in laboratories around the world; when a new gene is found, it is routine to assume that it arose by duplication and that it belongs to a known gene family whose member genes and functions may already have been documented in other systems, or can practicably be done in affordable model systems. Any new function of DNA or RNA is assumed to work by coding. Principles such as those we discuss below are being built into sophisticated theory to explain how the genomes that contain the genes that direct development have come about – in ways consistent with the core of Darwinian theory: descent with modification from common ancestry.

The principles are broad enough that they can be applied to life on the brief time scale of organismal embryological development, the expansive time scale of evolution, and on the ecological scale of interacting organisms in the evolutionary development of ecosystems (i.e., the adjustment of plant and animal systems to each other's presence over time). They account beautifully for the diverging tree-like relationships among tissues within an organism, and why this has the same topology (general shape characteristics) as the relationship among species. In fact these notions of how life works are often the unstated default assumptions when experimental studies are designed, and the explanation of their results inferred. And they are as applicable to basically any species on earth, as any useful generalizations about life should be.

Fundamental Principles of Life

We have written elsewhere at length about these principles (Weiss and Buchanan 2004, 2009a, 2009b; Buchanan *et al.* 2009: Weiss *et al.* 2011), but will present their essential points briefly here. That should suffice to account for their importance as we see it. From the evo-devo perspective, what evolves in populations by evolutionary principles is used, or nested, within the development of each individual.

1. Inheritance with Memory: Life Is One Continuous History

Inheritance with memory explains why offspring look like their parents. The memory largely resides in genes (DNA sequence itself), but there are environmental factors as well, including ones that cause epigenetic effects that modify DNA and alter its gene-usage without changing the sequence.

Also important is that inheritance is not just from parent to offspring, but among the billions of cells in the body as they descend by cell divisions from the initial fertilized egg. Because all cells ultimately descend from other cells, the fact of inheritance with memory is what connects life among organisms on the evolutionary scale, with life on the cellular scale in individual lifetimes. And it is important to remember that cells are more than DNA and that other substances within them, including organelles such as mitochondria or Golgi apparati, for example, which are not made anew, are divided and transmitted to descendant cells.

2. Like LEGOs, Life Is Built from Modular Elements

From the cellular components to the biosphere, life is organized as functional subunits with varying degrees of interdependence and specialization. All the molecular components of a cell, the cell's organelles and the cells themselves, the substructures of tissues, tissues and organs, organisms, species, and ecosystems are all units, or modules that

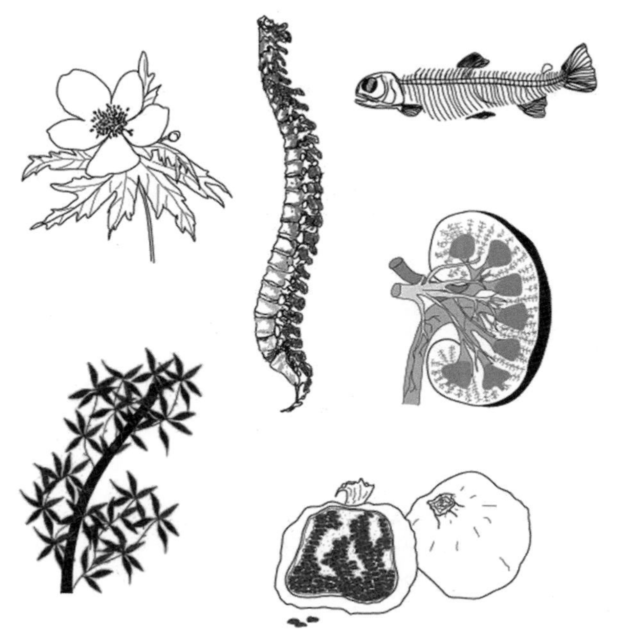

Figure 13.2. Nature is modular.

comprise the whole. Once an organism has the instructions for producing a given type of module, it can produce multiple copies of it, often with diversifying variation. A module can be small in one iteration and large in another; teeth, a trait of longstanding interest to anthropologists, come in many forms, even within the same individual. They have internal modularity as in cusps and roots. Vertebrae, ribs, hair, nephrons in the kidney, and so forth are all modular, the basic form made once, but tweaked during development as directed. New or complex traits can be made in this way, and modularity works during the development of an individual as well as the evolution of a lineage. Feathers can be made, and then variations are used and reused both in a single bird and in the evolution of birds. There isn't a separate set of genes for each hair or kidney nephron, and so on, but a separate iteration of the same developmental process.

3. Life is about Partial Sequestration (Units Isolated from Each Other)

The diversity of life at all levels and how it is produced depend on partial sequestration, the near isolation of cellular components, cells, tissues, organs, and organisms as they go about doing what they do. Units like cells have internal integrity because they are isolated from the outside. But this isolation is incomplete: they must monitor the outside world to determine how to change their behavior. For example, cell membranes are permeable, individuals are not isolates, as they need others to reproduce, organisms within ecosystems need each other as food, and so on. Without sequestration, basically at all levels from modular units in DNA to whole organisms and species, there would be no cellular or other differentiation, and life would not have changed from the primordial soup it was 4 billion years ago.

4. Coding: The Action Is about Interaction

DNA by itself is essentially an inert molecule. Things in life happen as a rule strictly through interaction with other things. Change within cells and organisms occurs because of switches, cellular responses to signaling. Signaling involves codes, such as the well-known genetic code, but there are many others, not all contained in genes. These codes determine how cells respond to signals, and change the way cells behave

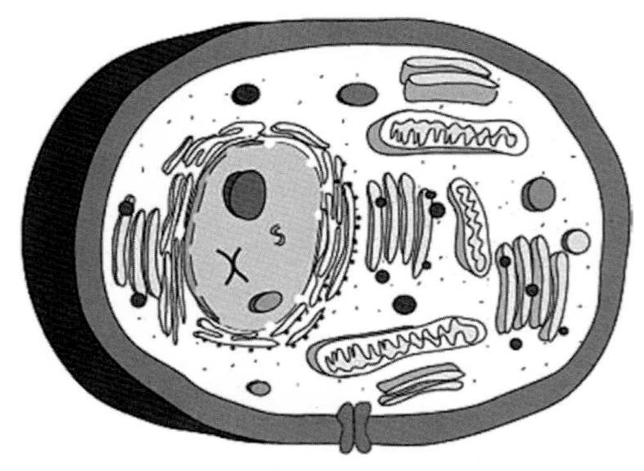

Figure 13.3. Sequestered organelles typify the internal structures of cells.

during development and the rest of life. Codes are important in ecosystems as well, as organisms respond to mating signals, be they hormones or visual displays, and, say, bees respond to flowers.

5. Contingency: Working Only from Today

Evolution has neither hindsight nor foresight. Neither does development, and neither do ecosystems. These are phenomena strictly rooted in the present. They are affected by yesterday only because yesterday sets the table for today, and they affect tomorrow only by how they respond to what is here today, regardless of how it got here or what it might be like tomorrow. That's how individuals in species evolve, and development is the same; whatever happens next is entirely contingent on the current stage of development. This is a cellular phenomenon – a cell responds to signals that tell it to express a gene now, for example, but it is also an ecological phenomenon, whether between cells, tissues, organs, or organisms. Life at all levels is an orchestration of steps and stages, all dependent on what is happening at every level now.

This can have long-term consequences, but that is because of local contingency and context, not foresight. Being rooted in the present is a constraint in the sense that evolution can't make a leg into a tail, for example, but it is also a strong reason for aspects of life that are predictable but would otherwise not be. The contingent nature of evolution, reflected in the development of each species, has often been referred to as "tinkering" (Bock and Goode 2006).

6. Chance: Change without Direction

Random unpredictable events play a role in the life of every organism, from the happenstance of which sperm fertilizes an egg, to DNA being altered by environmental factors such as X-rays, to which in a school of fish is scooped up by a shark. Chance effects can affect the germline and thus be inherited, or they can by random somatic mutations that affect only the development of a single individual. Chance can alter evolutionary trajectories, as well, when a mutation happens to change a morphological trait without harming – or benefitting – an organism. A central aspect of chance is that it makes prediction of the future imprecise or even impossible and, in particular, explains in part why an individual organism's traits and life history, such as disease, morphology, and behavior, are not usually predictable from its DNA alone, except in broad terms.

7. Adaptability in the Face of Changing Circumstances

Perhaps because life is in many ways so unpredictable, adaptability is a fundamental and essential fact of life. Organisms with built-in facultativeness that can respond to environmental change and contingencies in ways not prescribed in advance are those that are more likely to survive in the long term. This is sometimes called "plasticity" in the evo-devo literature (e.g., West-Eberhard 2003 and see the October 15, 2015 special issue of the journal Heredity devoted to plasticity) and can be as true of cells and organs, such as the brain and the liver, as it is of individual organisms or species.

There is certainly a lot of preprogramming contained ultimately in DNA, inherited from conception, but the success of animals such as primates heavily depends on their ability to sense their unpredictable circumstances and respond to them. One might say that we are programmed *not* to be programmed!

8. Cooperation: Getting Along as well as Getting Ahead

This final principle, cooperation, is probably the most likely to be unacceptable to some, but we believe it is probably more fundamental than any other because the others *require* cooperation. Cooperation happens at all levels, from interacting genes, cells, and tissues during development and then throughout life, to the interacting organisms of which stable ecosystems are made. At the molecular level, it should suffice here simply to note that gene expression, cellular differentiation, and the like occur only by proper *combinations* or *arrangements* of contributing molecules, signals, signal receptors, nutrients, physical structures, and the like. This is typically hierarchical: the actions at one level subsequently set the stage for the next level of comparably complex interactions. Most important functions require tens, hundreds, or more of these interacting units. It is their combined, and hence cooperative, interaction that fulfills the tasks of life, from conception on up in organisms, and from organisms to species, to ecosystems. Development and ecosystems, and thus evolution, all depend on cooperation among subcellular components, cells, organs, organisms, and species.

Geneticists have for practical and historical reasons considered genomic function to work in what is known as 'cis', that is, a gene and its chromosomally nearby regulatory regions were treated as largely independent functional units. But the chemical inertness of DNA by itself and the fundamentally interconnected nature of biological function means that genomic action is more of a 'trans' phenomenon. That is, systems of interactions among many, often hundreds of elements in the genome are responsible even for simple, and clearly for more complex, biological and developmental function. These are elaborate but not yet well-understand. However, it is clear that life, and both its development and evolution, are 'trans' phenomena.

Principles Applied to Evo-Devo in the Anthropological Context

These characteristics of life have been known for some time, even if details, especially at the cellular and molecular levels, are only gradually being revealed experimentally and in specific detail. We describe below how general principles of this sort might lead to a broader approach to anthropological questions, or even expand the kinds of questions that are being asked in the field. The principles have little relevance to the kinds of anthropological questions whose answer depends on judgment or cultural norms or definitions – for example, is this fossil a new species, how different does the DNA between two primates have to be before we conclude they are different species, or is race biological? As any student of anatomy knows very well, variation in details of development is found among people, or among individuals in any species.

But major differences, such as between clearly distinct species, can be addressed within the framework of these principles.

The Pace of Evolutionary Change

One of the major debates in evolutionary biology has been the degree to which Darwin's ideas of very slow gradualism are accurate, or when and if large changes can occur rapidly. In Chapter IV of *On the Origin of Species*, Darwin vigorously stated his dictum that *Natura non facit saltum*, Nature does not make leaps. Darwin did not see species arise suddenly with new, complex adaptive traits like upright posture or language. Instead, variation was mainly quantitative and change gradual. Each generation looked pretty much like its parents. This was important, because saltational change would be hard to explain: it would mean a birth of an individual with traits not present in its parents. A chimp never gave birth to a human or an elephant to a mouse! By contrast, variation is natural within species, and between parents, and is easy to understand (in principle) in their offspring. And, more important perhaps to Darwin, saltational change was ready-made for theological creationist explanations, and Darwin saw life as a continuous historical, *material*, process that could be explained without recourse to a chain of discrete events that would tempt a supernatural explanation.

Darwin's idea is that while species do not form between one generation and the next, accumulating divergence means that at some point in any given lineage, members alive today could not have mated successfully with their ancestors many generations ago, nor with distantly related contemporary populations. This is generally not directly testable, nor attributable to any fixed amount of heritable difference, and while this discussion in Chapter IV of the *Origin* was perhaps fundamental to his argument for evolution by natural selection, Darwin basically just asserted this as a fact, or perhaps more accurately, theoretical assumption. Most of the rest of his writings were about evidence for common ancestry and for detailed aspects of adaptation.

Indeed, historically, besides the problem of defining species in a definitive way, the assumption that species arose *because* of natural selection posed major problems even after genes had been discovered (gradually) during the first half of the 20th century. This is because it was not possible to understand the origins of complex structures in terms of classical "Mendelian" genes conferring two distinct states on traits but which didn't seem to change in viable ways from one generation to the next – in contrast to Darwin's continual if gradual change, that constancy of inherited units was one of Mendel's main points. If traits are determined by genes, but genes don't change (or when they do, the change is drastically dysfunctional), how can species evolve – which means change?

If species do not generally form among individuals within a population, however, the situation is very different *within* individuals. There, development clearly produces discrete modular structures like ribs, hair, teeth, and lungs that do not just vary quantitatively. A brain and a braincase are qualitatively different structures, but are within the same individual, made of cells with the same genome (and closely related in terms of cell lineages since the fertilized egg). However, despite having essentially the same genome, the cells in the separate structures of the same individual are as different as species, in the sense, for example, that one of these cell types cannot differentiate into the other.

And yet, by analogy with species, biologists were led to ask, if Darwinian gradualism was true on an evolutionary, generational scale, how could we explain the evolution of *homeotic* changes, changes in the number of structures, like teeth, from such genes, that not only clearly were found as differences between very closely related species but also *within* species? Thus, some individuals have more or fewer ribs or teeth or tooth-cusps, hairs, feathers, scales, and so on, than others – and these can even differ from left to right within the same person! Such apparent saltation, or rapid jumps, might suggest that evolution can after all, proceed in that way. Indeed, how can a single organism inheriting a single "genome" (whatever that was – the term itself didn't exist at the time) have many hairs, ribs, teeth, or cusps in the first place? The question was addressed in detail by William Bateson, who hypothesized rather unspecified processes of pattern formation during development (Bateson 1894). Bateson (Bateson 1913) and others doubted Mendelism as the basis of evolution and invoked a kind of built-in internal essentialism to explain changes in the number of repeated structures within individual organisms. Indeed, the genetic explanation of that phenomenon wasn't known for nearly a century, when evo-devo research developed a satisfactory understanding of the genetic basis of morphogenic processes. The answer is first the devo one: an organism can have multiple hairs not requiring a separate gene for each, but by spatially reusing the same genes.

Over the past generation or so, research has identified genes contributing to the development of distinct traits like the heart or teeth. And ideas that date back implicitly to Bateson, and to Alan Turing (Turing 1952) and others (e.g.,Wolpert *et al.* 1998), about simple patterning mechanisms (reaction-diffusion, morphogens, etc.) can account for distinct, organized structures that can be involved in homeotic change and thus evolve. These mechanisms involve interactions between the proteins coded by different, known genes, and the way these affect the expression levels of genes in specific cell types and stages during development. Good reading on this subject can be found in Webster and Goodwin (1996). The modular, repetitively organized nature of life at all levels helps account for many complex traits. Once a structure – a segment, hair, scale, kidney, tooth, and so on – has evolved, the same instructions can be invoked repeatedly, or modified along some axis in time or along the embryo or within the organ. Thus we have many hairs on our head and kidneys have many nephrons, and indeed this modularity extends below and above the level of such units. A number of recent studies have made it clear that homeotic change can be due to combinatorial expression patterns, in particular of closely related genes (e.g., Gilbert 2010; Sheth *et al.* 2012).

In fact like everything in evolution, developmental states build upon and diversify from each other over time. A secondary structure *evolves* only after its primary structure precursor has evolved, and in some descendant lineage of that structure. This also means that a secondary structure requires a primary structure from which to *develop* during embryological time. The result is nested modularity, and if you think about it, it is a thread that knits evo and devo together, and accounts, at least in principle, for the long-standing persistence of some traits, such as basic body form.

Once a basic body form is established, other forms can evolve from it, but it is difficult to undo the primary form without lethal damage to the embryo. As a result, trait states can become entrenched over very long evolutionary time. This is known as developmental *canalization*, an idea going back to C. H. Waddington and even earlier (Waddington 1942, 1957; Siegal and Bergman 2002). Similarly, there can be *morphological integration* by which various developmentally related but otherwise independent states can coevolve (Cheverud 1982; Ackermann and Cheverud 2000; Mezey *et al.* 2000; Ehrich *et al.* 2003; Klingenberg *et al.* 2004; Rolian *et al.* 2010).

In this sense, differences among species are the accumulation of the same sorts of developmental-genetic mechanisms that operate during development within individuals. Thus change in the number of vertebrae or teeth, or dental shape and cusp structure, the location of hair, and the relative length of digits are all modifications of repetitive structures that constitute putatively important adaptational differences among primates.

Saltation versus Gradualism in Light of Modern Evo-Devo

These various facts of development, brought to light by evo-devo research over the last few decades, are entirely consistent with classical Darwinism in the sense that things happen gradually, and that organisms and their traits (and the underlying genetic mechanisms, no matter how complex they are) have evolved historically by Darwinian descent with modification and divergence. One can debate the degree to which quasi-deterministic natural selection is responsible, but not the essential fact of evolution.

From this perspective, canalization of basic body plan presumably took millions of years, after which we're all today "just" variations on the evolving, diversifying theme. We know that some traits, like modular structures such as fingers and ribs, can change number, and we know the reasons. They involve "jumps" in the invocation of shared process, but only to proliferate or modify an existing structural "program." And this works by context-specific changes in the usage, but not the variation in genes. The numbers of repetitive traits can change rapidly, though starting with only one individual who carries and transmits a variant that affects the number or shape of a modular unit, and increasing gradually in frequency in the population. Thus, it is possible, in theory, to understand the genetic changes that account for differences in limb or digit length, or numbers of ribs or vertebrae between primate species. It is also possible to follow such changes over evolutionary time, and thus to account for much more significant changes between widely distant species.

In this light, there is no real controversy about the gradual nature of the evolution of even the processes that generate modular structures. As Darwin said, after long time periods, these differences can accumulate or lead to what later seem to be very different structures, like bat wings and gorilla arms, and changes such as additional or loss of legs in arthropods or snakes, or of rib numbers in mammals. They don't by themselves cause speciation, but they are generally correlated with it in that reproductive isolation is roughly correlated with amount of time since common ancestry, and so are morphological and behavioral traits. It is entirely consistent with the facts that these kinds of alterations arose as minor changes in one individual and proliferated gradually over many generations by the normal population genetic processes.

The Relevance of Complex Genetic Causation and the Conservation of Complex Structures

We know that most traits of interest to anthropology are the result of many contributing genes (and their regulation), along with a plethora of environmental inputs. The latter include culture, which adds enormous complexity to the challenge.

But even just considering genes, when traits are affected by many, even hundreds of different genes, each one of which varies in the population and when most serious variation is quickly purged by purifying selection, the causal landscape becomes difficult. Most individual causal effects are so small that they are difficult to identify, even if the basic underlying architecture (genes that are used to make this or that structure) can be characterized experimentally and are conserved among deep phylogenies.

The fact of polygenic – multiple-gene – causation with each cause generally having individually very small effects salvaged the seeming incompatibility between Mendelian inheritance as shown by Mendel's work and the gradual nature of evolution. If there are many genes contributing, each varying slightly (not in the grotesque way it was thought in the early 1900s that most mutations were), each inherited in a completely Mendelian way, then even if each factor hardly changed, there is enough variation in a population that selection on a resulting trait could move that trait gradually. It is clear from the kinds of traits in peas that Mendel worked with that evolution did not have to wait around for very very rare favorable mutations at the one or few genes that might be responsible for a trait.

The issue of individual variation is, in general, a relatively small one in this context. We are so alike as members of our species that our variation is slight relative to our similarity. Our genomes are not identical, but the genes they include are very similar. In other words, the trait variation among close relatives such as primates may generally only require small amounts of adjustment of essentially the same genes. The temptation to exaggerate the differences (or, for example, to exceptionalize humans beyond necessity, or to treat intergroup differences as more profound than they are) may be greater than is warranted.

If we focus too hard on or over-interpret the amount of variation, we can come away with a very misleading *gestalt* on evolution. This is even more the case regarding traits that require environmental interaction or years of life for differences to accumulate – experience that affects behavior or blood pressure, and so on. Similarly, we are so alike that even countless small differences interlock or have compensating effects such that they don't prevent the trait from having a kind of central tendency and coherent distribution among individuals in a population or species or in a population over time. Examples are the essentially normal distributions of height, blood pressure, IQ, and so on among people within and between populations.

It is easy to see that slight, gradual changes in genetic variation, nudged slowly by selection (and chance), lead the central tendency – average differences and/or variances – to change over long time periods, perhaps to the extent of no overlap between the species. No saltation is needed, and there is no "controversy" about how this happens.

For similar reasons, we can account for other patterns of difference. Since most contributing causal effects are small, and their future fates largely statistical (genetic drift, weak selection that depends on local context), and since most new mutational variants are unique at the DNA level (because they recur very rarely), the genetic basis of the "same" trait in different populations, or even among different individuals, can come to be quite different. With strong selection, such as related to malarial difference or UV light exposure, we see just this: different genetic adaptations in different human populations, due to different genotypes. There can be such differences within populations, but because of geographic distance and separation times, different genotypes can be associated with the same phenotype (such as malaria resistance, dark or light skin pigmentation, etc.). But mechanisms more or less fixed by earlier evolution, that canalize development, can persist for immense time periods even if things they are involved in, like basic body plans, change substantially.

We know these facts are important aspects of evolution. But they challenge some standard notions of how things evolve, that impair an ability to reconstruct the evolution of development with much confidence. This is the danger of assuming that phenotypes map tightly onto specific genotypes, in the face of the many-to-many causal reality (ignoring environmental and life-history factors). There are simply too many ways to get from here to there, and too many "there"s that "here" can lead to. If selection is weak, it is very hard to identify with much confidence at the individual gene level (even if we can assume it and make up *post hoc* stories about it). But that does not introduce any fundamental controversy about development or its evolution: they are genetic phenomena, but ones with many pathways and redundancies.

Perspectives on the Evolution of Development

When we try to put these ideas into practice in fields like anthropology, we face great challenges. We can do DNA sequencing on humans and other primates, and we can try to relate genetic variation to trait variation. But in the face of the kind of complexity we've just described, it is very hard to relate correlations of this sort to convincing inferences about causation. Because the principles we've discussed apply to all of life, we can use animal models like mice as tools for indirect, perhaps hopeful, inference about primates and humans, but this does not eliminate complexity. In fact, there are many problems in using and interpreting results from model systems, and often no easy solutions.

We can see this in the following way. Experimentally, we can take a model system, such as a particular inbred strain of laboratory mouse, in which variation among members of the strain is very minimal, and work out the means – the mechanism – by which this variation comes about. This is where our basic principles are invoked, and indeed they are, implicitly or explicitly, the basis of much of experimental design. The idea is that the variation, if adequately controlled in the design, is minor relative to the major mechanism being found. It isn't perfect, but if we are not mesmerized by similarities so much as to over-interpret them, models at least allow us to make educated guesses about similar phenomena in species, like primates, that are not ethically or even practicably useful as actual experimental subjects. At the same time, careful examination of experimental studies, as in mice, reveals that they often don't report variation among what are purportedly genetically identical (e.g., inbred) animals (e.g., Couzin-Frankel 2013).

Said in another way, we can learn much from experimental work of this sort, and it is exceedingly illuminating of the way complex traits develop. But they provide a *stereotypical* understanding of mechanism. We must be circumspect in how we extend this to the world outside the laboratory, in which variation is often what is most interesting. The mouse is, after all, not a primate! Outside the lab, every gene, in every individual, in every population, and in every species is subject to variation. And not only is everyone genotypically unique, but the functions of the genes involved, in their actual implementation, act probabilistically.

In the real world, this variation implies a modeling problem: modeling is about repeatable characteristics, but no two individuals are alike (it is an uncomfortable fact, often comfortably ignored, that no two laboratory mouse strains are alike either). Genetic causation is typically complex, affected by the environment as well as many genes. Indeed, the environment of a given gene is everything else in the genome, cell, organ, organism, species, and ecosystem. There are no *a priori* limits on where effects can come from. Not only are there innumerable causal components, but their variation is not uniform. It is an inherent aspect of the way life works that most variation has individually small effects, and only a subset of causes has major effects in any given case. And the latter vary from individual to individual in ways for which, again, there is no *a priori* theory. The take-home message is that we need more effort on developing cogent theory, not just additional data of the kinds we already amply have. This is a profound challenge today, and perhaps we have the tools to improve the nature and extensibility of our models.

Practical Issues: Sampling, Mapping, and Causal Inference

There are many practical problems that one faces even within an experimental context, and certainly when it comes to making inferences about primates and humans. Among them are that the same genetic variant even within species does not always have the same effect – and often can be harmful in some people but have no effect in others. Transgenic mice carrying the same engineered mutation vary one from another, even though an inbred strain is supposed to be genetically homogeneous (reported effects are often just of the clearest instance, assumed to be the most "correct"). Related to this is a notorious more general problem of the publication only of positive results. Every individual human, and doubtless every chimpanzee or rhesus monkey, is carrying a large number, well over 100 of, mutant genes, but with no discernible effect. Negative results are an essential part of triangulating these problems.

Laboratory mice produce generations that must include new mutations, but are assumed to be homogeneous not just within one developmental laboratory but among labs, and we know that the same engineered mutation can have dramatically different effects in different inbred strains.

There is also the sampling issue. To show genotype-phenotype correlations in natural populations, one must collect some sort of sample. We usually say we want "random" samples, but it is not clear in this connection what such a term means. If everyone is not the same, the sampling question is not just about minimizing the effects of measurement error, though the statistical analysis often treats the results as if that were the source of variation.

Theory and Its Discontents

Academic society loves "controversy." It is much more fun than normal science, unexciting if illuminating as that can be. Our culture rewards attention-seeking claims, and it is only natural for us, who want our lives to have real meaning, and our times to be important, to portray our current era as laden with controversy. There are always disagreements about nuances (the less difference, the more heated, perhaps) and there are always uncertainties. Indeed, "gradualism" and other similar generalizations about evo and devo are subjective terms that inherently leave room for discussion. The lure for public showmanship is strong and strongly rewarded, perhaps especially in so subjective and emotive an area as anthropology, where the topic is evolution and development in ourselves. Controversy where the issues are substantive *and* there is some way to resolve them, is at the core of good science. But we think that public bickering over minor points, or over points for which neither side has very strong evidence, is a disservice to what our science establishment should encourage.

The implications of the ultimately only weakly predictable details of physical traits can perhaps be understood in a subjective way. Compare this to notions of "free will." From a materialist point of view there is no such thing: we are all determined by the molecules and energy interactions of our bodies. Nobody in science seems to think that the appearance of a leg, jaw, or eye, or diabetes or cancer are mystical events, no matter how wondrous they may truly be as they emerge out of an assemblage of contributing components. Likewise, the limits in predicting behavior from genotypes may be inherent, in a way that could be defined by saying the individual has "free will." The physical traits of embryologic development are more mechanical and less vulnerable to these sorts of controversies, but development is a similar kind of phenomenon in that we see the net result of four-dimensional interactions among thousands of contributing components in billions of quasi-independent cells.

An overriding danger here, and the reason for any sense of "controversy," is the hunger for simplicity in the face of complexity. Careers are not built nor research grants funded upon uncertainties of the sort that a candid view will recognize in Nature. In the absence of a strongly precise theory of development and its evolution, we have no externally derived guide for interpreting results, even, often times, of experiments. We want to find the gene "for" this or that substantial effect, not a beach-load of individual sand grains. But the latter may be closer to the truth. When genes typically have small effects, are used in many different ways even within the same body but at different times and different ways during embryogenesis, and interact with other genes differently in these contexts, we face a real interpretive problem! But that does not mean that, properly understood, this is a "controversy." It is simply that we are far from knowing everything. And integrating the findings we've discussed here with higher-level ecological concepts is a vibrant, and not new, area as well (Gottlieb 1992; Gilbert *et al.* 1996; Oyama 2000; Gilbert 2001, 2003; Oyama *et al.* 2001; Gilbert and Bolker 2003).

There would be a great prize to be had if someone could identify important instances of *real* saltational evolution, of new, highly organized structures. There are always people reaching for claims, often (and usually properly) on the fringe of science. History has examples such as polyphenism in which different environments induce very different phenotypes in the same species. Homeotic changes, such as in the number of lumbar vertebrae or molars, are well known, though often the mechanism is not. We know of countless genetic variants that, by themselves, can generate discrete phenotypic differences in humans (eye color is a benign example, but the literature is rife with devastating developmental diseases due to single-gene changes). But none of the known instances poses any basic challenge to the general idea of how evolution works.

History shows that there is always the possibility that dogma will be shown to be wrong. We need to be open to new ideas, but also to have a healthy level of skepticism, so that our foundational understanding is both as correct as possible at any given time and not too vulnerable to opportunistic or misguided intrusion. At present, there are a great many unknowns about how genomes evolve and are used in development, but there is no real evo-devo anthropology controversy in the areas of development that we have discussed.

It is not controversial that science doesn't know everything, and that can be frustrating. But we shouldn't fret too much, or look for controversy where it's not to be found – because in fact we *do* know quite a lot!

References

Ackermann, R. R., and Cheverud, J. M. 2000. Phenotypic covariance structure in tamarins (genus Saguinus): A comparison of variation patterns using matrix correlation and common principal component analysis. *American Journal of Physical Anthropology* 111(4):489–501.

Bateson, W. 1894. *Materials for the study of variation, treated with special regard to discontinuity in the origin of species*. Reprinted 1992 with introductory essays. London: Macmillan (reprint Baltimore: Johns Hopkins Press).

Bateson, W. 1913. *Problems of genetics*. New Haven, CT: Yale University Press.

Bock, G., and Goode, J. (eds) 2006. *Tnkering: The microevolution of development*. London: Novartis Foundation.

Buchanan, A. V., Sholtis, S., Richtsmeier, J., and Weiss, K. M. 2009. What are genes "for" or where are traits "from"? What is the question? *BioEssays: News and Reviews in Molecular, Cellular and Developmental Biology* 31(2):198–208.

Cheverud, J. M. 1982. Phenotypic, genetic, and environmental morphological integration in the cranium. *Evolution* 36:1737–1747.

Couzin-Frankel, J. 2013. When mice mislead. *Science* 342(6161):922–923, 925.

Dunn, L. C. 1965. *A short history of genetics: The development of some of the main lines of thought: 1864–1939*. New York: McGraw-Hill.

Ehrich, T. H., Vaughn, T. T., Koreishi, S. F. *et al.* 2003. Pleiotropic effects on mandibular morphology I. Developmental morphological integration and differential dominance. *Journal of Experimental Zoology*. Part B, Molecular and Developmental Evolution 296(1):58–79.

Gilbert, S. F. 2001. Ecological developmental biology: Developmental biology meets the real world. *Developmental Biology* 233(1):1–12.

Gilbert, S. F. 2003. The morphogenesis of evolutionary developmental biology. *International Journal of Developmental Biology* 47(7–8):467–477.

Gilbert, S. F. 2010. *Developmental biology*. Sunderland, MA, Sinauer.

Gilbert, S. F., and Bolker, J. A. 2003. Ecological developmental biology: Preface to the symposium. *Evolution & Development* 5(1):3–8.

Gilbert, S. F., Opitz, J., and Raff, R. 1996. Resynthesizing evolutionary and developmental biology. *Developmental Biology* 173:357–372.

Gottlieb, G. 1992. *Individual development and evolution: The genesis of novel behavior*. New York: Oxford University Press.

Gould, S. J. 1981. *The mismeasure of man*. New York: Norton.

Gould, S. J. 2003. *The structure of evolutionary theory*. Cambridge, MA: Belknap Press.

Hall, B. K. 1999. *Evolutionary developmental biology*. Dordrecht, Holland: Klewer.

Kawasaki, K., Buchanan, A. V., and Weiss, K. M. 2007. Gene duplication and the evolution of vertebrate skeletal mineralization. *Cells Tissues Organs* 186(1):7–24.

Kawasaki, K., Buchanan, A. V., and Weiss, K. M. 2009. Biomineralization in humans: Making the hard choices in life. *Annual Review of Genetics* 43:119–142.

Kawasaki, K., Suzuki, T., and Weiss, K. M. 2004. Genetic basis for the evolution of vertebrate mineralized tissue. *Proceedings of the National Academy of Sciences, USA* 101(31):11356–11361.

Kawasaki, K., and Weiss, K. M. 2003. Mineralized tissue and vertebrate evolution: The secretory calcium-binding phosphoprotein gene cluster. *Proceedings of the National Academy of Sciences, USA* 100(7):4060–4065.

Kawasaki, K., and Weiss, K. M. 2006. Reading the palimpsests of life. *Evolutionary Anthropology* 15(6):207–211.

Klingenberg, C. P., Leamy, L. J., and Cheverud, J. M. 2004. Integration and modularity of quantitative trait locus effects on geometric shape in the mouse mandible. *Genetics* 166:1009–1921.

Mayr, E. 1982. *The growth of biological thought*. Cambridge, MA: Harvard University Press.

Mayr, E., and Provine, W. (eds) 1980. *The evolutionary synthesis: Perspectives on the unification of biology*. Cambridge, MA: Harvard University Press.

Mezey, J. G., Cheverud J. M., and Wagner, G. P. 2000. Is the genotype-phenotype map modular? A statistical approach using mouse quantitative trait loci data. *Genetics* 156(1):305–311.

Oyama, S. (ed.) 2000. *The ontogeny of information: Developmental systems and evolution*. Durham, NC: Duke University Press.

Oyama, S., Griffiths, P. E., and Gray, R. D. (eds) 2001. *Cycles of contingency: Developmental systems and evolution*. Cambridge, MA: MIT Press.

Provine, W. 1971. *The origins of theoretical population genetics*. Chicago: University of Chicago Press.

Rolian, C., Lieberman, D. E., and Hallgrímsson, B. 2010. The coevolution of human hands and feet. *Evolution* 64(6):1558–1568.

Sheth, R., Marcon, L., Bastida, M. F. *et al.* 2012. *Hox* genes regulate digit patterning by controlling the wavelength of a Turing-type mechanism. *Science* 338(6113):1476–1480.

Siegal, M. L., and Bergman, A. 2002. Waddington's canalization revisited: Developmental stability and evolution. *Proceedings of the National Academy of Sciences, USA* 99(16):10528–10532.

Turing, A. 1952. The chemical basis of morphogenesis. *Philosophical Transactions of the Royal Society of London, Series B* 237:37–72.

Waddington, C. H. 1942. Canalization of development and the inheritance of acquired characters. *Nature* 150:563–565.

Waddington, C. H. 1957. *The strategy of the genes: A discussion of some aspects of theoretical biology*. London: George Allen & Unwin.

Webster, G., and Goodwin, B. 1996. *Form and transformation: Generative and relational principles in biology*. Cambridge: Cambridge University Press.

Weiss, K. M., and Buchanan, A. V. 2004. *Genetics and the logic of evolution*. New York: Wiley-Liss.

Weiss, K. M., and Buchanan, A. V. 2009a. The cooperative genome: Organisms as social contracts. *International Journal of Developmental Biology* 53(5–6):753–763.

Weiss, K. M., and Buchanan, A. V. 2009b. *The mermaid's tale: Four billion years of cooperation in the making of living things*. Cambridge, MA: Harvard University Press.

Weiss, K. M., Buchanan, A. V., and Lambert, B. W. 2011. The red queen and her king: Cooperation at all levels of life. *American Journal of Physical Anthropology* 146 Suppl. 53:3–18.

West-Eberhard, M. J. 2003. *Developmental plasticity and evolution*. Oxford/New York: Oxford University Press.

Wolpert, L., Beddington, R., Brockes, J. *et al.* 1998. *Principles of development*. London: Current Biology.

Index

Note: Page numbers in *italics* refer to Figures; those in **bold** to Tables.

Developmental Approaches to Human Evolution, First Edition. Edited by Julia C. Boughner and Campbell Rolian.